# TELECOMMUNICATIONS:

## VOICE / DATA
## WITH FIBER OPTIC
## APPLICATIONS

# TELECOMMUNICATIONS:

## VOICE / DATA
## WITH FIBER OPTIC
## APPLICATIONS

*WAYNE TOMASI*

Mesa Community College

*VINCENT F. ALISOUSKAS*

Devry Institute of Technology

PRENTICE HALL   Englewood Cliffs, New Jersey 07632

Library of Congress Cataloging-in-Publication Data

Tomasi, Wayne.
    Telecommunications : Voice/data with fiber optic applications
    Wayne Tomasi, Vincent F. Alisouskas.
        p.    cm.
    Includes index.
    ISBN 0-13-902602-9
    1. Telecommunication.    2. Optical communications.    I. Alisouskas.
Vincent F.    II. Title.
TK5101.T64    1988
621.38—dc19                                              87–26869

Editorial/production supervision and
    interior design: *Margaret Lepera*
Cover design: *George Cornell*
Cover photo courtesy NASA
Manufacturing buyers: *Lorraine Fumoso and Peter Havens*

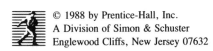
PRENTICE-HALL INTERNATIONAL (UK) LIMITED, *London*
PRENTICE-HALL OF AUSTRALIA PTY. LIMITED, *Sydney*
PRENTICE-HALL CANADA INC., *Toronto*
PRENTICE-HALL HISPANOAMERICANA, S.A., *Mexico*
PRENTICE-HALL OF INDIA PRIVATE LIMITED, *New Delhi*
PRENTICE-HALL OF JAPAN, INC., *Tokyo*
SIMON & SCHUSTER ASIA PTE. LTD., *Singapore*
EDITORA PRENTICE-HALL DO BRASIL, LTDA., *Rio de Janeiro*

# *CONTENTS*

**PREFACE** *v*

ONE   *THE TELEPHONE NETWORK* *1*

TWO  *COMMUNICATIONS SYSTEM OVERVIEW* *12*

THREE  *THE TELEPHONE CIRCUIT* *23*

FOUR  *TRANSMISSION CODES AND ERROR
DETECTION AND CORRECTION* *48*

FIVE  *UNIVERSAL ASYNCHRONOUS RECEIVER/
TRANSMITTER* *63*

SIX  *SERIAL INTERFACES* *71*

iv

SEVEN  *DATA TRANSMISSION WITH ANALOG CARRIERS*  *88*

EIGHT  *VOICE OR DATA TRANSMISSION WITH ANALOG CARRIERS*  *134*

NINE  *VOICE OR DATA TRANSMISSION WITH DIGITAL CARRIERS*  *154*

TEN  *ASYCHRONOUS PROTOCOL*  *214*

ELEVEN  *BINARY SYNCHRONOUS COMMUNICATIONS*  *225*

TWELVE  *SYNCHRONOUS DATA LINK CONTROL AND HIGH LEVEL DATA LINK CONTROL*  *266*

THIRTEEN  *PACKET SWITCHING AND LOCAL AREA NETWORKS*  *286*

FOURTEEN  *FIBER OPTIC COMMUNICATIONS*  *304*

INDEX  *345*

# *PREFACE*

Since 1973, technological advances in the digital area have transformed telecommunications into a highly dynamic field. Where previously telecommunications systems had only voice to accommodate, with the advent of the microprocessor chip and the accompanying low-cost computers, the need for transmission of digital information has arisen. Digital technology has also provided a means of building smaller, cheaper, and more reliable telecommunications equipment. Laser technology is providing a means of transmitting information on optic fibers. The introduction of satellites into space has expanded long-distance communications.

Along with these advances, numerous texts have been published about each new area. Because of the complexity of each area, many of these texts are highly specialized. Other texts presume either that the reader has already been in the communications field and wishes to update his or her knowledge, or that the reader is highly skilled in mathematics and can follow complex derivations that establish limits on equipment and techniques or derive the probabilities associated with error rates. These books are excellent texts for the audience intended.

However, the rapid advances in technology in this area have aroused the interest of many who have never been previously associated with this field. The unique terms and numerous acronyms used in communications, if not explained, can leave the newcomer bewildered. Also, just as the advances in digital technology have brought about the introduction to digital theory and computer programming at the high school level, the authors foresee the introduction of telecommunications systems at lower levels in our educational institutions. This is the goal of this book—to provide an understanding of existing systems to a newcomer in this area. Hopefully, this book is written so that a reader with a background in basic communications

(AM and FM theory), basic digital theory, and mathematics through trigonometry can easily become acquainted with this area. Numerous examples are provided for a better understanding of the theory.

Since the existing telephone lines are still the major media of all communication systems, Chapter 1 explains how direct calls are made (Direct Distance Dialing) and describes some of the associated hardware and terminology. It also describes the private-line services that are available to the subscriber. Chapter 2 is a general chapter which provides an overview of the various aspects involved in a communications system: whether it is two-point or multipoint; if multipoint, what configuration; whether it is synchronous or asynchronous; line protocol; and error detection. Many of these aspects are covered in detail in subsequent chapters. Chapter 3 deals with the telephone lines themselves and describes the specifications they must meet in order to conduct information. It also describes the various forms of distortion that may be produced in a transmitted message. If messages are to be transmitted in digital form, Chapter 4 describes the various codes that are used and error detection and correction schemes. Chapter 5 explains the operation of a UART, a device whose main purpose is to convert parallel digital information into serial format, and vice versa. Information coming from a data terminal is in parallel form. It must be converted to a serial format in order to be sent on a single transmission line. If the data terminal is receiving information, the reverse must be accomplished. Chapter 6 deals with the RS 232C interface signals between the UART and the modem. RS 449, RS 422, and RS 423, the anticipated replacements for the RS 232C, are also described. Chapter 7 deals with the various modulation schemes used to transform digital signals to analog form suitable for transmission on a voice line. These digital signals which have been converted to analog form can now be frequency-division multiplexed with other voice channels and the resultant signal transmitted on an analog line. This is the subject of Chapter 8. Since there are lines suitable for the transmission of digital information, Chapter 9 explains pulse code modulation and delta modulation—schemes for converting analog signals to digital pulses. It also explains how these channels are time-division multiplexed together prior to being transmitted on a digital carrier. Chapters 10 through 12 deal with line protocol. Chapter 10 describes asynchronous protocol, while Chapters 11 and 12 describe the synchronous protocols: Bisynch, SDLC, and HDLC. Bisynch is a character-oriented protocol, while SDLC and HDLC are bit-oriented protocols. Chapter 13 describes how different types of computers and terminals are connected to form local area networks, and packet switching and the associated standards governing their use. Chapter 14 covers the basic concepts of a fiber optic communications system. A detailed explanation is given for light-wave propagation through a guided fiber. Also, several light sources and detectors are discussed, contrasting their advantages and disadvantages.

*Wayne Tomasi*
*Vincent F. Alisouskas*

# TELECOMMUNICATIONS:

## VOICE / DATA
## WITH FIBER OPTIC
## APPLICATIONS

# ONE

# *THE TELEPHONE NETWORK*

The American Telephone and Telegraph Company (AT&T) described the Bell System as "the world's most complicated machine." A telephone call made from any telephone in the United States to any other telephone must use this machine. The characteristic of the Bell System that made it unique from other giant corporations is that every piece of equipment, technique, or procedure, new or old, must have been capable of working with the rest of the system. This is equivalent to interfacing every IBM computer with every other IBM computer or manufacturing parts for 1983 Buicks that would also work in 1952 Chevrolets. The Bell System, prior to January 1, 1983, serviced 83% of the telephones in the United States. It was the world's largest corporation, employing over 1,000,000 people. Its assets were three times those of General Motors; its revenues 20 times those of Bank of America. What was previously the Bell System, in conjunction with nearly 2000 independent telephone companies, make up the Public Telephone Network (PTN). The PTN was designed for voice communications. When the high-volume data communications boom hit in the early 1970s, the use of the PTN was an attractive alternative to constructing separate facilities (at tremendous cost) for data circuits. Gradually, communications systems are being developed that use fiber optic links and satellite transponders which are designed for data transmission. It will be many years before these systems provide any real relief to the already overloaded PTN.

The following explanation is limited to telecommunications using the PTN. *Telco* includes all of the telephone companies that make up the PTN. Telco offers two general categories of service: *direct distance dialing* (DDD) and *private line*. DDD originally included only those switches and facilities required to complete a long-distance telephone call without the assistance of a Telco operator. The DDD now includes the entire public switched network; that is, any service associated with a telephone number. Private-line services are dedicated to a single user.

## DDD NETWORK

The DDD network, commonly referred to as the *dial-up* or *switched network*, imposes several limitations that must be overcome before it can be used for data communications. A basic understanding of the electrical operation of the telephone network would be helpful. The DDD network can be divided into four main sections: instruments, dial switches, local loops, and trunk circuits. An *instrument* is the device used to originate and receive signals, such as a telephone set (telset). The instrument is often referred to as *station equipment* and the location of the instrument as the *station*. A *dial switch* is a programmed matrix that provides a temporary signal path. A *local loop* is the dedicated transmission path between an instrument and the nearest dial switch. A *trunk circuit* is a transmission path between two dial switches. The dial switches are located in Telco central offices and are categorized as local, tandem, or toll. A *local* dial switch serves a limited area. The size of the area is determined by how many telephone numbers are required or desired in a given geographical area. Telco designates these areas as branch area exchanges. A branch exchange is a dial switch. The *subscriber* is the operator or user of the instrument: if you have a home telephone, you are a subscriber. A subscriber is the customer of Telco: the person placing the call. Telco refers to this person as either the talker or the listener, depending on their role at a particular time during a conversation.

A telephone number consists of seven digits. The first three digits make up the prefix while the last four constitute the extension number. Each prefix can accommodate 10,000 telephone numbers (0000 to 9999). The capacity of a dial switch is determined by how many prefixes it serves. The local dial switch provides a two-wire cable (local loop) for each telephone number it serves (Figure 1–1). One wire is designated the *tip* and the other the *ring*. The station end of the loop is terminated in a telephone set. The dial switch applies -48 V dc on the tip and a ground on the ring of each loop. This dc voltage is used for supervisory signaling and to provide a talk battery for the telset microphone. On-hook, off-hook, and dial pulsing are examples of supervisory signaling.

When a subscriber goes off-hook (lifts the handset off the teleset cradle), a switch hook is released, completing a dc short between the tip and the ring of the loop through the telset microphone. The dial switch senses a dc current in the loop and recognizes this as an off-hook condition. This procedure is referred to as a *loop start operation*: the loop is completed to indicate an off-hook condition. The dial switch responds with an audible dial tone. On hearing the dial tone, the subscriber dials the destination telephone number. The originating and destination telephone numbers are referred to as the *calling* and the *called numbers*, respectively.

Dialing is accomplished with switch closures (dial pulses) or touch-tone signal-

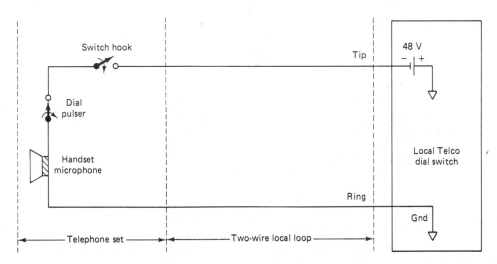

**FIGURE 1–1** Simplified two-wire loop showing telset hookup to a local dial switch (loop start operation).

ing. Dial pulsing is the interruption of the dc loop current by a telset dialing mechanism. Eight tone oscillators are contained in each telset equipped with a touch-tone pad. In touch-tone signaling, depending on the digit depressed, the telset outputs two of the eight tone frequencies.

After the entire called number has been dialed, the dial switch searches for a signal path through the switching matrix to the loop associated with the number called. Once the dial switch identifies a signal path and locates the destination loop, it tests the loop for an off-hook (busy) condition. If the destination loop is busy, the dial switch signals the calling loop with a busy signal. A station busy signal is a 60-ppm buzz. If the destination loop is on-hook (idle), the dial switch applies a 20-Hz 110-V ac ringing signal to it. A typical ringing cycle is 2 seconds on, 4 seconds off. When the destination telephone is answered (goes off-hook), the dial switch terminates the ringing signal and completes the transmission path between the two loops through the matrix. The signal path through the dial switch will be maintained as long as both loops remain closed. When either instrument goes on-hook, the signal path is interrupted.

What if the calling and the called telephone numbers are not served by the same dial switch? Generally, a community is served by only one local telephone company. The community is divided into zones; each zone is served by a different dial switch. The number of zones established in a given community is determined by the number of stations served and their density. If a subscriber in one zone wishes to call a station in another zone, a minimum of two dial switches are required. The calling station receives off-hook supervision and dial pulses are outputted as previously described. The dial switch in the calling zone recognizes that the prefix of the destination telephone number is served by a different dial switch. There are two ways that the serving dial switch can complete the call. It can locate a direct trunk (interoffice) circuit to the dial switch in the destination zone or it can route the call through a tandem switch. A *tandem switch* is a switcher's switch. It is a switching matrix used to interconnect dial switches. Trunk circuits that terminate in tandem switches are called *tandem trunks*. Normally, direct trunk circuits are provided only between adjacent zones. If a call must pass through more than one zone, a tandem switch must be used (Figure 1–2). If no direct trunks between the originating and the terminating dial switches exist and a common tandem switch is not available, the call is classified as a *toll call* and cannot be completed as dialed. Toll calls involve an additional charge and the dialed number must be preceded by a "1."

The telephone number prefix identifies which particular dial switch serves a station. With three digits, 1000 (000 to 999) prefixes can be generated for dial switches. A single metropolitan exchange alone may serve 20 to 30 prefixes. In the United States, there are over 20,000 local dial switches. It is obvious that

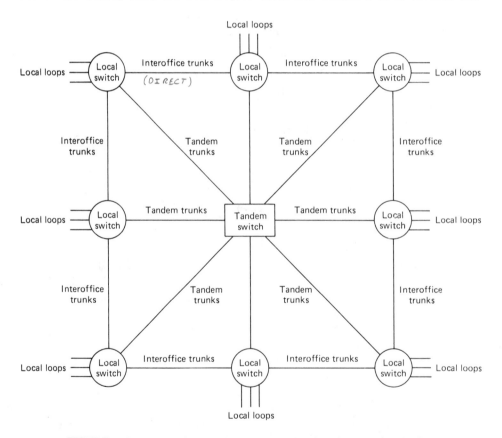

**FIGURE 1–2  Community telephone system showing the use of a tandem switch to facilitate interzone calling.**

further encoding is required to differentiate between the same prefix and extension in two different parts of the country. In the United States, an additional three-digit area code is assigned to each telephone number. The area code precedes the prefix but needs to be included only when calls are destined outside the area of the originating station. When a dialed telephone number is preceded by a "1," it is routed from the local dial switch to a toll switch by way of a *toll-connecting trunk*. Telco's present toll-switching plan includes five ranks or classes of switching centers. From highest to lowest classification, they are the regional center, sectional center, primary center, toll center, and end office. Local dial switches are classified as end (central) offices. All toll switches are capable of functioning as tandem switches to other toll switches.

The Telco switching plan includes a switching hierarchy that allows a certain degree of route selection when establishing a long-distance call (Figure 1–3). The choice is not offered the subscriber, but rather, the toll switches, using software translation, select the best route available at the time the call is made. The best route is not necessarily the shortest route. It is the route requiring the fewest number of dial switches. If a call cannot be completed because the necessary trunk circuits are not available, the local dial switch signals the calling station with an equipment busy signal. An equipment busy signal is similar to a station busy signal except

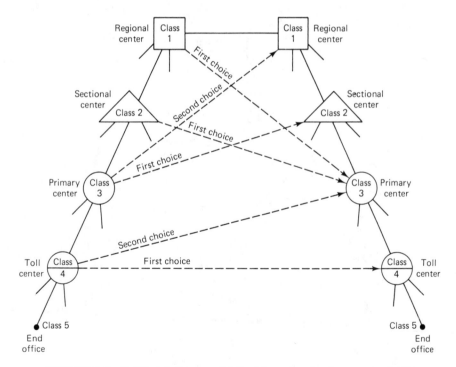

**FIGURE 1–3  Public telephone switching hierarchy showing some possible route choices to complete a toll call between two end offices.**

that it repeats at a 120-ppm rate. The worst-case condition encountered when completing a long-distance call is when seven tandem toll (intertoll) trunks are required. Based on Telco statistics, the probability of this occurring is 1 in 100,000. Because software translations in the automatic switching machines permit the use of alternate routes, and each route includes many different trunk circuits, the probability of using the same facilities on identical calls is unlikely. This is an obvious disadvantage when using the PTN for data transmission because inconsistencies in the transmission parameters are introduced from call to call. Telco guarantees the transmission parameters of each local loop and trunk circuit to exceed the minimum requirements of a basic voice-grade (VG) communications channel. However, two to nine separate facilities in tandem may be required to complete a telephone call (see Figure 1–4). Since transmission impairments are additive, it is quite possible that the overall transmission parameters of a telephone connection, established through the public switched network, may be substandard. Because transmission paths vary from call to call, it is difficult to compensate for line impairments. Subscribers to the DDD network lease a dedicated loop from their station to the nearest Telco dial switch. Any additional facilities required for the subscribers to complete a call are theirs only temporarily. The subscriber uses these facilities only for the duration of the call and then they are made available for other users of the network. These temporary

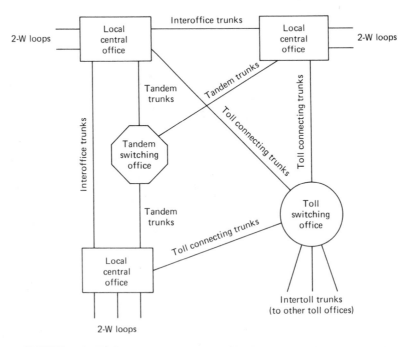

**FIGURE 1–4   Typical switching layout showing the relationship between local, tandem, and toll switches in the public telephone network.**

facilities are called *common usage trunks*—they are shared by all of the subscribers of the network.

The switching transients associated with the dial switches are another disadvantage of using the public switched network for data transmission. The older dial switches were electromechanical machines. Mechanical relay contacts were used to establish a signal path. The contact closures in the switching machines induced static interference that bled into adjacent signal paths. The static electricity caused impulse noise which produced transmission errors in the data signals. Telco is rapidly converting to *Electronic Switching Systems* (ESS). ESS machines are by no means perfectly quiet, but they are a tremendous improvement over the older electromechanical machines.

In order for a toll call to be completed, the dialed phone number must be transferred from switch to switch. Ultimately, the switching matrix at the destination end office requires the prefix and extension numbers to establish the final connection. The transmission paths between switching machines are very often carrier systems: microwave links, coaxial cables, or digital T-carriers. Microwave links and coaxial cables use analog carriers and are ac coupled. Therefore, the traditional dc supervisory and dial pulsing techniques cannot be used. An alternate method of transferring supervisory signals using a *single-frequency (SF) tone* has been devised. An idle trunk circuit has a 2600-Hz SF tone present in both directions. An off-hook indication at either end is indicated by the removal of the SF tone. The receiving switch acknowledges the off-hook indication by removing the SF in the opposite direction. Two methods are available for dial pulsing. The SF tone can be pulsed on and off to represent the dialed number or a signaling method called *multifrequency (MF) signaling* can be used. MF is a two-of-six code similar to touch tone. However, the MF tone frequencies are higher and are transmitted at a faster rate. Touch tone and MF are not compatible. Digital T-carriers use a completely different method for transferring supervisory information. This method is explained in Chapter 9.

**Hybrids and echo suppressors.** Local loops (dial-up lines) consist of a two-wire pair (signal and ground). However, to achieve satisfactory long-distance performance, the telephone company found it necessary to provide four-wire circuits (see Figure 1–5). When two-wire loops are switched onto four-wire facilities as in a long-distance telephone call, an impedance-matching device called a *hybrid* or *terminating set* (Figure 1–6) is used to affect the interface.

The hybrid coil transfers the signal from the two- to the four-wire line. The

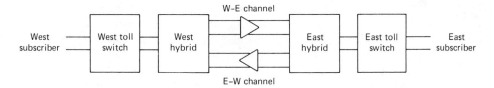

**FIGURE 1–5 Simplified diagram of a long-distance telephone connection.**

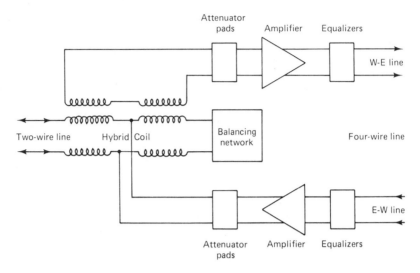

FIGURE 1–6   **Two-wire to four-wire terminating set (hybrid).**

balancing network compensates for impedance variations in the two-wire portion of the circuit. The amplifiers and attenuators adjust the signal voltages to required levels. Signals traveling W-E enter the terminating set from the two-wire loop and are inductively coupled into the W-E portion of the four-wire trunk. Signals received from the E-W portion of the four-wire trunk are applied to the center taps of the hybrid coils. If the impedance of the two-wire loop and the balancing network are properly matched, all currents produced by the E-W signal in the upper half of the hybrid coil will be equal and in opposite directions. Therefore, any voltages induced in the secondaries will cancel. This prevents any received signal from being returned to the sender.

If the impedance of the two-wire loop and the balancing network are not matched, the voltages induced in the secondaries of the hybrid coil do not cancel completely. This imbalance causes a portion of the received signal to be returned to the sender on the W-E portion of the four-wire trunk. The returned portion of the signal is heard as an echo by the talker and, if the round-trip delay of this signal exceeds 45 ms. the echo can become quite annoying. Delays of this magnitude occur when the distance between the two telephones exceeds 1500 miles. *Echo suppressors* (Figure 1–7) are inserted at one end of a four-wire trunk circuit to eliminate this echo. The speech detector in the echo suppressor senses the presence and the direction of the signal. It then enables the appropriate amplifier and disables the amplifier for the opposite direction. With an echo suppressor in the circuit, the two parties cannot talk simultaneously. If the conversation is changing rapidly, the people talking may hear the echo suppressor turning on and off.

The telephone lines are not only used for voice but also for the transmission of digital information. When used in this manner (Chapter 9), simultaneous transmis-

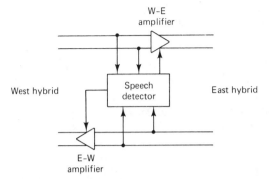

**FIGURE 1–7  Echo suppressor.**

sion of digital information in both directions is often desired. To accommodate this form of transmission, the telephone company has equipped the echo suppressors with a disabling mechanism. A single-frequency tone in the range 2010 to 2240 Hz, if applied for a period of 400 ms or longer, will disable the echo suppressor—both amplifiers will be enabled. The echo suppressor will remain disabled as long as signals in the 300- to 3000-Hz band are transmitted in either direction. If this signal is interrupted for 50 ms or more, the echo suppressor will automatically become active again.

## PRIVATE-LINE SERVICE

In addition to subscriptions to the public switched telephone network, Telco offers a comprehensive assortment of private-line services. Private-line subscribers lease those facilities required for a complete circuit. These facilities are hard-wired together in the Telco offices and are available only to one subscriber. Private-line circuits are *dedicated*, *private*, *leased* facilities. A dedicated circuit can be designed to meet voice-grade requirements from station to station—the end-to-end transmission parameters are fixed at the time the circuit is installed and will remain relatively constant. Circuit impairments will also remain relatively constant and can be compensated for by the subscriber. Private-line circuits afford several advantages over conventional dial-up circuits:

1. Availability
2. Improved performance
3. Greater reliability
4. Lower cost

Since private-line circuits are leased on a 24-hour basis, they are always available to the subscriber. Because the transmission parameters on a private-line circuit are guaranteed end to end, the overall performance is improved and a more reliable communications link is established. Heavy-usage private-line circuits are more eco-

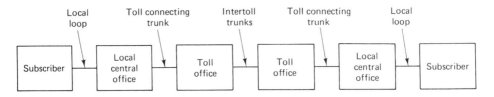

**FIGURE 1–8** Simplified private-line circuit layout showing dedicated loops and trunks with hard-wired cross-connects in each Telco office (two-point circuit).

nomical than dial-ups. However, dial-ups are more cost-effective for a person using the lines only a small percentage of the time.

Private-line circuit arrangements differ from dial-ups only in the fact that their circuits are permanently connected and dial switches and common-usage trunks are unnecessary. The terms "local loop" and "trunk" have a slightly different meaning to private-line subscribers. On a private-line circuit, a local loop is a transmission path between an instrument and the nearest Telco office; a trunk circuit is a transmission path between two Telco offices (see Figure 1–8). The only change in the definition is the substitution of "Telco office" for "dial switch."

Examples of private-line offerings:

1. Foreign exchange (FX)
2. Full data (FD)
3. Full period (FP)
4. Digital data service (DDS)

FX circuits differ from conventional DDD subscriptions only in the fact that subscribers, instead of leasing a dedicated loop to the nearest dial switch, lease a loop to a dial switch of their choice. This facilitates toll-free interzone calling to specific zones. FD circuits are four-wire, dedicated data circuits capable of full-duplex operation at a data rate of 9600 bps (bits per second). A local bank's system of automatic teller machines is an example of an FD circuit. FP circuits are four-wire, dedicated voice circuits. The hoot-and-holler (yell-down) circuits used by auto dismantlers (previously called "junkyards") to locate used auto parts is an example of an FP circuit. DDS circuits provide two-point, full-duplex operation at synchronous data rates of 2.4, 4.8, 9.6, or 56 kbps. DDS is intended to provide a communications medium for the transfer of digital data from station to station. Conventional digital-to-analog data modems are replaced by digital-to-digital communications service units (CSUs). DDS circuits are guaranteed to average 99.5% error-free seconds at 56 kbps, and even better performance is achieved at the lower bit rates.

# QUESTIONS

1. Identify the four main sections that make up the DDD network.
2. What determines the capacity of a dial switch?
3. List three examples of supervisory signals.
4. Explain loop start operation.
5. Identify two methods by which dialing may be accomplished with a telephone set.
6. What is a tandem switch called?
7. How many unique telephone numbers can a three-digit prefix accommodate?
8. Identify when the three-digit area code must be included when dialing a number.
9. How does a local dial switch identify a toll call?
10. List the five classes of Telco's present switching plan.
11. The best route between an originating and a destination station is always the shortest route. (T, F)
12. Explain the difference between an equipment busy signal and a station busy signal.
13. What is the predominant disadvantage of using the DDD network for data transmission?
14. Identify two methods of transferring supervisory signals over analog carrier systems.
15. Local loops are two-wire facilities.   (T, F)
16. Identify the Telco device that performs two-wire to four-wire conversion.
17. What is the purpose of an echo suppressor?
18. When are echo suppressors required?
19. Private-line circuits are dedicated to a single user.   (T, F)
20. Identify four advantages of a private-line circuit compared to a dial-up circuit.

# TWO

## *COMMUNICATIONS SYSTEM OVERVIEW*

Communications is the transmission of information from one place to another. The transmission path may be short, as when two people are talking face to face with each other or when a computer is outputting information to a printer located in the same room. *Telecommunications* is long-distance communications. The original source information is in either analog (voice) or digital form. Voice frequencies are either transmitted directly on a voice band communications channel or used to modulate a carrier frequency and then the modulated waveform is transmitted. Conversational voice frequencies are normally in the range 300 to 3000 Hz. The voice channel over which these frequencies are transmitted has an allocated bandwidth of 4 kHz (0 to 4 kHz). Voice source information can also be transmitted in digital format. *Codecs* (coders/decoders) are integrated circuits which are capable of converting voice frequencies to a series of digital pulses or taking the digital pulses and converting them back to voice frequencies. Codecs or similar circuits perform the necessary conversions at the transmit and receive ends and the digital pulses themselves are transmitted on special lines called digital T-carriers. Digital source information can take on one of two forms. It can be raw digital data such as the op codes, addresses, and data of a computer program, or it can be a computer-contained message which is stored and/or transmitted in a digital code such as ASCII or EBCDIC. Digital information is characterized by having all information represented by a sequence of two voltage levels which represent logic 1's and 0's. In general, *digital communications* is the transmission of information in digital form (1's and 0's). The source information may be either digital or analog. *Data communications* is the transmission of information that was originally digital in nature. This information can be transmitted as either digital or analog signals. Figure 2–1 illustrates the

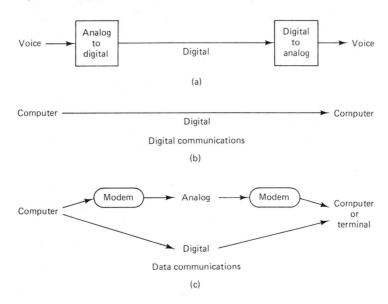

**FIGURE 2–1 Digital and data communications.**

distinction between these two forms of communications. These definitions are very broad and, as will be seen in Chapters 7 and 9, the distinction is not that exclusive.

Both the computer system and the data communications system can be very simple or very complex. The data communications system may simply be a link between a computer and a remote terminal, or it may be a link between a central computer and many terminals. Banks are good examples of such a system where all branch locations in a city tie into one main computer. These communications systems are broadly categorized as *two-point* or *multipoint* systems. A two-point system can be between a computer and a printer and/or a CRT. In such a two-point system, the flow of information would always be unidirectional. A two-point system could also be between two computers that have keyboards, printers, and displays as peripheral equipment. In this system, bidirectional transfer of information is possible. The computer which initiates information transfer is called the *master*, while the other computer is called the *slave*. Obviously, these roles are interchangeable depending on which computer initiates the call. In such a system, a problem of *contention* arises when both computers attempt to initiate a call simultaneously. Should this be possible, built-in delays resolve the problem and give one priority over the other.

Multipoint links take on many variations; the network selected is determined by system requirements. A few of these variations are depicted in Figure 2–2. The *star network* has the advantage of ready access by remote sites to the central computer. The associated disadvantage is the tariff paid for the separate lines for each terminal. If the line traffic between the remote terminals and the central computer is small, there may not be sufficient justification for such a system. But then, the ready

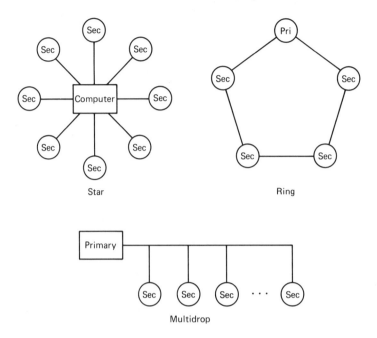

Star

Ring

Multidrop

**FIGURE 2–2   Multipoint system configurations.**

access may be an overriding factor. In a *ring network*, the information is passed around in a loop. Each terminal extracts and inserts its own information. In this type of configuration, if one element in the loop goes out of commission, the entire system is down. Backup configurations are possible where information can be circulated either clockwise or counterclockwise in the loop. Probably the most used configuration is the *multidrop*. Here, all of the terminals are connected to the same main line. In this system, the main computer is identified as the *primary* (host, CPU) and the connecting terminals are called *secondaries* (remotes, tributaries). The primary is so called because it controls all movement of information. A secondary cannot transmit or receive information unless it is allowed to do so by the primary. The primary can communicate with all of the secondaries, but each secondary can communicate only with the primary. If one secondary wishes to send a message to another secondary, it must send the information to the primary, which would then relay the information to its intended destination. In an expanded system, one station can be a secondary in one data link and serve as a primary in a different data link. In Figure 2–3, station D is a secondary in a link for which A is the primary. Station D is also the primary for the secondary stations E, F, and G. In this text, a multipoint system will be used synonymously with a multidrop system.

The types of communications possible between any two points are classified as:

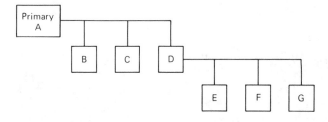

FIGURE 2–3   Station D acting as a primary and a secondary.

1. *Simplex*. Information can be sent only in one direction. If errors are present in a transmission, the receive end has no way of asking for a retransmission. UPI and AP have such a system where the central news-gathering facility transmits the news to many regional offices. These offices are equipped with read-only printers (ROPs). The printers can display received information but cannot transmit. Simplex systems require two-wire lines (signal and reference).

2. *Half-duplex* (HDX). Information can be transmitted in either direction but not simultaneously. Communicating with a walkie-talkie is an example of such a system. These transceivers have a push-to-send or a send/receive button. When in the send mode, the receiver is disabled; when in the receive mode, the transmitter is disabled. HDX systems also require only two-wire lines.

3. *Full duplex* (FDX) *or duplex*. Information can be transmitted in both directions between the same two points simultaneously. Such systems are generally four-wire systems—a signal and a reference line for information transmission in each direction. FDX operation can be achieved with two-wire lines through the use of frequency-division multiplexing. Two-wire FDX operation is explained in Chapter 7.

4. *Full/full duplex* (F/FDX). In a F/FDX system, the primary has the capability of transmitting to one secondary while receiving from a different secondary simultaneously. This classification is restricted in use to a multipoint system.

These four terms—simplex, half-duplex, full duplex, and full/full duplex— are used not only to describe a system but also to describe the mode of operation. In the multipoint system of Figure 2–2, the primary and the secondaries all have four-wire connections. However, the primary is capable of operating in the F/FDX mode while the secondaries operate only in the HDX mode.

## HISTORY OF DATA COMMUNICATIONS

It is highly likely that data communications began long before recorded time in the form of smoke signals or tom-tom drums, although it is improbable that these signals were binary coded. If we limit the scope of data communications to methods that use electrical signals to transmit binary-coded information, then data communications began in 1837 with the invention of the *telegraph* and the development of the *Morse code* by Samuel F. B. Morse. With telegraph, dots and dashes (analogous to binary 1's and 0's) are transmitted across a wire using electromechanical induction. Various combinations of these dots and dashes were used to represent binary codes for letters, numbers, and punctuation. Actually, the first telegraph was invented in England by Sir Charles Wheatstone and Sir William Cooke, but their contraption required six different wires for a single telegraph line. In 1840, Morse secured an American patent for the telegraph and in 1844 the first telegraph line was established between Baltimore and Washington, D.C. In 1849, the first slow-speed telegraph printer was invented, but it was not until 1860 that high-speed (15 bps) printers were available. In 1850, the Western Union Telegraph Company was formed in Rochester, New York, for the purpose of carrying coded messages from one person to another.

In 1874, Emile Baudot invented a telegraph *multiplexer*, which allowed signals from up to six different telegraph machines to be transmitted simultaneously over a single wire. The telephone was invented in 1876 by Alexander Graham Bell and, consequently, very little new evolved in telegraph until 1899, when Marconi succeeded in sending radio telegraph messages. Telegraph was the only means of sending information across large spans of water until 1920, when the first commercial radio stations were installed.

Bell Laboratories developed the first special-purpose computer in 1940 using electromechanical relays. The first general-purpose computer was an automatic sequence-controlled calculator developed jointly by Harvard University and International Business Machines Corporation (IBM). The UNIVAC computer, built in 1951 by Remington Rand Corporation (now Sperry Rand), was the first mass-produced electronic computer. Since 1951, the number of mainframe computers, small business computers, personal computers, and computer terminals has increased exponentially, creating a situation where more and more people have the need to exchange digital information with each other. Consequently, the need for data communications has also increased exponentially.

Until 1968, the AT&T operating tariff allowed only equipment furnished by AT&T to be connected to AT&T lines. In 1968, a landmark Supreme Court decision, the Carterfone decision, allowed non-Bell companies to interconnect to the vast AT&T communications network. This decision started the *interconnect industry*, which has led to competitive data communications offerings by a large number of independent companies.

## DATA COMMUNICATIONS OVERVIEW

A simplified block diagram of a data communications link is shown in Figure 2–4. Although only one secondary is shown, it represents a typical secondary in a multipoint system. One of the functions of the host computer is to store the applications programs

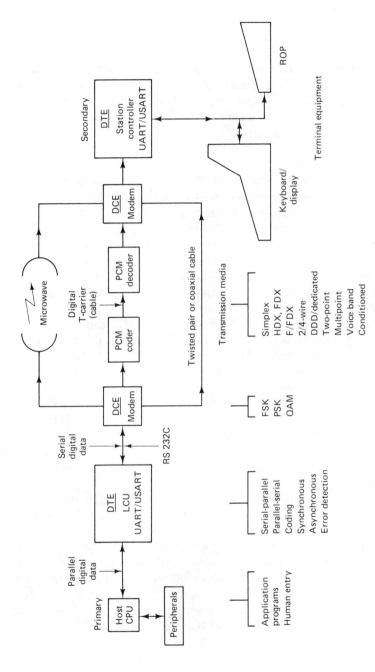

**FIGURE 2–4  Simplified block diagram of a data communications link.**

for the various secondaries. The secondaries may have similar or dissimilar functions. The same control code received from two different secondaries may warrant two different actions on the part of the host computer. If the system so requires, the host computer must also provide storage for the centralized data and the software for database management. Data information is normally stored in auxiliary memories to which the computer has ready access. The peripherals connected to the host computer allow for human entry and intervention. As needs change, application programs may require additions, deletions, or changes. Data link problems also arise which are beyond the range of the computer's programmed capabilities. In such cases, human entry is required to make the necessary changes to resolve the existing problems.

The end equipment which either generates the digital information for transmission or uses the received digital data can be computers, printers, keyboards, CRTs, and so on. This equipment generally manipulates digital information internally in word units—the number of bits that make up a word in a particular piece of equipment are transferred in parallel. Digital data, when transmitted, are in serial form. Parallel transmission of an 8-bit word would require eight pairs of transmission lines—not at all cost-effective. *Data terminal equipment* (DTE) is a general phrase encompassing all of the circuitry necessary to perform parallel-to-serial and serial-to-parallel conversions for transmission and reception respectively and for data link management. The *UART* (Universal Asynchronous Receiver/Transmitter) and *USART* (Universal Synchronous/Asynchronous Receiver/Transmitter) are the devices that perform the parallel-to-serial and serial-to-parallel conversions (Chapter 5). The primary DTE includes a *line control unit* (LCU or LinCo) which controls the flow of information in a multipoint data link system. A *station controller* (STACO) is the corresponding unit at the secondaries. If there is software associated with the LCU, it is then called a *front-end processor* (FEP). At one time, the DTE was the last piece of equipment that belonged to the subscriber in a data link system. Between the DTEs, starting with the modems, was communications equipment owned and maintained by Telco. Recent judgments have removed modems from the realm of exclusive Telco property.

*Data communications equipment* (DCE) accepts the serial data stream from the DTE and converts it to some form of analog signal suitable for transmission on voice-grade lines. At the receive end, the DCE performs the reverse function of converting the received analog signal to a serial digital data stream. The simplest form of DCE is a *modem* (modulator/demodulator) or *data set*. At the transmit end, the modem can be considered a form of digital-to-analog converter, while at the receive end, it can be considered a form of analog-to-digital converter. The most common forms of modulation by modems are *frequency shift keying* (FSK), *phase shift keying* (PSK), and *quadrature amplitude modulation* (QAM) (Chapter 7).

If the cables and signal levels used to interconnect the DTE and DCE were left unregulated, the variations generated would probably be proportional to the number of manufacturers. Electronics Industries Association (EIA), an organization of manufacturers concerned with establishing industry standards, have agreed on the RS 232C as the standard interface between the DTE and the modem. This is a 25-pin cable whose pins have designated functions and specified signal levels (Chapter 6). The RS 232C is anticipated to be replaced by an updated standard.

All information that leaves the modem is in analog form suitable for transmission over a voice-grade line. The alternatives for the actual transmission of these signals, at this point, are identical to voice which occupies the same frequency range:

1. This information can be transmitted directly on twisted pairs or coaxial cables. The twisted pairs may be dial-up or dedicated lines.

2. This information could further amplitude-modulate other carrier frequencies to produce single sidebands. These single sidebands would be frequency-division multiplexed with other sidebands similarly derived and then transmitted on a microwave link (Chapter 8).

3. This information could be converted to a different form of digital signal through a process of pulse-code modulation (PCM) and then transmitted on T-carriers (Chapter 9).

Whichever the method, the reverse process must be accomplished at the receiver to reconstruct the original data stream.

For successful transfer of digital information over an analog link, three types of synchronization must be accomplished:

1. *Carrier synchronization* between the modems so that the received signal may be properly demodulated.

2. *Bit synchronization*, so that the receive DTE knows when to sample the incoming bit stream. Errors would be introduced if either a bit was sampled twice (sampling rate too fast) or not at all (sampling rate too slow).

3. *Character synchronization*. It is one thing to identify the value of each bit and another to associate it properly with the word to which it belongs. There must be some means of identifying where each word starts and ends.

   The transmitted data and the modems are each classified as either synchronous or asynchronous.

   (a) *Synchronous data* implies the inclusion of unique characters in the transmission for the purpose of establishing character synchronization. These characters are transmitted as the first part of any message.

   (b) *Asynchronous data* implies that character synchronization is achieved by framing each transmitted character with a start bit and stop bit(s).

   (c) *If a modem is synchronous*, it extracts clocking information from the received message and its own internal clock is synchronized to this signal.

   (d) *If the modems are operating asynchronously*, the internal clocks of the transmit and the receive modems are operating independently of one another.

Generally, synchronous data use synchronous modems and asynchronous data use asynchronous modems. Synchronous modems are used where higher bit transmission rates are required. It is unlikely that synchronous data would ever be transmitted using asynchronous modems. However, asynchronous data are sometimes transmitted using synchronous modems. Such transmissions are termed *isochronous*.

Thus far, we have been concerned with the hardware involved in communications and the types of electrical signals used. In data communications, several other aspects must be examined. The first is line protocol. *Line protocol* is a set of rules governing the transmission of digital information. In a data communications system, some of the aspects which these rules must clarify are:

1. How modem synchronization is to be achieved.
2. How the primary will select a particular secondary.
3. How the primary will tell the secondary that it is about to receive information.
4. What the secondary should do if it is not ready to receive this information.
5. What should be done if there is an error in transmission.
6. How the primary gives a particular secondary permission to transmit.
7. What form of error detection should be used. In addition, *message protocol* must be established. If the transmitted information is strictly text, message protocol is unnecessary. However, if the primary must supply message forms for completion by the secondary, agreement must be reached on:
   (a) What signals will be used to control cursor movement on the CRT.
   (b) How the secondary operator can be prevented from overwriting the transmitted form.
   (c) If the same form is to be used for many items, how only operator-entered information can be transmitted back to the primary and the basic form retained by the secondary for reuse (Chapters 10, 11, and 12).

## Information Capacity

The *information capacity* of a communications system represents the number of independent symbols that can be carried through the system in a given unit of time. The most basic symbol is the *binary digit* (bit). Therefore, it is often convenient to express the information capacity of a system in *bits per second* (bps). In 1928, R. Hartley of Bell Telephone Laboratories developed a useful relationship among bandwidth, transmission time, and information capacity. Simply stated, *Hartley's law* is

$$C \approx B \times T$$

where $C$ = information capacity
       $B$ = bandwidth
       $T$ = transmission time

From the above equation it can be seen that the information capacity is a linear function and is directly proportional to both the system bandwidth and the transmission time. If either the bandwidth or the transmission time is changed, the information capacity will change proportionally.

In 1948, C. E. Shannon (also of Bell Telephone Laboratories) published a paper in the *Bell System Technical Journal* relating the information capacity of a communications channel to bandwidth, transmission time, and signal-to-noise ratio. Mathematically stated, the *Shannon limit for information capacity* is

$$C = B \log_2 \left( 1 + \frac{S}{N} \right)$$

where  $C$ = information capacity (bps)
   $B$ = bandwidth
   $\dfrac{S}{N}$ = signal power-to-noise ratio

For a standard voice band communications channel with a signal-to-noise ratio of 1000 (30 dB) and a bandwidth of 2.7 kHz, the Shannon limit for information capacity is

$$C = 2700 \log_2 (1 + 1000)$$

$$= 26.9 \text{ kbps}$$

Shannon's formula is often misunderstood. The results of the preceding example indicate that 26.9 kbps can be transferred through a 2.7-kHz channel. This may be true, but it cannot be done with a binary system. To achieve an information transmission rate of 26.9 kbps through a 2.7-kHz channel, each symbol transmitted must contain more than one bit of information. Therefore, to achieve the Shannon limit for information capacity, digital transmission systems that have more than two output conditions must be used. Several such systems are described in the following chapters. These systems include both analog and digital modulation techniques and the transmission of both digital and analog signals.

# QUESTIONS

1. What are the lines called on which digital data are transmitted directly?
2. What is the difference between a codec and a modem?
3. ASCII is a form of raw digital data.  (T, F)
4. Modems are synonymous with data sets.  (T, F)
5. In digital communications, the source information can be either digital or analog.  (T, F)
6. In data communications, the source information can be either digital or analog.  (T, F)
7. Contention can be a problem in a multipoint system.  (T, F)

8. In a two-point communications system, one (the same) end is always identified as the master. (T, F)

9. Secondaries cannot communicate directly with each other in a multipoint system. (T, F)

10. Applications programs may be stored at either the primary or the secondary. (T, F)

11. A tributary is an alternate name for the primary. (T, F)

12. Explain the difference between simplex, HDX, FDX, and F/FDX.

13. What is an ROP?

14. What is the name of the interface between the DTE and the DCE?

15. Bits are transferred in parallel between the DTE and the DCE. (T, F)

16. The UART would be found in the DTE. (T, F)

17. The modem would be found in the DTE. (T, F)

18. What is meant by a "synchronous modem"?

19. What is meant by "asynchronous data"?

20. Digital source information could be converted to analog information and then back to digital information before transmission. (T, F)

21. Voice source information is always in analog form from source to destination. (T, F)

22. What are the three types of synchronization required in data communications?

23. Define line protocol.

24. Line protocol does not concern itself at all with the message contents. (T, F)

25. Identify three types of modulation performed by modems.

26. Define information capacity.

27. Define Hartley's law.

28. Describe the relationships among information capacity, bandwidth, transmission time, and signal-to-noise ratio.

# THREE

## *THE TELEPHONE CIRCUIT*

A *telephone circuit* consists of two or more facilities, interconnected in tandem, to provide a transmission path between a source and a destination. The interconnects may be temporary, as in a dial-up circuit, or permanent private-line circuits. The facilities may be cable pairs or carrier systems, and the information may be transferred on a coaxial, metallic, microwave, optic fiber, or satellite communications system. The information transferred is called the *message* and the circuit used is called the *message channel*. Telco offers a wide assortment of message channels ranging from a basic 4-kHz voice band circuit to wideband (30-MHz) microwave channels that are capable of transferring high-resolution video signals. The following discussion will be limited to a basic voice band circuit. In Telco terminology, the word *message* originally denoted speech information. This definition has been extended to include any standard voice frequency signal. Thus a message channel may include the transmission of speech, supervisory signaling, or voice band data.

## THE LOCAL LOOP

The local loop is the only Telco facility required by all voice band data circuits. It is the primary cause of attenuation distortion and phase distortion. A local loop is a metallic transmission line (cable pair), consisting of two insulated conductors twisted together. The insulating material may be wood pulp or polyethylene plastic, while the wire conductor is usually copper and, in some instances, aluminum. Wire pairs are stranded together into units. Adjacent wire pairs within a unit are twisted with different pitch (twist length). This reduces the undesired effects of inductive

23

coupling between pairs and helps to eliminate crosstalk. Units are cabled together into cores and then placed inside a plastic sheath. Depending on the insulating material used, sheaths contain between 6 and 900 pairs of wire. Sheaths are connected together and strung between distribution frames within Telco central offices and junction boxes located in manholes, back alleys, or telephone equipment rooms within large building complexes. The length of a subscriber loop depends on the station location relative to a local central telephone office.

The transmission characteristics of a cable pair depend on the wire diameter, conductor spacing, dielectric constant of the insulator, and the conductivity of the wire. These physical properties, in turn, determine the inductance, resistance, capacitance, and conductance of the line. The resistance and inductance are distributed along the length of the wire, while the conductance and capacitance exist between the two wires. If the insulation is good, the effect of conductance is generally negligible.

The electrical characteristics of a cable (Figure 3–1) are uniformly distributed along its length and are appropriately referred to as *distributed parameters*. Since it is cumbersome working with distributed parameters, it is common practice to lump them into discrete values per unit length (i.e., millihenrys per 1000 ft). The amount of attenuation and phase delay experienced by a signal propagating down a line is a function of the *frequency* of the signal and the *electrical characteristics* of the cable pair. Figure 3–2 illustrates the effect of frequency on attenuation for a given length of line. On this unloaded wire, a 3000-Hz signal suffers 6 dB more attenuation than a 500-Hz signal on the same line. The cable acts as a low-pass filter for the signal. Extensive studies of attenuation on cable pairs have shown that a reduction in attenuation can be achieved if inductors are added in series with the wire. This is called *loading*. Loaded cable is identified by the addition of the letters H, D, or B and an inductance value to the wire gauge number. H, D, and B indicate that the loading coils are separated by 6000, 4500, and 3000 ft, respectively. Generally, the amount of series inductance added is either 44, 88, or 135 mH. A cable pair with the designation 26H88 is made of 26-gauge wire with 88 mH of series inductance added every 6000 ft. If a loaded cable is used, a 3000-Hz signal will suffer only 1.5 dB more loss than a 500-Hz signal. Note that the loss versus frequency characteristics for a loaded cable are relatively flat up to approximately 2000 Hz.

The low-pass filter characteristics of a cable also affect the phase distortion versus frequency characteristics of a signal. The amount of *phase distortion* is proportional to the *length* and *gauge* of the wire. Loading a cable also affects the phase characteristics of a line. The telephone company must often add gain and delay

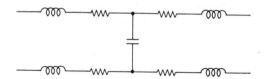

FIGURE 3–1  Electrical characteristics of a transmission line.

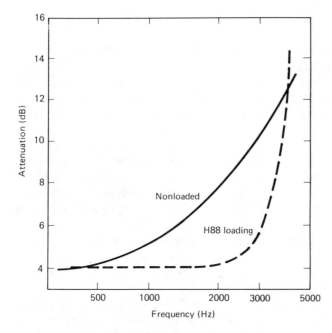

**FIGURE 3–2** **Frequency versus atten-**
**uation for 12,000 ft of 26-gauge wire.**

equalizers to a circuit in order to achieve the minimum requirements. Equalizers introduce discontinuities or ripples in the bandpass characteristics of a circuit. Automatic equalizers in modems are sensitive to this condition and, very often, an overequalized circuit causes as many problems to a data signal as an underequalized circuit.

## TRANSMISSION PARAMETERS

Transmission parameters are divided into three broad categories:

1. *Bandwidth parameters*, which include:
   (a) Attenuation distortion
   (b) Envelope delay distortion
2. *Interface parameters*, which include:
   (a) Terminal impedance
   (b) In-band and out-of-band signal power
   (c) Test signal power
   (d) Ground isolation
3. *Facility parameters*, which include:
   (a) Noise measurements
   (b) Frequency and phase distortion
   (c) Amplitude distortion
   (d) Nonlinear distortion

## Bandwidth Parameters

The only transmission parameters with limits specified by the Federal Communications Commission (FCC) are attenuation distortion and envelope delay. *Attenuation distortion* is the difference in circuit gain experienced at a particular frequency with respect to the circuit gain of a reference frequency. This characteristic is also called *frequency response, differential gain*, and *1004-Hz deviation. Envelope delay* is an indirect method of evaluating the phase delay characteristics of a circuit. FCC tariff number 260 specifies the limits for attenuation distortion and envelope delay distortion. The limits are prescribed by line conditioning requirements. Through line conditioning, the attenuation and delay characteristics of a circuit are artificially altered to meet prescribed limits. Line conditioning is available only to private-line subscribers for an additional monthly charge. The basic voice band, 3002 channel satisfies the minimum line conditioning requirements.

Telco offers two types of line conditioning, C type and D type. *C-type conditioning* specifies the maximum limits for attenuation distortion and envelope delay distortion. *D-type conditioning* sets the minimum requirement for signal-to-noise ratio (*S/N*) and deals with nonlinear distortion.

D-type conditioning is referred to as high-performance conditioning and has two categories, D1 and D2. Limits imposed by D1 and D2 conditioning are identical. D1 conditioning is available for two-point circuits, while D2 conditioning is available for multipoint arrangements. D-type conditioned circuits must meet the following specifications:

1. Signal-to-C-notched noise: $\geq$ 28 dB
2. Nonlinear distortion:
   (a) Signal to second order: $\geq$ 35 dB
   (b) Signal to third order:  $\geq$ 40 dB

D-type conditioning is mandatory when the data transmission rate is 9600 bps. If a facility is assigned by Telco for use as a 9600-bps circuit, and it does not meet the minimum requirements of D-type conditioning, it is so identified by Telco data technical support personnel and never considered for that purpose again. A different facility is sought. A circuit cannot be upgraded to meet D-type conditioning requirements by adding corrective devices.

C-type conditioning pertains to line impairments for which compensation can be made, to a certain degree, by filters and equalizers. This is accomplished with Telco-provided equipment. When a circuit is turned up for service with a particular C-type conditioning, it must meet the minimum requirements for that type of conditioning. The subscriber may include devices within the station equipment that compensate for minor long-term variations in the bandwidth requirements.

There are five classifications of C-type conditioning:

1. C1 and C2 pertain to two-point and multipoint circuits.
2. C3 conditioning is for access lines and trunk circuits associated with private switched networks.
3. C4 pertains to two-point circuits and multipoint arrangements that have a maximum of four station locations.
4. C5 specifications pertain only to two-point circuits.

Private switched networks are telephone systems dedicated to a single customer, usually with a large number of stations. An example is a large corporation with offices and complexes at different geographical locations with an on-premise private branch exchange (PBX) at each location. A PBX is a low-capacity dial switch where the subscribers are generally limited to stations within the same building complex. Common-usage access lines and trunk circuits are required to interconnect the PBXs. They are common only to the subscribers of the private network and not to the entire public telephone network.

Table 3–1 shows the various limits prescribed by the different types of

**TABLE 3–1   Bandwidth Parameter Limits**

| Channel conditioning | Attenuation distortion (frequency response) relative to 1004 Hz | | Envelope delay distortion | |
|---|---|---|---|---|
| | Frequency range (Hz) | Variation (dB) | Frequency range (Hz) | Variation (μs) |
| Basic | 500–2500 | +2 to −8 | 800–2600 | 1750 |
| | 300–3000 | +3 to −12 | | |
| C1 | 1000–2400 | +1 to −3 | 1000–2400 | 1000 |
| | 300–2700 | +2 to −6 | 800–2600 | 1750 |
| | 300–3000 | +3 to −12 | | |
| C2 | 500–2800 | +1 to −3 | 1000–2600 | 500 |
| | 300–3000 | +2 to −6 | 600–2600 | 1500 |
| | | | 500–2800 | 3000 |
| C3 (access line) | 500–2800 | +0.5 to −1.5 | 1000–2600 | 110 |
| | 300–3000 | +0.8 to −3 | 600–2600 | 300 |
| | | | 500–2800 | 650 |
| C3 (trunk) | 500–2800 | +0.5 to −1 | 1000–2600 | 80 |
| | 300–3000 | −0.8 to −2 | 600–2600 | 260 |
| | | | 500–2800 | 500 |
| C4 | 500–3000 | +2 to −3 | 1000–2600 | 300 |
| | 300–3200 | +2 to −6 | 800–2800 | 500 |
| | | | 600–3000 | 1500 |
| | | | 500–3000 | 3000 |
| C5 | 500–2800 | +0.5 to −1.5 | 1000–2600 | 100 |
| | 300–3000 | +1 to −3 | 600–2600 | 300 |
| | | | 500–2800 | 600 |

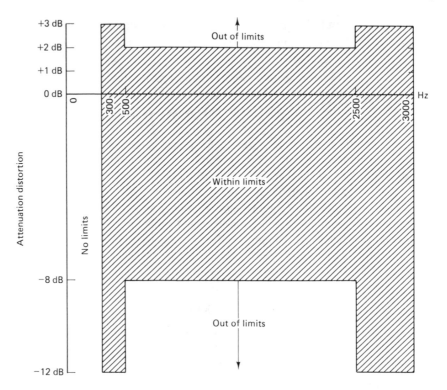

**FIGURE 3–3   Graphic presentation of the limits for attenuation distortion
in a basic 3002 channel.**

C-conditioning on attenuation distortion and envelope delay. Figures 3–3 through
3–7 show a graphic presentation of several of the bandwidth parameter limits.

**Attenuation distortion and C-type conditioning.**   The attenuation distortion
limits for a basic 3002 channel require the circuit gain at any frequency between
500 and 2500 Hz to be not greater than 2 dB above the circuit gain at 1004 Hz
and not more than 8 dB below the circuit gain at 1004 Hz (Figure 3–3). For attenuation
distortion, the circuit gain at 1004 Hz is always used as the reference. Also, within
the frequency bands from 300 to 499 Hz and from 2501 to 3000 Hz, the circuit
gain cannot be greater than 3dB above or more than 12 dB below the gain at
1004 Hz.

**EXAMPLE**

A 1004-Hz test tone is transmitted at 0 dBm and received at −16 dBm. The circuit
gain is −16 dB (this is actually a loss of 16 dB: negative dB values are negative
gains, or losses). Frequencies from 500 to 2500 Hz must be received at a minimum
level of −24 dBm and at a maximum level of −14 dBm. Frequencies within the

bands of 300 to 499 Hz and 2501 to 3000 Hz must be received at signal strength levels between −13 and −28 dBm inclusive. If the same transmission levels were applied to a circuit with C2 conditioning, frequencies between 500 and 2800 Hz must be received at signal levels ranging from −15 to −19 dBm. Frequencies within the bands of 300 to 499 Hz and 2801 to 3000 Hz must be received at signal levels ranging from −14 to −22 dBm. Table 3–1 shows that the higher the classification of conditioning imposed on a circuit, the flatter the frequency response, therefore, a better-quality circuit.

**Envelope delay.**    A linear phase versus frequency relationship is a requirement for error-free data transmission. This relationship is difficult to measure because of the difficulty in establishing a phase reference. Envelope delay is an alternate method of evaluating the phase versus frequency relationship of a circuit.

The time delay encountered by a signal as it propagates from source to destination is called *propagation time* or *phase delay*. All frequencies in the usable voice band (300 to 3000 Hz) do not experience the same time delay in a circuit. Therefore, a complex frequency spectrum, such as the output from a modem, does not possess the same phase versus frequency characteristics when it is received as when it was transmitted. This condition represents a possible impairment to a data signal. The *absolute phase delay* is the actual time required for a particular frequency to be propagated from source to destination. The difference between the absolute delays of different frequencies is phase distortion. A graph of phase delay versus frequency for a typical circuit is nonlinear.

By definition, envelope delay is the first derivative of phase with respect to frequency:

$$\text{envelope delay} = \frac{d\phi(\omega)}{d\omega}$$

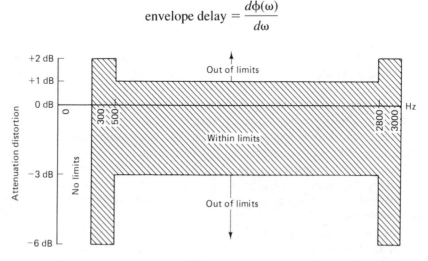

**FIGURE 3–4   Graphic presentation of the limits for attenuation distortion on a channel with C2 conditioning.**

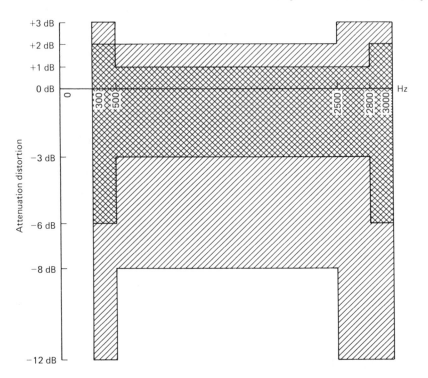

**FIGURE 3–5    Overlay of Figures 3–3 and 3–4 to demonstrate the more stringent requirements imposed by C2 conditioning.**

In actuality, envelope delay only closely approximates $d\phi(\omega)/d\omega$. Envelope delay measurements do not evaluate true phase versus frequency characteristics, but rather, the phase of a wave that is the resultant of a narrow band of frequencies. It is a common misconception to confuse true phase distortion (also called delay distortion) with envelope delay distortion (EDD). *Envelope delay* is the time required to propagate a change in an AM envelope through a transmission medium. To measure envelope delay, a narrowband amplitude-modulated carrier, whose frequency is varied over the usable voice band, is transmitted. (The AM modulation rate is typically between 25 and 100 Hz.) At the receiver, phase variations of the low-frequency envelopes are measured. The phase difference at the different carrier frequencies is *envelope delay distortion*. The carrier frequency that produces the minimum envelope delay is established as the reference and is normalized to zero. Therefore, EDD measurements yield only positive values and indicate the relative envelope delays of the various carrier frequencies with respect to the reference frequency. The reference frequency of a typical voice band circuit is approximately 1800 Hz.

EDD measurements do not yield true phase delays, nor do they determine the relative relationships between true phase delays. EDD measurements are used

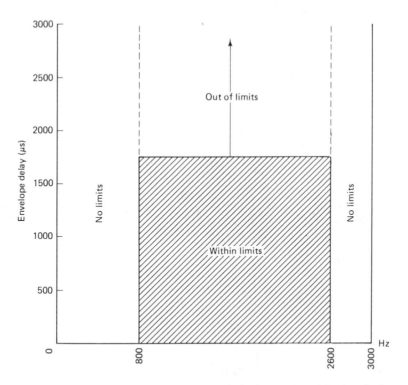

**FIGURE 3–6** Graphic presentation of the limits for envelope delay in a basic 3002 channel.

to determine a close approximation of the relative phase delay characteristics of a circuit.

The EDD limit of a basic 3002 channel is 1750 μs between 800 and 2600 Hz, as shown on Table 3–1. This indicates that the maximum difference in envelope delay between any two carrier frequencies in this range cannot exceed 1750 μs.

**EXAMPLE**

An EDD test on a basic 3002 channel indicated that an 1800-Hz carrier experienced the minimum absolute delay of 400 μs. Therefore, it is the reference. The absolute envelope delay experienced by any frequency within the 800- to 2600-Hz band cannot exceed 2150 (400 + 1750) μs.

The absolute time delay encountered by a signal between any two points in the continental United States will never exceed 100 ms. This is not sufficient to cause any problems. Consequently, relative rather than absolute values of envelope delay are measured. For the previous example, as long as EDD tests yielded relative values less than +1750 μs, the circuit is within limits.

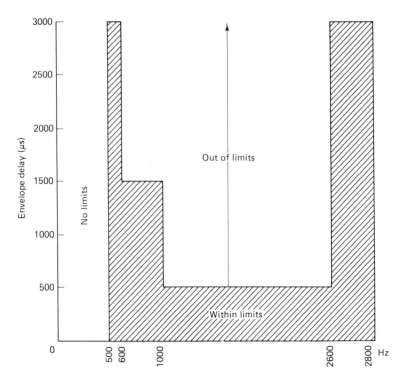

**FIGURE 3–7    Graphic presentation of the limits for envelope delay on a channel with C2 conditioning.**

## Interface Parameters

The two primary considerations of the interface parameters are:

1. Electrical protection of the telephone network and its personnel
2. Standardization of design arrangements

These considerations are summarized below. Station equipment impedances should be 600 Ω resistive over the usable voice band, and the station equipment should be isolated from ground by a minimum of 20 MΩ dc and 50 kΩ ac. The basic voice-grade telephone circuit is a 3002 channel; it has a usable bandwidth of 300 to 3000 Hz. The circuit gain at 3000 Hz is 3 dB below the specified in-band signal power. The gain at 4 kHz must be at least 15 dB below the gain at 3 kHz. The maximum transmitted signal power for a private-line circuit is 0 dBm. The transmitted signal power for dial-up circuits is established for each loop so that the signal is received at the central Telco office at −dBm.

## Facility Parameters

Facility parameters represent potential impairments to a data signal. These impairments are caused by Telco equipment and the limits specified pertain to all voice band circuits regardless of conditioning.

**1004-Hz variation.**   Telco has established 1004 Hz as the standard test-tone frequency. The frequency 1004 Hz was selected because of its relative location in the passband of a standard voice band circuit. The purpose of this test tone is to simulate the combined signal power of a standard voice band transmission. The 1004-Hz channel loss for a private-line circuit is 16 dB. A 1004-Hz test tone applied to the transmit local loop at 0 dBm should appear at the output of the destination loop at −16 dBm. Long-term variations in the gain of the transmission facilities should not exceed ±4 dB; the received signal power must be within the limits of −12 to −20 dBm.

**Noise.**   *Noise* can be generally defined as any undesired energy present in the usable passband of a communications channel. The noise is either correlated or uncorrelated. *Correlation* implies a relationship between the signal and the noise. Uncorrelated noise is energy present in the absence of a signal such as thermal noise. Correlated noise is unwanted energy which is present as a direct result of the signal, such as nonlinear distortion.

**Noise weighting.**   Signal interference by noise is categorized in terms of annoyance or intelligibility. Noise may be annoying to the listener but not to the degree that the conversation cannot be understood. Western Electric Company conducted experiments in which groups of listeners were asked to rate the annoyance caused by 14 different audible frequencies between 180 and 3500 Hz. These frequencies were presented to the listeners on a standard 500-type telephone (the old black dialer that Grandma had). The listeners first compared the annoyance of each frequency to the annoyance of a reference frequency of 1000 Hz in the absence of speech power. Then the same experiments were repeated with speech present. The results of the two tests were averaged and smoothed to produce the *C-message weighting curve*. This curve is shown in Figure 3–8.

C-message noise is significant because it indicates the passband characteristics that should be considered when conducting noise tests on voice band communications circuits. The frequencies from 600 to 3000 Hz proved to be the most annoying and, as the frequency decreased below 600 Hz, the annoyance factor also decreased. Because of these results, C-message filters were developed that have a response similar to the C-message weighting curve. These filters increase the attenuation of the noise power at frequencies below 600 Hz and above 3000 Hz, but they have a relatively flat response to noise power within the 600- to 3000-Hz passband. This response makes the measured strength of the noise frequencies proportional to the amount of annoyance they produce. These filters are inserted in noise-measuring

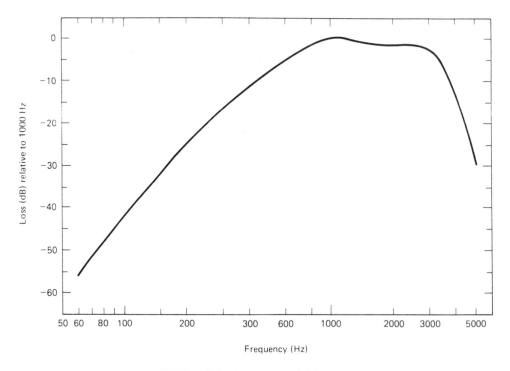

**FIGURE 3–8** C-message weighting curve.

sets at a point just before where the noise power is measured. The noise-measuring set then evaluates the noise in a manner similar to the human ear. The human ear cannot appreciate the true rms power of sound if the duration of the sound is 200 ms or less. Therefore, a noise-measuring set for voice circuits includes a 200-ms time constant that prevents them from reacting to short bursts of noise power. What significance does C-message weighting have on data circuits? Typical voice band data modems exhibit an output frequency spectrum that concentrates most of the transmitted signal power in the 600- to 3000-Hz band, Since a C-message filter has a relatively flat frequency response over this range, noise measurements with this type of filter are valid for data applications also.

**C-message noise.**   C-message noise measurements determine the average continuous rms noise power. This noise is commonly referred to as background, white, thermal, or Gaussian noise. This noise is inherently present in a circuit due to the electrical makeup of the circuit. Since white noise is additive, its magnitude is dependent, in part, on the electrical length of the circuit. C-message noise measurements are the terminated rms power readings at the receive end of a circuit with the transmit end terminated in the characteristic impedance of the telephone line (see Figure 3–9). There is a disadvantage in measuring noise this way. The overall circuit characteristics, in the absence of a signal, are not necessarily the same as

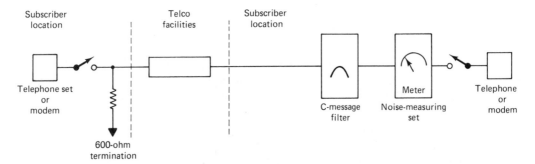

**FIGURE 3–9   Terminated C-message noise test.**

when a signal is present. The use of companders and automatic gain devices in the circuit causes this difference. Companders are devices that exhibit frequency-dependent gain characteristics (companders are explained in detail in Chapter 9). For this reason C-notched noise measurements were developed.

C-notched noise measurements differ from standard C-message noise measurements only in the fact that a holding tone (usually, 1004 or 2804 Hz) is applied to the transmit end of the circuit during the test. The holding tone ensures that the circuit operation simulates a loaded voice or data transmission. "Loaded" is a communications term that indicates the presence of a signal power comparable to the power of an actual message transmission. The holding tone is filtered (notched out) in the noise-measuring set prior to the C-message filter. The bandpass characteristics of the notch filter are such that only the holding tone is removed. The noise power in the usable passband is measured with a standard C-message noise-measuring set. This test (Figure 3–10) assures that the noise readings obtained actually reflect the loaded circuit characteristics of a normal voice band transmission.

The physical makeup of a private-line data circuit may require the use of several trunk circuits in tandem. Each individual trunk may be an analog, digital, companded, or a noncompanded facility. Various combinations of these facilities may be used to configure a circuit. Telco has established realistic C-notched noise requirements for each type of facility for various trunk lengths. These requirements

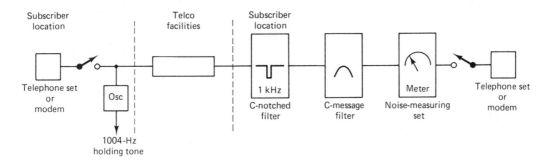

**FIGURE 3–10   C-notched noise test.**

assist Telco in evaluating the performance of a facility and in expediting trouble-isolation procedures. A subscriber to the telephone network need only be concerned with the overall (end-to-end) C-notched noise requirement. Standard private-line data circuits, operating at less than 9600 bps, require a minimum signal-to-C notched noise ratio of 24 dB. Data circuits operating at 9600 bps require a high-performance line with a minimum signal-to-C-notched noise ratio of 28 dB.

      **Impulse noise.**    *Impulse noise* is characterized by high-amplitude peaks (impulses) of short duration in the total noise spectrum. The significance of impulse noise hits on data transmission has been a controversial topic. Telco has accepted the fact that the absolute magnitude of the impulse hit is not as important as the magnitude of the hit relative to the signal amplitude. Empirically, it has been determined that an impulse hit will not produce transmission errors in a data signal unless it comes within 6 dB of the signal level (see Figure 3–11). Hit counters are designed to register a maximum of seven counts per second. This produces a "dead" time of 143 ms between counts when additional impulse hits are not registered. Contemporary high-speed data formats transfer data in block form and whether one hit or many occur during a single block is unimportant. Any error within a block necessitates retransmission of the entire block. Counting additional impulses

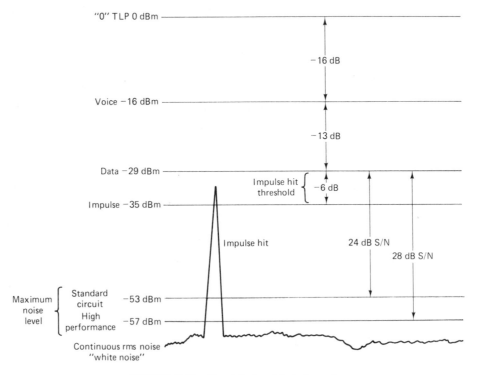

**FIGURE 3–11  C-notched and impulse noise.**

during the time of a single block does not correlate well with data transmission performance.

Impulse tests are performed by placing a 2804-Hz holding tone on the circuit to ensure loaded circuit characteristics. The counter records the number of hits in a prescribed time interval (usually 15 minutes). An impulse hit is typically less than 4 ms and never more than 10 ms in duration. Telco's limit for recordable impulse hits is 15 within a 15-minute time interval. This does not limit the number of hits to one per minute but rather limits the average to one per minute.

**Gain hits and dropouts.** A *gain hit* is a sudden, random change in the gain of a circuit. Gain hits are classified as temporary variations in the gain exceeding $\pm$ 3 dB, lasting more than 4 ms and returning to the original value within 200 ms. The primary cause of a gain hit is that of transients caused by switching radio facilities in the normal course of a day. Atmospheric fades produce the necessity for switching radio facilities. A *dropout* is a decrease in circuit gain of more than 12 dB that lasts longer than 4 ms. Dropouts are characteristics of a temporary open-circuit condition and are caused by deep radio fades or Telco maintenance activities. Dropouts occur at a rate of approximately one per hour. Gain hits and dropouts (see Figure 3–12) are detected by monitoring the receive level of a 1004-Hz test tone.

**Phase hits.** *Phase hits* (see Figure 3–13) are sudden, random changes in the phase of a transmitted signal. Phase hits are classified as temporary variations in the phase of a signal that last longer than 4 ms. Generally, phase hits are not recorded unless they exceed $\pm 20°$ peak. Phase hits, like gain hits, are caused by transients produced when radio facilities are switched.

**Phase-jitter.** *Phase jitter* (Figure 3–13) is a form of incidental phase modulation—a continuous, uncontrolled variation in the zero crossings of a signal. Generally,

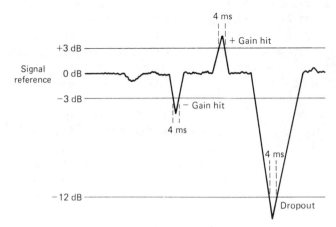

**FIGURE 3–12   Gain hits and drop-outs.**

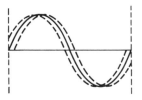

Phase jitter
continuous changes < 10 p-p

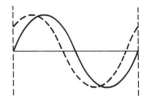

Phase hit
rapid changes > 20 p

**FIGURE 3–13   Phase distortion.**

phase jitter occurs at less than a 300-Hz rate, and its primary cause is a low-frequency ac ripple in Telco power plant supplies. The number of power supplies required in a circuit is directly proportional to the number of trunk circuits and Telco offices that make up a message channel. Each trunk facility has a separate phase jitter specification. The maximum end-to-end phase jitter allowed is 10° peak-to-peak regardless of the number of radio links, cable facilities, or digital carrier spans used in the circuit. Limiting the number of trunk circuits is a primary design consideration for a data circuit. Phase jitter is measured by observing the zero crossings of a 1004-Hz test tone.

**Single-frequency interference.**    *Single-frequency interference* is the presence of one or more continuous, unwanted tones within a message channel. The tones are called spurious tones and are often caused by crosstalk or cross-modulation between different channels in a carrier system. Spurious tones are measured by terminating the transmit end of a circuit, then searching through the channel spectrum with a frequency-selective voltmeter or observing the channel passband with a spectrum analyzer. Spurious tones can produce the same undesired circuit behavior as white noise.

**Frequency shifts.**    Analog carrier systems used by telephone companies operate single-sideband suppressed carrier (SSBSC) and therefore require coherent demodulation. In *coherent demodulation*, the frequency of the suppressed carrier must be recovered and reproduced exactly by the receiver. If this is not done, the demodulated signal will be offset in frequency by the difference between the transmit and the receive carrier frequencies. Frequency shift is measured by transmitting a 1004-Hz test tone and then measuring the frequency of the tone at the receiver.

**Phase intercept distortion.**    *Phase intercept distortion* occurs in coherent SSBSC systems when the received carrier is not reinserted with the exact phase relationship to the received signal as the transmit carrier possessed. This impairment causes a constant phase shift to all frequencies. This impairment is of little concern with data modems that use frequency shift keying, differential phase shift keying, or quadrature amplitude modulation. Since these are the more common methods of modulation, the telephone company has not set any limits on phase intercept distortion.

**Nonlinear distortion.** Nonlinear distortion is an example of correlated noise. The noise caused by nonlinear distortion is in the form of additional tones present because of the nonlinear amplification of a signal. No signal, no noise! Nonlinear distortion produces distorted sine waves. Two classifications of nonlinear distortion are:

1. *Harmonic distortion*: unwanted multiples of the transmitted frequencies
2. *Intermodulation distortion*: cross-products (sums and differences) of the input frequencies

Harmonic and intermodulation distortion, if of sufficient amplitude, can cause a data signal to be destroyed. The degree of circuit nonlinearity can be measured using either harmonic or intermodulation distortion tests.

Harmonic distortion is measured by applying a single-frequency test tone to a data channel. At the receive end, the power of the fundamental, second, and third harmonic frequencies is measured. Harmonic distortion is classified as second, third, *n*th order or as total harmonic distortion. Generally, harmonics above the third extend beyond the passband of a voice band channel and are of insufficient amplitude to be important. The actual amount of nonlinearity in a circuit is determined by comparing the power of the fundamental to the combined powers of the second and third harmonics. Harmonic distortion tests use a single-frequency (704-Hz) source (see Figure 3–14); therefore, no cross-product frequencies are produced.

Although simple harmonic distortion tests provide an accurate measurement of the nonlinear characteristics of an analog message channel, they are inadequate for digital (T-carrier) facilities. For this reason, a more refined method was developed which uses a multifrequency test tone signal. Four test frequencies are used (see Figure 3-15): two designated the A band (A1 = 856 Hz, A2 = 863 Hz), and two designated the B band (B1 = 1374 Hz, B2 = 1385 Hz). The four frequencies are transmitted with equal power, and the total combined power is equal to that of a composite data signal. The nonlinear amplification of the circuit produces multiples of each frequency and also their cross-products. For reasons beyond the scope of this text, the following second- and third-order products were selected for measurement: B + A, B − A, and 2B − A. The combined signal power of the four A and B band frequencies is compared with the second-order products, and then compared with the third-order products. Harmonic and intermodulation distortion tests

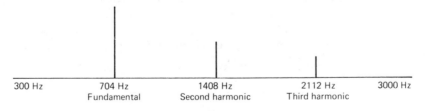

| 300 Hz | 704 Hz | 1408 Hz | 2112 Hz | 3000 Hz |
| | Fundamental | Second harmonic | Third harmonic | |

**FIGURE 3–14 Harmonic distortion.**

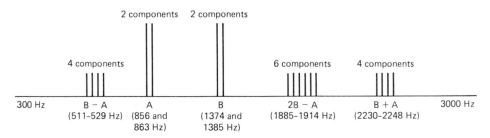

FIGURE 3–15  Nonlinear distortion.

do not directly determine the amount of interference caused by nonlinear circuit gain. They serve as a figure of merit only when evaluating circuit parameters.

**Peak-to-average ratio.**   The difficulties encountered in measuring true phase distortion or envelope delay distortion led to the development of peak-to-average (PAR) tests. A signal containing a series of distinctly shaped pulses with a high peak-to-average ratio (hence, the name) is transmitted. Delay distortion in a circuit has a tendency to spread the pulses and reduce the peak-to-average ratio. The received peak-to-average ratio is converted to a number between 0 and 100. The higher the number, the less the phase distortion. Peak-to-average tests do not indicate the exact amount of phase distortion present in a circuit; they only provide a figure of merit that can indicate the presence of a problem. Peak-to-average tests are less sensitive to attenuation distortion than envelope delay distortion tests and are easier to accomplish.

## SUMMARY

Tables 3–2 and 3–3 summarize the interface parameter limits and facility parameter limits discussed in this chapter.

### Abbreviations Associated with Signal and Noise Level Measurements

1. *dB (decibel)*. Experiments indicate that a listener cannot give a reliable estimate of the loudness of a sound, but he can distinguish the difference in loudness between two sounds. The ear's sensitivity to a change in sound power follows a logarithmic rather than linear scale, and the dB has become the unit of this change.

$$dB = 20 \log \frac{V_1}{V_2} \qquad dB = 20 \log \frac{I_1}{I_2} \qquad dB = 10 \log \frac{P_1}{P_2}$$

If the larger value is assigned to the numerator, the dB value will be positive.

**TABLE 3–2   Summary of Interface Parameter Limits**

| Parameter | Limit |
|---|---|
| 1. Recommended impedance of terminal equipment | 600 Ω resistive ± 10% |
| 2. Recommended isolation to ground of terminal equipment | At least 20 MΩ dc<br>At least 50 kΩ ac<br>At least 1500 V rms breakdown voltage at 60 Hz |
| 3. Data transmit signal power | 0 dBm (3-s average) |
| 4. In-band transmitted signal power | 2450- to 2750-Hz band should not exceed signal power in 800- to 2450-Hz band |
| 5. Out-of-band transmitted signal power:<br>*Above voice band*: | |
| (a) 3995–4005 Hz | At least 18 dB below maximum allowed in-band signal power |
| (b) 4–10 kHz band | Less than −16 dBm |
| (c) 10–25 kHz band | Less than −24 dBm |
| (d) 25–40 kHz band | Less than −36 dBm |
| (e) Above 40 kHz | Less than −50 dBm |

*Below voice band*:

(f) Rms current per conductor as specified by Telco, but never greater than 0.35 A.

(g) Magnitude of peak conductor-to-ground voltage not to exceed 70 V.

(h) Conductor-to-conductor voltage shall be such that conductor-to-ground voltage is not exceeded. For an ungrounded signal source, the conductor-to-conductor limit is the same as the conductor-to-ground limit.

(i) Total weighted rms voltage in band from 50 to 300 Hz, not to exceed 100 V. Weighting factors for each frequency component ($f$) are: $f^2/10^4$ for $f$ between 50 and 100 Hz, and $f^{3.3}/10^{6.6}$ for $f$ between 101 and 300 Hz.

6. Maximum test signal power: same as transmitted data power

---

dB yields a relative value—the relative size of the numerator compared to the denominator.

2. *dBm*.
   (a) dBm is the dB in reference to 1 mW.
   (b) dBm is a measure of absolute power.
   (c) A 10 dBm signal is equal to 10 mW.
   (d) A 20 dBm signal is equal to 100 mW.

3. *TLP* (*transmission level point*). Knowing the signal power at any point in a system is relatively worthless. A signal at a particular point can be 10 dBm. Is this good or bad? This could be answered if it is known what the signal strength should be at that point. The TLP does just that. The reference for TLP is 0 dBm. A −15 dBm TLP indicates that, at this specified point, the signal should be −15 dBm. A 0 TLP is a TLP where the signal power should be 0 dBm. The TLP says nothing about the signal itself.

4. *dBmO*. dBmO is a power measurement adjusted to 0 dBm that indicates

**TABLE 3–3   Summary of Facility Parameter Limits**

| Parameter | Limit |
|---|---|
| 1. 1004-Hz loss variation | Not more than ± dB long-term |
| 2. C-message noise | Maximum rms noise at modem receiver (nominal −16 dBm point) |

| Facility miles | dBm | dBrncO |
|---|---|---|
| 0–50 | −61 | 32 |
| 51–100 | −59 | 34 |
| 101–400 | −58 | 35 |
| 401–1000 | −55 | 38 |
| 1001–1500 | −54 | 39 |
| 1501–2500 | −52 | 41 |
| 2501–4000 | −50 | 43 |
| 4001–8000 | −47 | 46 |
| 8001–16000 | −44 | 49 |

| Parameter | Limit |
|---|---|
| 3. C-notched noise | (minimum values) |
| (a) Standard voice band channel | 24-dB signal-to-C-notched noise |
| (b) High-performance line | 28-dB signal-to-C-notched noise |
| 4. Single-frequency interference | At least 3 dB below C-message noise limits |
| 5. Impulse noise: | |

| Threshold with respect to 1004-Hz holding tone | Maximum counts above threshold allowed in 15 minutes |
|---|---|
| 0 dB | 15 |
| +4 dB | 9 |
| +8 dB | 5 |

| Parameter | Limit |
|---|---|
| 6. Frequency shift | ±5 Hz end to end |
| 7. Phase intercept distortion | No limits |
| 8. Phase jitter | No more than 10° peak to peak (end-to-end requirement) |
| 9. Nonlinear distortion (D-conditioned circuits only) | |
| Signal-to-second order | At least 35 db |
| Signal-to-third order | At least 40 dB |
| 10. Peak-average ratio | Reading of 50 minimum end to end with standard PAR meter |
| 11. Phase hits | 8 or less in any 15-minute period greater than ±20° peak |
| 12. Gain hits | 8 or less in any 15-minute period greater than ±3 dB |
| 13. Dropouts | 2 or less in any 15-minute period greater than 12 dB |

what the power would be if it were measured at 0 TLP. This value compares the actual signal at a point to what that signal should be at that point (TLP). A +4 dBm signal measured at a −16 dBm TLP is +20 dBmO. It is 20 dB stronger than the reference. −2 dBm at a −10 dBm TLP is equal to +8 dBmO.

5. *rn* (*reference noise*). This value is the dB value used as noise reference. This value is *always* −90 dBm or 1 pW. This value is taken as the reference because the signal value at any point should never be less than this.

6. *dBrnc*. This is the dB value of noise with respect to reference noise with C-message weighting. The larger this value is, the worse the condition because of a larger amount of noise.

7. *dBrncO*. This is the amount of noise corrected to 0 TLP. The amount of noise at a point does not really indicate how detrimental its effect is on a given signal. 0.3 dBm of noise with a 20 dBm signal is insignificant, while the same amount of noise with a 0.3 dBm signal strength may completely obscure the signal. Noise of 34 dBrnc at +7 dBm TLP yields a value of 27 dBrncO. dBrncO relates noise power readings (dBrnc) to 0 TLP. This unit establishes a common reference point throughout the system.

8. *DLP* (*data level point*). This parameter is equivalent to the TLP as far as its function is concerned. It is used as a reference for data transmission. TLP is used as a reference for voice transmission. Whatever the TLP is, the DLP is always 13 dB below that value for the same point. If the TLP is −12 dBm, the DLP is −25 dBm for that same point.

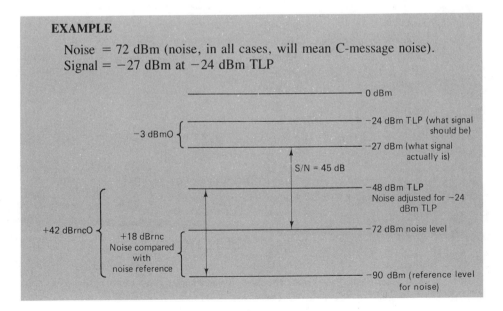

**EXAMPLE**

Noise = 72 dBm (noise, in all cases, will mean C-message noise).
Signal = −27 dBm at −24 dBm TLP

Find:

| | | |
|---|---|---|
| Signal _____ dBmO | answer −3 |
| Noise _____ dBrnc | answer +18 |
| Noise _____ dBrncO | answer +42 |
| S/N _____ dB | answer +45 |

9. *dBrn.* This value is used only when there is excessive low-frequency noise and that noise is exceeding the levels specified in the interface parameters. Its value is measured with a device that has a 3-kHz flat filter (instead of a C-message filter) (see Figure 3–16). In normal circuits, dBrn is typically 1.5 dB above what would be read with a C-message filter.

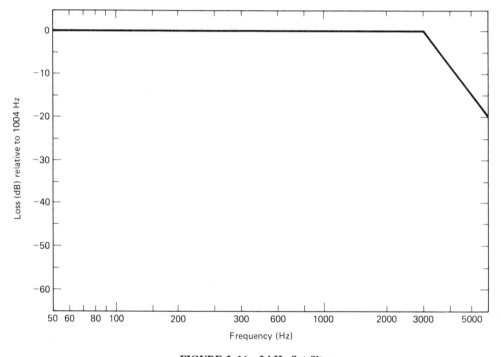

**FIGURE 3–16    3-kHz flat filter.**

# QUESTIONS

**1.** Higher frequencies generally suffer greater attenuation than lower frequencies when traveling through a given cable of a fixed length.   (T, F)

2. A loaded telephone cable identification number indicates three characteristics of that cable. What are they?

3. A metallic cable acts as what type of filter?

4. A cable is loaded by adding _____ in series with the wire.

5. 1004-Hz deviation is another name for envelope delay.   (T, F)

6. Dial-up lines may be conditioned.   (T, F)

7. What type of line has minimum conditioning?

8. D-type conditioning may be achieved by adding inductors in series to the line.   (T, F)

9. D-type conditioning is mandatory if the transmission rate is _____ bps.

10. Which frequency is used as the reference for attenuation distortion measurements when determining the type of conditioning?

11. Which frequency is used as the reference for envelope delay measurements when determining the type of conditioning?

12. All frequencies take the same time to propagate down a particular cable of fixed length. (T, F)

13. In determining the amount of phase distortion, (*relative, absolute*) phase delay is the more important parameter.

14. A C4 conditioned line must meet (*more, less*) stringent requirements than a C2 conditioned line.

15. Phase distortion is caused by the fact that it takes different frequencies different times to propagate down a line.   (T, F)

16. Phase distortion is synonymous with delay distortion.   (T, F)

17. Delay distortion is synonymous with EDD.   (T, F)

18. The maximum allowed transmitted signal power for a dial-up circuit is always 2 mW. (T, F)

19. The maximum allowed transmitted signal power for a private-line circuit is _____ mW.

20. All voice band frequencies suffer the same attenuation when passing through a C-message filter.   (T, F)

21. C-message filters are used in circuits to improve the signal-to-noise ratio.   (T, F)

22. What is the difference in measuring C-message noise and C-notched noise?

23. For a private line, what is the minimum signal-to-C-notched noise for:
   **(a)** Transmission at less than 9600 bps?
   **(b)** 9600-bps transmission?

24. Empirical determinations are the results of mathematical proofs of theory.   (T, F)

25. Three impulse hits on a message block would necessitate a different action than a single impulse hit on the same block.   (T, F)

26. What is the difference between a gain hit and an impulse hit?

27. Name two types of nonlinear distortion.

28. dB is a (*relative, absolute*) value.

29. dBm is a (*relative, absolute*) value.

30. What unit of measurement indicates the actual signal strength relative to what the signal strength should be?

**31.** What is the magnitude of noise, in dBm, that is used as the reference?

**32.** What is the unit of noise measurement that is corrected to 0 TLP?

**33.** Data transmission relates a signal strength to a DLP which is always _____ dB below the TLP?

**34.** What is the noise unit that is used to measure low-frequency noise?

# PROBLEMS

**1.** A cable is identified as 16D44. This indicates that _____ -mH inductors have been added in series every _____ feet to a _____ -gauge wire.

**2.** A 1800-Hz signal takes 2 ms to propagate down a line. A 600-Hz signal takes 2.5 ms to travel down the same line. What is the absolute phase delay of the 600-Hz signal? What is the relative phase delay of this same signal?

**3.** A line has C4 conditioning. A 1004-Hz signal is received at −10 dBm and the propagation delay for a 1800-Hz signal is 5 ms. A signal frequency of 400 Hz can take no more than _____ ms to travel down this line and its received signal strength must be between _____ and _____ dBm.

**4.** Draw a graphic representation for attenuation distortion and envelope delay distortion for a channel with C4 conditioning.

**5.** Frequencies of 250 Hz and 1 kHz are applied to a C-message filter. Their difference in amplitude would be (*greater*, *the same*, *less*) at the output than at the input.

**6.** Dedicated lines are used to transmit data at 4800 bps. The received signal level is 500 μW. What is the maximum amount of C-notched noise allowed to satisfy D-type conditioning?

**7.** A 2000-Hz carrier and a frequency of 300 Hz is transmitted. Identify the type of nonlinear distortion if:
   **(a)** 1500 Hz is present at the output
   **(b)** 1700 Hz is present at the output

**8.** A C-message noise measurement taken at a −22 dBm TLP indicates −72 dBm of noise. A test tone is measured at the same TLP and measures −25 dBm. Determine the following signal/noise levels.
   **(a)** Signal power relative to TLP (dBmO)
   **(b)** C-message noise relative to reference noise (dBrnc)
   **(c)** C-message noise relative to reference noise at a 0 TLP (dBrncO)
   **(d)** Overall signal-to-noise ratio (dB)

**9.** A C-message noise measurement taken at −20 dBm TLP indicates a corrected reading of 43 dBrncO. A test tone transmitted at data level is used to determine a signal-to-noise of 31 dB. Determine the following signal/noise levels.
   **(a)** Signal power relative to TLP (dBmO)
   **(b)** C-message noise relative to reference noise (dBrnc)
   **(c)** Actual signal power of test tone (dBm)
   **(d)** Actual C-message noise power

**10.** A test-tone signal power of −62 dBm is measured at −61 dBm TLP. The C-message noise is measured at the same TLP and measures $-10$ dBrnc. Determine the following signal-to-noise levels.

(a) C-message noise relative to reference noise at a 0 TLP (dBrncO)

(b) Actual C-message noise power (dBm)

(c) Signal power relative to TLP (dBmO)

(d) Overall signal-to-noise ratio (dB)

# FOUR

## *TRANSMISSION CODES AND ERROR DETECTION AND CORRECTION*

## TRANSMISSION CODES

*Transmission codes* are used for encoding alpha/numeric characters and symbols (punctuation, etc.) and are consequently often called *character sets*, *character languages*, or *character codes*. Essentially, three types of characters are used in data communications codes: *data link control* characters, which are used to facilitate the orderly flow of data from the source to the destination; *graphic control* characters, which involve the syntax or presentation of the data at the receive terminal; and *alpha/numeric* characters, which are used to represent the various symbols used for letters, numbers, and punctuation in the English language.

The first data communications code that saw widespread usage was the Morse code. The Morse code used three unequal-length symbols (dot, dash, and space) to encode alpha/numeric characters, punctuation marks, and an interrogation word.

The Morse code is inadequate for use in modern digital computer equipment because all characters do not have the same number of symbols or take the same length of time to send, and each Morse code operator transmits code at a different rate. Also, with Morse code, there is an insufficient selection of graphic and data link control characters to facilitate the transmission and presentation of the data typically used in contemporary computer applications.

The three most common character sets presently used for character encoding are the Baudot code, the American Standard Code for Information Interchange (ASCII), and the Extended Binary-Coded Decimal Interchange Code (EBCDIC).

## Baudot Code

The *Baudot code* (sometimes called the *Telex code*)was the first fixed-length character code. The Baudot code was developed by a French postal engineer, Thomas Murray, in 1875 and named after Emile Baudot, an early pioneer in telegraph printing. The Baudot code is a 5-bit character code that is used primarily for low-speed teletype equipment such as the TWX/Telex system. With a 5-bit code there are only $2^5$ or 32 codes possible, which is insufficient to represent the 26 letters of the alphabet, the 10 digits, and the various punctuation marks and control characters. Therefore, the Baudot code uses *figure* shift and *letter* shift characters to expand its capabilities to 58 characters. The latest version of the Baudot code is recommended by the CCITT as the International Alphabet No. 2. The Baudot code is still used by Western

**TABLE 4-1   Baudot Code**

| Character shift | | Binary code |
| --- | --- | --- |
| *Letter* | *Figure* | *Bit:  4  3  2  1  0* |
| A | — | 1 1 0 0 0 |
| B | ? | 1 0 0 1 1 |
| C | : | 0 1 1 1 0 |
| D | $ | 1 0 0 1 0 |
| E | 3 | 1 0 0 0 0 |
| F | ! | 1 0 1 1 0 |
| G | & | 0 1 0 1 1 |
| H | # | 0 0 1 0 1 |
| I | 8 | 0 1 1 0 0 |
| J | ' | 1 1 0 1 0 |
| K | ( | 1 1 1 1 0 |
| L | ) | 0 1 0 0 1 |
| M | . | 0 0 1 1 1 |
| N | , | 0 0 1 1 0 |
| O | 9 | 0 0 0 1 1 |
| P | 0 | 0 1 1 0 1 |
| Q | 1 | 1 1 1 0 1 |
| R | 4 | 0 1 0 1 0 |
| S | bel | 1 0 1 0 0 |
| T | 5 | 0 0 0 0 1 |
| U | 7 | 1 1 1 0 0 |
| V | ; | 0 1 1 1 1 |
| W | 2 | 1 1 0 0 1 |
| X | / | 1 0 1 1 1 |
| Y | 6 | 1 0 1 0 1 |
| Z | " | 1 0 0 0 1 |
| Figure shift | | 1 1 1 1 1 |
| Letter shift | | 1 1 0 1 1 |
| Space | | 0 0 1 0 0 |
| Line feed (LF) | | 0 1 0 0 0 |
| Blank (null) | | 0 0 0 0 0 |

Union Company for their TWX and Telex teletype systems. The AP and UPI news services also use the Baudot code for sending news information around the world. The most recent version of the Baudot code is shown in Table 4-1 (p. 49).

## ASCII

ASCII is a 7-bit code and is shown in Table 4–2. For the ASCII codes 5EH and 5FH, many ASCII tables show UP ARROW and LEFT ARROW, respectively. Often, two hexadecimal characters are used to express this code with an assumed 0 for the most significant bit. The ASCII code for A expressed in hexadecimal is 41 H. The trailing H signifies hexadecimal. ASCII characters can be divided into three groups:

1. **Data Link Control Characters**

   SOH Start of Heading
   STX Start of Text
   ETX End of Text
   EOT End of Transmission
   ENQ Enquiry
   ACK Positive Acknowledgment
   DLE Data Link Escape

**TABLE 4–2 ASCII**

| | | Bits 6, 5, 4 | | | | | | | | |
|---|---|---|---|---|---|---|---|---|---|---|
| Bits | | 000 | 001 | 010 | 011 | 100 | 101 | 110 | 111 | |
| 3, 2, 1, 0 | Hex | 0 | 1 | 2 | 3 | 4 | 5 | 6 | 7 | Hex |
| 0000 | 0 | NUL | DLE | SP | 0 | @ | P | ` | p | |
| 0001 | 1 | SOH | DC1 | ! | 1 | A | Q | a | q | |
| 0010 | 2 | STX | DC2 | " | 2 | B | R | b | r | |
| 0011 | 3 | ETX | DC3 | # | 3 | C | S | c | s | |
| 0100 | 4 | EOT | DC4 | $ | 4 | D | T | d | t | |
| 0101 | 5 | ENQ | NAK | % | 5 | E | U | e | u | |
| 0110 | 6 | ACK | SYN | & | 6 | F | V | f | v | |
| 0111 | 7 | BEL | ETB | ' | 7 | G | W | g | w | |
| 1000 | 8 | BS | CAN | ( | 8 | H | X | h | x | |
| 1001 | 9 | HT | EM | ) | 9 | I | Y | i | y | |
| 1010 | A | LF | SUB | * | : | J | Z | j | z | |
| 1011 | B | VT | ESC | + | ; | K | [ | k | { | |
| 1100 | C | FF | FS | , | < | L | \ | l | ¦ | |
| 1101 | D | CR | GS | - | = | M | ] | m | } | |
| 1110 | E | SO | RS | . | > | N | ∧ | n | ~ | |
| 1111 | F | SI | US | / | ? | O | — | o | DEL | |

SYN Synchronous Idle

NAK Negative Acknowledgment

ETB End of Transmission Block

The specific meaning and functional use of the link control characters are explained and demonstrated in Chapters 10 and 11.

2. **Graphic Control Characters**

| | | | |
|---|---|---|---|
| BEL | Bell | DC3 | Device Control 3 |
| BS | Back Space | DC4 | Device Control 4 |
| HT | Horizontal Tab | EM | End of Medium |
| LF | Line Feed | ESC | Escape |
| VT | Vertical Tab | FS | File Separator |
| FF | Form Feed | GS | Group Separator |
| CR | Carriage Return | RS | Record Separator |
| SO | Shift Out | US | Unit Separator |
| SI | Shift In | DEL | Delete |
| DC1 | Device Control 1 | CAN | Cancel |
| DC2 | Device Control 2 | | |

These characters are transmitted to identify where information will be printed or displayed or to signal the operator.

3. **Alphanumeric Characters**

These are the remaining characters that were not included in the first two groups. They are the numbers, letters, punctuation marks, and so on, that make up the message that is displayed on the CRT or printed.

## EBCDIC

EBCDIC is an 8-bit code devised by IBM. This code is shown in Table 4–3. EBCDIC characters can be divided into the same three major categories as ASCII. The following are abbreviations that differ from ASCII:

| | | | |
|---|---|---|---|
| BYP | Bypass | PF | Punch Off |
| CC | Unit Backspace | PN | Punch On |
| DS | Digit Select | PRE | Prefix |
| FF | Form Feed | RES | Restore |
| IFS | Interchange File Separator | SI | Shift In |
| IGS | Interchange Group Separator | SM | Start Message |
| IL | Idle | SMM | Repeat |

TABLE 4-3   EBCDIC

| Bit position<br>0 1 2 3 →<br>4 5 6 7 ↓ | 0000 | 0001 | 0010 | 0011 | 0100 | 0101 | 0110 | 0111 | 1000 | 1001 | 1010 | 1011 | 1100 | 1101 | 1110 | 1111 |
|---|---|---|---|---|---|---|---|---|---|---|---|---|---|---|---|---|
| 0000 | NUL | DLE | DS |  | SP | & | - |  |  |  | SMM | VT | FF | CR | SO | SI |
| 0001 | SOH | DC1 | SOS |  |  |  | / |  | a | j |  |  | A | J |  | 1 |
| 0010 | STX | DC2 | FS | SYN |  |  |  |  | b | k | s |  | B | K | S | 2 |
| 0011 | ETX | DC3 |  |  |  |  |  |  | c | l | t |  | C | L | T | 3 |
| 0100 | PF | RES | BYP | PN |  |  |  |  | d | m | u |  | D | M | U | 4 |
| 0101 | HT | NL | LF | RS |  |  |  |  | e | n | v |  | E | N | V | 5 |
| 0110 | LC | BS | EOB | UC |  |  |  |  | f | o | w |  | F | O | W | 6 |
| 0111 | DEL | IL | PRE | EOT |  |  |  |  | g | p | x |  | G | P | X | 7 |
| 1000 |  | CAN |  |  |  |  |  |  | h | q | y |  | H | Q | Y | 8 |
| 1001 |  | EM |  |  |  |  |  |  | i | r | z |  | I | R | Z | 9 |
| 1010 | SMM | CC | SM |  | ¢ | ! |  | : |  |  |  |  |  |  |  |  |
| 1011 | VT |  |  |  | . | $ | , | # |  |  |  |  |  |  |  |  |
| 1100 | FF | IFS | DC4 |  | < | * | % | @ |  |  |  |  |  |  |  |  |
| 1101 | CR | IGS | ENQ | NAK | ( | ) | \| | ' |  |  |  |  |  |  |  |  |
| 1110 | SO | IRS | ACK |  | + | ; | > | = |  |  |  |  |  |  |  |  |
| 1111 | SI | IUS | BEL | SUB | \| | ¬ | ? | " |  |  |  |  |  |  |  | □ |

| | | | |
|---|---|---|---|
| IRS | Interchange Record Separator | SO | Shift Out |
| IUS | Interchange Unit Separator | SOS | Start of Significance |
| LC | Lower Case | SUB | Substitute |
| NL | New Line | UC | Upper Case |
| NUL | Null | | |

Caution should be used in using EBCDIC since IBM identifies the least significant bit as bit 7 and the most significant bit as bit 0.

> **EXAMPLE**
> EBCDIC for $n = 10010101 = 95H$.

## ERROR DETECTION

### Redundancy

*Redundancy* involves transmitting each character twice. If the same character is not received twice in succession, a transmission error has occurred. The same concept can be used for messages. If the same sequence of characters is not received twice in succession, in exactly the same order, a transmission error has occurred.

### Exact-Count Encoding

With *exact-count encoding*, the number of 1's in each character is the same. An example of an exact-count encoding scheme is the ARQ code shown in Table 4–4. With the ARQ code, each character has three 1's in it, and therefore a simple count of the number of 1's received can determine if a transmission error has occurred.

### Parity

The most common method used for error detection is through the use of parity. A single bit called the *parity bit* is added to every word so that the total number of 1's in a word, together with the parity bit, is odd for odd parity and even for even parity.

> **EXAMPLE**
> Assume that odd parity is desired and that ASCII is the code used. The code for A is P100 0001. P represents the parity bit. Since odd parity is desired, P should take on a value to make the total number of 1's odd. Therefore, P takes on a value of 1 and 1100 0001 would be transmitted.

**TABLE 4–4  ARQ Exact-Count Code**

| | Binary code | | | | | | Character | |
|---|---|---|---|---|---|---|---|---|
| Bit: | 1 2 3 4 5 6 7 | | | | | | *Letter* | *Figure* |
| | 0 0 0 1 1 1 0 | | | | | | Letter shift | |
| | 0 1 0 0 1 1 0 | | | | | | Figure shift | |
| | 0 0 1 1 0 1 0 | | | | | | A | — |
| | 0 0 1 1 0 0 1 | | | | | | B | ? |
| | 1 0 0 1 1 0 0 | | | | | | C | : |
| | 0 0 1 1 1 0 0 | | | | | | D | (WRU) |
| | 0 1 1 1 0 0 0 | | | | | | E | 3 |
| | 0 0 1 0 0 1 1 | | | | | | F | % |
| | 1 1 0 0 0 0 1 | | | | | | G | @ |
| | 1 0 1 0 0 1 0 | | | | | | H | £ |
| | 1 1 1 0 0 0 0 | | | | | | I | 8 |
| | 0 1 0 0 0 1 1 | | | | | | J | (bell) |
| | 0 0 0 1 0 1 1 | | | | | | K | ( |
| | 1 1 0 0 0 1 0 | | | | | | L | ) |
| | 1 0 1 0 0 0 1 | | | | | | M | . |
| | 1 0 1 0 1 0 0 | | | | | | N | , |
| | 1 0 0 0 1 1 0 | | | | | | O | 9 |
| | 1 0 0 1 0 1 0 | | | | | | P | 0 |
| | 0 0 0 1 1 0 1 | | | | | | Q | 1 |
| | 1 1 0 0 1 0 0 | | | | | | R | 4 |
| | 0 1 0 1 0 1 0 | | | | | | S | ' |
| | 1 0 0 0 1 0 1 | | | | | | T | 5 |
| | 0 1 1 0 0 1 0 | | | | | | U | 7 |
| | 1 0 0 1 0 0 1 | | | | | | V | = |
| | 0 1 0 0 1 0 1 | | | | | | W | 2 |
| | 0 0 1 0 1 1 0 | | | | | | X | / |
| | 0 0 1 0 1 0 1 | | | | | | Y | 6 |
| | 0 1 1 0 0 0 1 | | | | | | Z | + |
| | 0 0 0 0 1 1 1 | | | | | | | (blank) |
| | 1 1 0 1 0 0 0 | | | | | | | (space) |
| | 1 0 1 1 0 0 0 | | | | | | | (line feed) |
| | 1 0 0 0 0 1 1 | | | | | | | (carriage return) |

The parity bit is inserted at the transmitter and parity is checked at the receiver. The receiver should know the type of parity used. If odd parity is expected, and a word, together with its parity bit, have an even number of 1's, at least a 1-bit error has occurred. A limitation with parity is that if there is a 2-bit error in a word, this error will not be detected. There are many circuits that can be used for parity generators and parity checkers. The least complex of these is shown in Figure 4–1.

If there are $n$ bits in the word, $n - 1$ XOR gates must be used for this circuit to function as a parity generator. No matter how many bits there are in the word, the output of this circuit will always be a 1 if there is an odd number of 1's in the word; the output will always be 0 if there is an even number of 1's in the

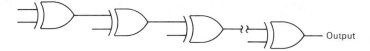

**FIGURE 4–1    Parity generator or parity checker.**

word. Therefore, if even parity is desired, the output of this circuit can be used directly for the parity bit. If odd parity is desired, the output of this circuit must be passed through an inverter to generate the proper parity bit. To be used as a parity checker, the word plus the parity bit are applied to the inputs of this circuit. If no errors are detected, the output will always be low for even parity and always high for odd parity. What is done if a parity error is detected is dependent on the system. In some systems, automatic request for retransmission is initiated on the detection of a single parity error. Sometimes the message can be understood despite the error. Some systems will not request retransmission but instead, display an unused symbol (ℑ) for the character that had the parity error. This is called *symbol substitution*. If the received message is ''Four scℑre and seven...,'' the missing character can be guessed. However, if the message is ''Pay Joe $?00,000.00...,'' it would be wise to request retransmission.

## Order of the Arrow

A problem arises in the interpretation of a string of 1's and 0's in serial transmission. The least significant bit is always transmitted first. There is no disagreement in identifying A as the first letter of ABC. However, if the ASCII codes for these letters were shown in the same sequence with a trailing parity bit:

P100 0001    P100 0010    P100 0011
  b      a      d      c      f      e

the transmission bit sequence would be incorrect. *a* to *b* would be transmitted first, then *c* to *d*, then *e* to *f*. These bits can be rearranged to show the proper transmission sequence as follows:

*Case 1*

P100 0011    P100 0010    P100 0001

Now the rightmost bit can be identified as the first bit transmitted and it is also the least significant bit of the first character. Similarly, the leftmost bit is the last bit transmitted and is the parity bit of the last character. Case 1 demonstrates the order of the arrow to the right—the rightmost bit is transmitted first and is the least significant bit of the first character. If the order of the arrow is to the left, the leftmost bit is transmitted first and is the least significant bit of the first character. This is shown in case 2.

*Case 2*

$$1000 \ 001P \qquad 0100 \ 001P \qquad 1100 \ 001P$$
$$\ \ 1 \qquad\ \ 4 \qquad\ \ 2 \qquad\ \ 4 \qquad\ \ 3 \qquad\ \ 4$$

If the order of the arrow is to the left, to interpret the bit pattern correctly, the bits must be taken in groups of 8, reversed, and then interpreted. If this were accomplished, 41H, 42H, and 43H would be obtained, which are the correct ASCII codes for A, B, and C.

Datascopes are used to monitor data lines. The first bit received would appear farthest left on the CRT. This means that the bit pattern received would be displayed with the order of the arrow to the left. In order to interpret a datascope presentation, a bit pattern with the order of the arrow to the left must be understood.

## Vertical Redundancy Check and Longitudinal Redundancy Check

If two parity bits are used, a single bit error can be detected and corrected. *VRC* (vertical redundancy check) is simply inserting and checking the parity bit for each word as previously described. A group of words is called a *block*. The specific number of words in a block is described more fully in Chapter 11. Additional

---

**EXAMPLE**

Odd parity is used for the VRC, even parity for the LRC.

```
                                    L
                T  H  E  S  C  A  T  R
                         P           C

       LSB      0  0  1  0  1  1  0  1
                0  0  0  0  1  0  0  1
                1  0  1  0  0  0  1  1
                0  1  0  0  0  0  0  1
                1  0  0  0  0  0  1  0
                0  0  0  1  0  0  0  1
       MSB      1  1  1  0  1  1  1  0

       VRC      0  1  0  0  0  1  0   * = 0


    01000011  00100000  01000101  11001000  01010100
        C        SP         E         H         T
    00101111  01010100  11000001
      LRC         T         A
```

The bits making up the LRC are also called the *block check sequence*. This sequence can represent an ASCII character. In this case, 010 1111 is the ASCII code for /, which is called the *block check character* for this particular block.

*\* is computed as a VRC bit. The serial transmitted bit stream is shown below with the order of the arrow to the right.*

parity bits are computed based on the first bit of every word, on the second bit of every word, and so on. These parity bits make up the *LRC* (longitudinal redundancy check).

For the example above, assume that bit 2 of the letter E was received in error as a 0. The VRC would detect a parity error for this character but, by itself, could not specify which bit was wrong. The LRC would also detect an error in the second bit of some word but, by itself, could not identify which word had the error. Together, however, they identify the bit at the intersection of bit 2 and the letter E as erroneous. An inversion of this bit will correct the error.

## Cyclic Redundancy Check

The *CRC* (cyclic redundancy check) is normally used with transmissions using EBCDIC as the character code. Basically, a generating function $G(x)$ is "divided into" a data string of 1's and 0's, the quotient is discarded, and the remainder is transmitted as the CRC. The number of bits in the CRC is equal to the highest exponent of the generating function. The generating function used with EBCDIC is

$$G(x) = x^{16} + x^{12} + x^5 + 1 \qquad \text{where } x^0 = 1$$

This function is also referred to as CRC-16 (because of the high-order exponent) and is 16 bits or two bytes long. These two bytes of CRC are also called the *block check sequence* (BCS) or the *frame check sequence* (FCS). The corresponding EBCDIC characters are called the *block check characters* (BCCs) or the *frame check characters* (FCCs).

The exponents identify the bit positions containing a 1. Therefore, the $G(x)$ shown above could be written as

$$1 \quad 0 \quad 0 \quad 0 \quad 1 \quad 0 \quad 0 \quad 0 \quad 0 \quad 0 \quad 1 \quad 0 \quad 0 \quad 0 \quad 0 \quad 1$$
$$b_{16} \qquad \quad b_{12} \qquad \qquad \qquad b_5 \qquad \qquad b_0$$

The "division" is not accomplished in the usual manner. In the subtract portion of the division, the remainder is obtained through an XOR operation rather than by an honest subtraction. Therefore, my $G(x)$ will "divide" into another number with an equal amount of digits one time regardless of whether $G(x)$ is larger or smaller.

**EXAMPLE**

Assume the message

$$M(x) = x^7 + x^5 + x^4 + x^2 + x + 1$$
$$= 1 \ 0 \ 1 \ 1 \ 0 \ 1 \ 1 \ 1$$

The least significant bit is on the *left*.

For simplicity, a $G(x)$ of $x^5 + x^4 + x + 1$ will be used. $M(x)$ is first multiplied by the number of bits in the CRC [same as the highest exponent of $G(x)$], producing

$$x^5 (x^7 + x^5 + x^4 + x^2 + x + 1) = x^{12} + x^{10} + x^9 + x^7 + x^6 + x^5$$

$$= 1\ 0\ 1\ 1\ 0\ 1\ 1\ 1\ 0\ 0\ 0\ 0\ 0$$

which is now divided by $G(x)$.

```
                    (discarded)     1 1 0 1 0 1 1 1
         1 1 0 0 1 1
                  )1 0 1 1 0 1 1 1 0 0 0 0 0
                   1 1 0 0 1 1
                     1 1 1 1 0 1
                     1 1 0 0 1 1
                       1 1 1 0 1 0
                       1 1 0 0 1 1
                         1 0 0 1 0 0
                         1 1 0 0 1 1
                           1 0 1 1 1 0
                           1 1 0 0 1 1
                             1 1 1 0 1 0
                             1 1 0 0 1 1
                               0 1 0 0 1 = CRC
```

The CRC is appended to the message producing a transmitted bit stream of

$$1\ 0\ 1\ 1\ 0\ 1\ 1\ 1 \quad 0\ 1\ 0\ 0\ 1$$

At the receiver, this stream is "divided" by the same $G(x)$ as used by the transmitter and if the remainder is zero, there was no error.

```
                                    1 0 1 0 1 1 1
     1 1 0 0 1 1)1 0 1 1 0 1 1 1 0 1 0 0 1
                1 1 0 0 1 1
                  1 1 1 1 0 1
                  1 1 0 0 1 1
                    1 1 1 0 1 0
                    1 1 0 0 1 1
                      1 0 0 1 1 0
                      1 1 0 0 1 1
                        1 0 1 0 1 0
                        1 1 0 0 1 1
                          1 1 0 0 1 1
                          1 1 0 0 1 1
                          0 0 0 0 0 0     no error
```

The circuit used to generate the CRC is shown in Figure 4-2.

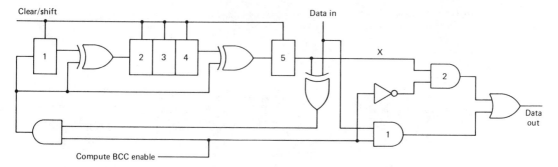

**FIGURE 4–2   CRC generating circuit.**

**Operation of the CRC generating circuit.**   Initially, all of the flip-flops (numbered boxes) of the shift register are cleared and the Compute BCC enable is high for as long as the data stream is applied. During this time, data are outputted by way of AND gate 1.

When the last data bit has passed through, the Compute BCC enable is made low. The BCC characters are contained in the shift register. With Compute BCC enable low, the circuit acts as a 5-bit shift register and, on succeeding clock pulses, the BCS is outputted by way of AND gate 2.

---

**EXAMPLE**

The flip-flops are identified by their number. The same $G(x)$ and data stream are used as in the preceding example.

At the transmitter:

| | Flip-flops: | 1 | 2 3 4 | 5 | Data in |
|---|---|---|---|---|---|
| | Initial clear: | 0 | 0 0 0 | 0 | 1 |
| After shift 1: | | 1 | 1 0 0 | 1 | 0 |
| After shift 2: | | 1 | 0 1 0 | 1 | 1 |
| After shift 3: | | 0 | 1 0 1 | 0 | 1 |
| After shift 4: | | 1 | 1 1 0 | 0 | 0 |
| After shift 5: | | 0 | 1 1 1 | 0 | 1 |
| After shift 6: | | 1 | 1 1 1 | 0 | 1 |
| After shift 7: | | 1 | 0 1 1 | 0 | 1 |
| After shift 8: | | 1 | 0 0 1 | 0 | |

CRC in the shift register.
Rightmost bit is transmitted first.

At the receiver:

| Flip-flops: | 1 | 2 3 4 | 5 | Data in Incl, CRC |
|---|---|---|---|---|
| Initial clear: | 0 | 0 0 0 | 0 | 1 |
| After shift 1: | 1 | 1 0 0 | 1 | 0 |
| After shift 2: | 1 | 0 1 0 | 1 | 1 |
| After shift 3: | 0 | 1 0 1 | 0 | 1 |
| After shift 4: | 1 | 1 1 0 | 0 | 0 |
| After shift 5: | 0 | 1 1 1 | 0 | 1 |
| After shift 6: | 1 | 1 1 1 | 0 | 1 |
| After shift 7: | 1 | 0 1 1 | 0 | 1 |
| After shift 8: | 1 | 0 0 1 | 0 | 0 (CRC) |
| After shift 9: | 0 | 1 0 0 | 1 | 1 |
| After shift 10: | 0 | 0 1 0 | 0 | 0 |
| After shift 11: | 0 | 0 0 1 | 0 | 0 |
| After shift 12: | 0 | 0 0 0 | 1 | 1 |
| After shift 13: | 0 | 0 0 0 | 0 | |

All zeros. No error.

*Variations*

1. Some circuits have an inverter located at point X on the circuit diagram. This causes the 1's complement of the CRC to be transmitted. An inversion would also be required at the receiver.
2. Some circuits initially preset the flip-flops. At the receiver, after the CRC has been received, a constant character (not 0's) is produced.

## Hamming Code

The Hamming code is one form of a forward error-correcting code (FEC). This means that the receiver has the capacity to detect and correct errors in the received transmission. The number of bits in the Hamming code is dependent on the number of bits in the data string. If $m$ represents the number of bits in the data string and $n$ represents the number of bits in the Hamming code, then $n$ is the smallest number such that

$$2^n \geq m + n + 1$$

The Hamming bits can be placed anywhere in the data string. Once these positions are decided on, they must remain fixed and must be known by the transmitter and the receiver.

## EXAMPLE

Assume that the data string consists of the 12 bits

```
1 0 1 1 0 0 0 1 0 0 1 0
```

An $n = 4$ is insufficient. An $n = 5$ must be used. An H is arbitrarily placed in the Hamming bit positions.

```
17 16 15 14 13 12 11 10  9  8  7  6  5  4  3  2  1
 H  1  0  1  H  1  0  0  H  H  0  1  0  H  0  1  0
```

The bit positions must be identified as shown.

To determine the value of the Hamming bits, express bit positions as a 5-bit number ($n = 5$) and XOR all positions containing a 1.

```
 2    0 0 0 1 0
 6    0 0 1 1 0
      0 0 1 0 0
12    0 1 1 0 0
      0 1 0 0 0
14    0 1 1 1 0
      0 0 1 1 0
16    1 0 0 0 0
      1 0 1 1 0    Hamming bits
```

*at the transmitter*

The Hamming bits will replace the H's left to right, respectively.

The encoded transmitted data string is now

```
1 1 0 1 0 1 0 0 1 1 0 1 0 0 0 1 0
```

Assume that these data are received with an error in bit 14:

```
1 1 0 0 0 1 0 0 1 1 0 1 0 0 0 1 0
```

The Hamming bits are extracted and XORed with all bit positions containing a 1.

```
      1 0 1 1 0
 2    0 0 0 1 0
      1 0 1 0 0
 6    0 0 1 1 0
      1 0 0 1 0
12    0 1 1 0 0
      1 1 1 1 0
16    1 0 0 0 0
      0 1 1 1 0 = 14    position of error
```

If the error occurred in the Hamming bit or if there was more than a 1-bit error, this coding scheme, as described, is not able to detect and correct. Additional coding bits would be required.

Any scheme for error detection and correction involves the transmission of additional bits. The purpose is to eliminate the need of retransmission. But the inclusion of additional bits for error detecting and correcting increases the transmission time. Before any such scheme for FEC is implemented, the frequency of errors and the corresponding time for retransmission must be compared with the increased transmission time caused by the insertion of additional bits for FEC.

# QUESTIONS

1. What is the purpose of the parity bits?
2. Parity is checked at the transmitter.   (T, F)
3. Name the three functional usages of the ASCII characters.
4. The LRC is made up of the individual parity bits for each word.   (T, F)
5. What bit is the most significant bit in EBCDIC?
6. If EBCDIC is the transmission code, what is used for error detection?
7. What is a redundancy code? What is it used for?
8. Describe how an exact-count code achieves error detection.
9. How does the ARQ code achieve error detection?

# PROBLEMS

1. Given the following ASCII codes with the order of the arrow to the right. What are the corresponding hexadecimal values and the corresponding ASCII characters?
   **(a)** 1010111     **(b)** 0000010     **(c)** 0010110     **(d)** 0101111
2. Given the following ASCII codes with the order of the arrow to the left. What are the corresponding hexadecimal values and the corresponding ASCII characters?
   **(a)** 101110     **(b)** 0101000     **(c)** 1011011     **(d)** 0010000
3. For Problem 1, if odd parity is used, what is the parity bit?
4. For Problem 2, if even parity is used, what is the parity bit?
5. **(a)** Compute the VRC and LRC for MARY. Use odd parity to compute the LRC and even parity to compute the VRC.
   **(b)** What ASCII character is represented by the LRC?
6. Compute the LRC (odd parity) and VRC (even parity) for your last name.
7. How many Hamming bits are required for each ASCII word?
8. **(a)** For the ASCII A (1000001), insert the required number of Hamming bits into every other location starting at the right:

   <pre>
   1 0 0 0 0 0 1
    X X     etc.
   </pre>

   **(b)** What are the Hamming bits used?
   **(c)** If the fifth bit from the right was received in error, show how the Hamming code would identify the location of the error.

# FIVE

## *UNIVERSAL ASYNCHRONOUS RECEIVER/ TRANSMITTER*

Computers move words internally in parallel. An 8-bit computer will move 8 bits at a time between registers or between memory and registers. If information stored in the computer is to be transmitted, this information must be converted to serial form. Similarly, if serial information is received, it must be converted to parallel form before it can be handled by the computer. The UART is a chip that performs these functions for asynchronous data—data in which the characters are framed by start and stop bits. The UART is functionally divided into the transmit and receive sections.

The transmit section:

1. Converts parallel information to a serial format
2. Inserts the appropriate parity bit if parity is used
3. Inserts a start bit
4. Inserts the required number of stop bits

The receive section:

1. Converts serial information to a parallel format
2. Checks for parity error
3. Checks to see that received information is properly transferred to the computer or terminal

Examples of UARTs currently on the market are (number of pins enclosed in parentheses):

National's INS 8250 (40)

Signetic's SCN2681 DART (Dual Asynchronous Receiver/Transmitter) (24, 28, and 40)

Intersil's IM 6402/3 (Universal Asynchronous Receiver/Transmitter) (40)

In addition, as the various manufacturers produce their family of microprocessor chips, they also produce *peripheral chips* that are compatible with the microprocessor. One of these chips is some form of UART, although the specific name will vary with the manufacturer. Examples of these are:

Intel's 8256 (40) for the 8085 microprocessor

Motorola's MC8650 ACIA (Asynchronous Communications Interface Adapter) (24) for the MC6800 microprocessor

Texas Instrument's TMS 9902 ACC (Asynchronous Communications Controller) (18) for the 9900 family of microprocessors

Zilog's Z8031 Z-ASCC (Asynchronous Serial Communications Controller) (40) for the Z8000 microprocessor

Motorola's MC68681 DUART (Dual Asynchronous Receiver/Transmitter) (40) for the MC68000 family of microprocessors

Although the performance characteristics of these chips vary, they all perform basically the functions described previously. The operation of Motorola's ACIA will be discussed in detail as a sample UART.

Figure 5–1 shows how the pins of the ACIA are functionally connected. The pins on the right side of the schematic are connected to the modem by way of the RS 232C interface. Motorola's designations, if different from RS 232C, are shown in parentheses. The purpose of these signals is explained in the following chapter. The pins on the left are connected to the MC6800 microprocessor. For the MC6800 to transfer data to and from the ACIA, the ACIA must first be selected. The MC6800 accomplishes this by applying a high to CS0 and CS1 and applying a low to $\overline{CS2}$. The E signal is actually a clocking signal which clocks the data to and from the data bus buffers. There are four 8-bit ACIA registers with which the MC6800 work: the *control register*, the *status register*, the *transmit data register*, and the *receive data register*. These registers are selected (addressed) by the MC6800 by applying the appropriate signals on RS and R/$\overline{W}$.

Initially, a *master reset* signal is applied to the ACIA by entering 1's into bits 0 and 1 of the control register. This resets all of the status register bits and outputs a high on $\overline{RTS}$ and $\overline{IRQ}$. The bits of the control register must be programmed before transmitting or receiving. For transmission, it identifies the bits to be transmitted: for reception, it identifies the expected sequence of bits to be received.

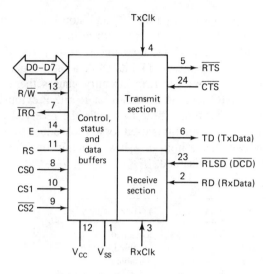

**FIGURE 5–1**  ACIA pin assignments.

## CONTROL REGISTER

Bits 0 and 1 determine what the input clock frequency is divided by to provide a clock for the transmit and receive shift registers. Divide by 16 is most commonly used.

| Bits: | 1 | 0 | |
|-------|---|---|------------|
|       | 0 | 0 | Divide by 1 |
|       | 0 | 1 | Divide by 16 |
|       | 1 | 0 | Divide by 64 |

Bits 2, 3, and 4 identify how many bits will make up a data word; if parity is to be used; if used, whether parity will be odd or even; and the number of stop bits.

| Bits: | 4 3 2 | Data bits | Type of parity | Number of stop bits |
|-------|-------|-----------|----------------|---------------------|
|       | 0 0 0 | 7         | Even           | 2                   |
|       | 0 0 1 | 7         | Odd            | 2                   |
|       | 0 1 0 | 7         | Even           | 1                   |
|       | 0 1 1 | 7         | Odd            | 1                   |
|       | 1 0 0 | 8         | None           | 2                   |
|       | 1 0 1 | 8         | None           | 1                   |
|       | 1 1 0 | 8         | Even           | 1                   |
|       | 1 1 1 | 8         | Odd            | 1                   |

With the ACIA, the sum of the data bits, parity bit, and stop bits will total either 9 or 10 bits. Other UARTs permit 5 to 8 data bits with no restrictions on whether 1 or 2 stop bits are used. With these UARTs, the same sum may vary from 6 bits (5 data bits, no parity bit, and 1 stop bit) to 11 bits (8 data bits, 1 parity bit, and 2 stop bits). In these UARTS, internal *steering* circuitry causes these bits to be transmitted sequentially. All UARTs insert a leading start bit. Previously, 2 stop bits were used to provide slower terminals time to handle the received data. With the speed of current terminals, only 1 stop bit is used for asynchronous transmissions.

Bits 5 and 6 determine whether a high or a low is output on $\overline{\text{RTS}}$ and if the transmit interrupt will be enabled or disabled.

| Bits: | 6 | 5 | $\overline{\text{RTS}}$ | Transmit interrupt |
|-------|---|---|-----|--------------------|
| | 0 | 0 | Low | Disabled |
| | 0 | 1 | Low | Enabled |
| | 1 | 0 | High | Disabled |
| | 1 | 1 | Low | Disabled; break level transmitted on TD |

If bit 7 of the control register is a 1, the microprocessor will be interrupted if:

1. The receive data register is full.
2. There is an overrun.
3. A low-to-high transition is detected on $\overline{\text{RLSD}}$ (loss of received carrier by the modem).

## STATUS REGISTER

The information in the status register denotes the status of the operation in progress. This information is obtained by the microprocessor by reading the status register. The following explanation of the status register bits of the ACIA is not complete. Only the details necessary for the understanding of a UART's operations are presented.

*Bit 0* (RDRF): *Receive Data Register Full.* This bit is set (high or a 1) after received data have been transferred from the *receive shift register* to the receive data register. It is cleared when the microprocessor reads the receive data register.

*Bit 1* (TDRE): *Transmit Data Register Empty.* This bit is set when the data have been transferred from the transmit data register to the *transmit shift register*. It is cleared when the microprocessor writes a new word into the transmit data register.

*Bit 2* ($\overline{\text{RLSD}}$). This bit will be set if the modem is not detecting a carrier.

*Bit 3* ($\overline{\text{CTS}}$). This bit is reset (low or 0) if there is an active clear-to-send signal from the modem.

*Bit 4* (FE): *Frame Error.* This bit is set if the stop bit is not detected.

*Bit 5*(OVRN): *Receiver Overrun.* This bit is set if the ACIA transfers received data from the receive shift register to the receive data register before the microprocessor has read the previous contents of the receive data register. Basically, it indicates that one or more pieces of information have been lost. This bit is reset when the microprocessor reads from the receive data register.

*Bit 6* (PE): *Parity Error.* This bit is set if a parity error is detected in the received information.

*Bit 7* ($\overline{\text{IRQ}}$). This bit is set if there is an active signal on the $\overline{\text{IRQ}}$ output pin to the microprocessor.

## TRANSMIT SECTION

Figure 5–2 shows a block diagram of the ACIA transmitter section. A control word is first written into the control register. As an example, let us assume that this control word is 10101101. Since bits 6 and 5 are 0 and 1, respectively, an active-low signal is supplied by the $\overline{\text{RTS}}$ output to the modem. After an appropriate delay, the modem will supply an active low to the $\overline{\text{CTS}}$ input of the ACIA. Data are ready to be output. The microprocessor reads the status register, and if TDRE (bit

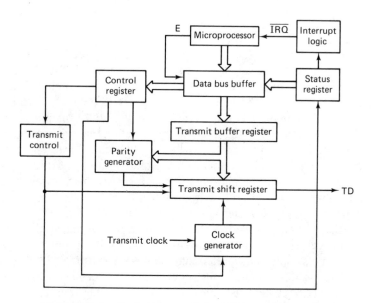

**FIGURE 5–2   Block diagram of the ACIA transmit section.**

FIGURE 5–3 **Transmitted sequence of information.**

1) is a 1, it sends a word to the transmit data register. This word is strobed into the transmit data register on the high-to-low transition of the signal on the E pin. This causes the TDRE bit of the status register to be reset. Internal logic computes the odd-parity bit required and moves the data together with the start bit, parity bit, and 1 stop bit into the transmit shift register. The data are shifted out on the TD line at a rate of $\frac{1}{16}$ of the clock frequency applied to the transmit clock input pin. When the data were moved to the transmit shift register, the TDRE bit of the control register went high. Again, since bits 6 and 5 of the control register are 0 and 1, respectively, when TDRE goes high, an interrupt signal is automatically sent to the microprocessor on the IRQ output. The microprocessor responds by sending the next word to the transmit data register even though the first word may not have been completely shifted out of the transmit shift register. This process is repeated until all of the information has been transmitted. The use of two registers in the transmit section is called *double buffering*. Its purpose is to speed up the time of transmission. While the first word is being shifted out, the microprocessor can obtain the second word from its memory and send it to the transmit data register. In this way, the second word is ready to be loaded into the transmit shift register after the first word has been shifted out. The sequence in which the information is shifted out is shown in Figure 5–3. The start bit is always low; the stop bit(s) are always high.

## RECEIVE SECTION

Figure 5–4 shows a block diagram of the receive section of the ACIA. Incoming serial digital data are applied to the RD pin of the ACIA. Normally, the input to this line is held high by the modem if there is no incoming data. These highs are called *idle line 1's*. The receive clock input to the ACIA is 16 times the input bit rate. The first input bit is a start bit and the ACIA monitors the RD line for a high-to-low transition. Eight clock cycles after the occurrence of this transition, this input is again checked for a low. If a low is still present, a valid start bit has been detected; otherwise, the ACIA would ascribe the high-to-low transition to noise and would start monitoring this line for a new high-to-low transition. Checking the RD line for a high-to-low transition and then rechecking this same input for a low eight clock cycles later is called *start bit validation*.

Eight clock cycles after the initial high-to-low transition of the start bit places us at the center of the start bit. With this as a starting point, every 16 clock cycles the data are sampled and shifted into the receive shift register. The input information is therefore sampled at the center of each bit. The received parity bit is then checked, and if an error is detected, bit 6 (PE) of the status register is set. If the required number of stop bits is not detected, bit 4 (FE) of the status register is set. After

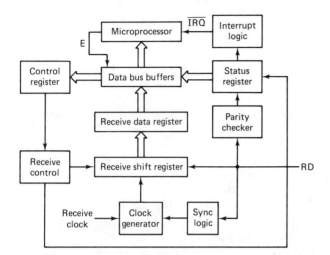

**FIGURE 5–4    Block diagram of the receiver section of the ACIA.**

the expected number of data bits have been received, the information is shifted in parallel from the receive shift register to the receive data register and bit 0 (RDRF) of the status register is set. If bit 7 of the control register had been set during initialization, an *interrupt* to the microprocessor is automatically generated by causing IRQ of the ACIA to go low. The microprocessor executes an interrupt service program and reads the status register to determine the cause of the interrupt. If the microprocessor finds the RDRF bit set, it reads the receive data register. This action clears the RDRF bit of the status register.

The receive section is also double buffered, allowing the next byte of information to be shifted into the receive shift register while the previously received byte is waiting to be read by the microprocessor from the receive data register. After the second and subsequent bytes have been placed in the receive shift register and before the ACIA moves this information to the receive data register, it checks the RDRF bit of the status register. If this bit is still set, the previous word had not yet been read by the microprocessor. The ACIA will overwrite the previous word with the current word in the receive shift register, allow the RDRF bit of the status register to remain set, and sets bit 5 (OVRN) of the status register. This condition would also generate an automatic interrupt to the microprocessor had bit 7 of the control register been set.

# QUESTIONS

1. List the functions of the ACIA.
2. The ACIA can operate in the half-duplex mode.   (T, F)
3. The ACIA can operate in the full-duplex mode.   (T, F)
4. What must bits 4, 3, and 2 of the control register be to transmit ASCII with odd parity and 1 stop bit?

5. Which bit of the status register is set if a received word is entered into the receive data register before the previous word had been read by the microprocessor? What is the name of this error?

6. The $\overline{\text{IRQ}}$ output of the ACIA will automatically output an active signal to the microprocessor during transmit operations if bit(s) _____ of the _____ register are _____ (respectively).

7. The $\overline{\text{IRQ}}$ output of the ACIA will automatically output an active signal to the microprocessor during receive operations if bit(s) _____ of the _____ register are _____ (respectively).

8. If the $\overline{\text{IRQ}}$ output of the ACIA is disabled, how does the microprocessor find out when it should send the next word to the ACIA in transmit operations and when it should read the next word from the ACIA in receive operations?

9. The parity bit is generated in the receive section of the ACIA.   (T, F)

10. In transmission, what is not optional? (*Presence of a start bit, number of data bit, presence of a parity bit, type of parity, number of stop bits*)

11. Identify two reasons why it is advantageous to divide the receive input clock frequency by a value other than 1.

12. If you wanted to increase the accuracy of the receive bit sample, the receiver input clock should be divided by a (*larger, smaller*) number.

13. The TxCLK used at the transmitter must be synchronized with the RxCLK at the receiver (these obviously are not on the same ACIA).   (T, F)

14. Which character codes can be used with the ACIA?

15. The ACIA is used for the transmission and reception of (*synchronous, asynchronous*) data.

16. All ACIA transmissions contain a parity bit.   (T, F)

## PROBLEMS

1. The following ASCII message is received by the ACIA. Even parity is used with the order of the arrow to the left and 2 stop bits.

```
000100001110010000011100011001111011110110110111101101011
011010111101001011011001110111101110011111000000010111
011100010110111101101101011110111110001001111101000010011
0110000011
```

   (a) Translate the received bit pattern into English.
   (b) Find the parity error.

2. The following message is received by the ACIA. Even parity is used with the order of the arrow to the left and 2 stop bits. Idle line 1's are present. *1 + 7e2*

```
111111001000010111111010101000111111111000110010111111
1111101010001111010010011111111111111111100110001110
```

   (a) Translate the message into English.
   (b) Identify three errors in this transmission.

# Six

## *SERIAL INTERFACES*

### RS 232C INTERFACE

In an effort to standardize the interface between data terminal equipment and data communication equipment, Electronic Industries Association (EIA) agreed on the RS (recommended standard) 232C specifications. These specifications identify the electrical, mechanical, and functional description of this interface. The RS 232C is a 25-wire cable. The wires are used for the trnsmission of control signals or of digital data. Figure 6–1 identifies the voltages and terminology associated with the control and data lines. The drivers for these voltages are limited to a range of + 5 to + 15V for a space or an ON condition; − 5 to − 15 V for a mark or an OFF condition. The difference between 3 and 5 V (+ or −) is allowed for a noise margin. The driver voltages can be expected on-line provided that the line is terminated. If a line is not terminated, the line voltages should not exceed ± 25 V.

The maximum physical length is normally identified as 50 ft. The terminal capacitance is specified as 2500 pF, including cable capacitance. For the type of cable intended, the cable capacitance is approximately 50 pF per foot. From this, the 50-ft maximum cable length is derived. In most instances where a cable connection is identified as RS 232C compatible, the compatibility is restricted to the mechanical connectors and the line voltages specified by RS 232C. The maximum impedance at the terminating end is 7000 Ω. The functional use of each wire does not always conform to the function specified by EIA. EIA identifies each pin by a two- or three-letter designator (Table 6–1). The functional use of each line is not immediately obvious from these letter designators. Signal direction on the RS 232C is shown in Table 6–2, and typical modem specifications are listed in Table 6–3.

**Data Lines**

| Logic 1 | Logic 0 |
|---------|---------|
| Mark | Space |
| Voltage of − 3 V or more | Voltage of + 3 V or more |
| Negative | Positive |

**Control Lines**

| OFF | ON |
|-----|-----|
| Low | High |
| Voltage of − 3 V or more | Voltage of + 3 V or more |
| Negative | Positive |

FIGURE 6–1   Data and control line voltages and terminology.

### EXAMPLE: SIGNALING SEQUENCE ON THE RS 232C INTERFACE

The primary wishes to transmit a message to the secondary. Assume that the primary and the secondary DSR and DTR are ON. The primary modem is operating in the switched carrier mode.

The primary causes the RTS to go ON. When the modem detects this circuit ON, it starts to transmit a carrier to the secondary. After the secondary has received the carrier for 4 ms, SQ turns ON, validating the reception of the carrier, and the secondary modem starts training on this carrier. After 41 ms the secondary has completed training and RLSD goes ON. Forty-eight milliseconds after the primary modem has received the RTS it turns CTS ON to notify the primary that it can start transmitting data. At this time, the primary starts sending serial digital data to the data set on TD. These data bits phase-shift modulate the carrier and the resultant signal is transmitted. At the receive end, the modem demodulates the incoming signal converting the information back to serial digital data and sends these data to the secondary DTE on the RD line. Since the RLSD is ON, the secondary DTE accepts the information on the RD line as data. When the primary completes data transmission, it returns RTS back to OFF. When the transmit modem detects this change on RTS, it causes CTS to go OFF and stops transmitting the carrier. At the receive end, 2 ms after the carrier is lost, SQ goes OFF, followed almost immediately by RLSD going OFF. The internal mark generator of the secondary modem will now output constant marks on the RD line. This last state of the RD line is called the *mark hold condition*. While a carrier is being received by the receiving station but no data are present on the carrier, the mark generator circuit of the receive modem is disconnected from the RD line and the carrier is converted to constant marks on the RD line. As far as the receive DTE is concerned, by simply observing the constant marks on the RD line, it cannot tell how they originated. The constant marks derived from the carrier produce an *idle line 1 condition*.

Demodulation at the receive end is not instantaneous. Since the transmitter stops transmitting a carrier after the last bit of data has been sent, it is possible for the RLSD at the receive end to go OFF prior to the time that the modem has completed demodulating the last piece of received information. If the RLSD is OFF, anything placed on the RD line by the modem will not be accepted as data by the DTE. Hence the need of the 2-ms RLSD OFF delay—to allow the last piece of demodulated information to get to the DTE after the carrier is turned OFF.

There are many different types of modems. They differ in their transmission

**TABLE 6–1   RS 232C Interface**

| Pin number | EIA nomenclature | Common or alternate nomenclature |
|---|---|---|
| 1 | Protective Ground (AA) | |
| 2 | Transmitted Data (BA) | Send Data (TD, SD) |
| 3 | Received Data (BB) | (RD) |
| 4 | Request to Send (CA) | (RS, RTS) |
| 5 | Clear to Send (CB) | (CS, CTS) |
| 6 | Data Set Ready (CC) | (DSR), Modem Ready (MR) |
| 7 | Signal Ground (AB) (Common Return) | |
| 8 | Received Line Signal Detector (CF) | (RLSD), Carrier on Detect (COD) |
| 9 | Reserved for Data Set Testing | + V (+ 10 V dc) |
| 10 | Reserved for Data Set Testing | − V (− 10 V dc) |
| 11 | Unassigned | Equalizer Mode (QM)—Synchronous Modems |
| | | Originate Mode (OM)—Asynchronous Modems |
| 12 | Secondary Received Line Signal Detector (SCF) | Local Mode (LM)—Asynchronous Modems |
| 13 | Secondary Clear to Send (SCB) | |
| 14 | Secondary Transmitted Data (SBA) | New Synch (NS) |
| 15 | Transmission Signal Element Timing (DB) (DCE Source) | Serial Clock Transmit (SCT) |
| 16 | Secondary Received Data (SBB) | Divided Clock Transmit (DCT) |
| 17 | Receiver Signal Element Timing (DD) (DCE Source) | Serial Clock Receive (SCR) |
| 18 | Unassigned | Divided Clock Receive (DCR) |
| 19 | Secondary Request to Send (SCA) | |
| 20 | Data Terminal Ready (CD) | (DTR) |
| 21 | Signal Quality Detector (CG) | (SQ) |
| 22 | Ring Indicator (CE) | (RI) |
| 23 | Data Signal Rate Selector (CH) (DTE/ DCE Source) | (SS) |
| 24 | Transmit Signal Element Timing (DTE Source) (DA) | Serial Clock Transmit External (SCTE) |
| 25 | Unassigned | |

**TABLE 6–2   Signal Direction on RS 232C Interface**

| | | | |
|---|---|---|---|
| | 2 | Transmitted Data | → |
| | 3 | Received Data | ← |
| | 4 | Request to Send | → |
| | 5 | Clear to Send | ← |
| | 6 | Data Set Ready | ← |
| | 8 | Received Line Signal Detector | ← |
| D | 12 | Secondary Received Line Signal Detector | ← |
| T | 13 | Secondary Clear to Send | ← |
| E | 14 | Secondary Transmitted Data | → |
| | 15 | Transmission Signal Element Timing (DCE Source) | ← |
| | 16 | Secondary Received Data | ← |
| | 17 | Receiver Signal Element Timing (DCE Source) | ← |
| | 19 | Secondary Request to Send | → |
| | 20 | Data Terminal Ready | → |
| | 21 | Signal Quality Detector | ← |
| | 22 | Ring Indicator | ← |
| | 23 | Data Signal Rate Selector (DTE/DCE Source) | → |
| | 24 | Transmit Signal Element Timing (DTE Source) | → |

(DCE column marked on right side)

**TABLE 6–3   Condensation of Typical Modem Specifications**

Data Rate:           4800 bps
Modulation:          8-PSK
Operation:           Synchronous, binary, serial
Line Requirements:   Basic four-wire, private line (type 3002)
Operating Modes:     Simplex, half-duplex, duplex
Request to Send (RTS)/Clear to Send (CTS) delay:
   Switched carrier: 48.5 ms
   Continuous carrier with switched RTS option: 8 ms
Signal Quality Detector (SQ):
   Turn on if data carrier is present for 4 ms or longer.
   Turn off if data carrier is absent for 2 ms or longer.
Receive Line Signal Detect (RLSD):
   Turn on 41 ms after SQ on.
   Turn off less than 1 ms after SQ off. (1-second holdover option available.)
Protective Ground (1): Connected to the frame and to external ground at either the DCE or DTE but not both.
Signal Ground (7): Reference for control and data line voltages. Generally connected to protective ground to eliminate longitudinal power line noise.
Data Set Ready (6): Local data set is capable of transmitting and receiving data, and the data set is not in the TEST mode. An ON condition does not imply that a channel has been established to the remote site.

Request to Send (4): With *switched carrier operation*, an ON condition indicates to the local data set transmitter of the intent of the DTE to transmit data. With *continuous carrier*, the *carrier* is transmitted continuously. RTS may be strapped ON at the modem, or the RTS/CTS option may be exercised to provide a time delay for the transmitting data terminal.

Received Line Signal Detector (8): This circuit turns ON 45 ms after the data carrier has been received (41 ms after SQ turns ON). The 41 ms allows the adaptive equalizer at the receive site to "train" itself and get into synchronization with the received carrier. This circuit must be ON before data can be received on RD. The 1-second holdover option is to prevent loss of synchronization at the receive modem even though the reception of the carrier is interrupted for 1 second or less. During this time, any demodulated signals are passed on to the data terminal on RD even though they may be invalid. RLSD ON enables the RD pin.

Clear to Send (5): With switched carrier operation, this circuit will go ON 48.5 ms after RTS is received by the modem. This delay is built into the modem and should be greater than the time for SQ ON plus the time for the RLSD ON at the receive modem. When this circuit goes ON, the modem indicates to the DTE that it is ready to transmit data. With continuous carrier operation, if RTS is strapped ON, the modem will supply a constant ON on this line. With continuous carrier and the switched RTS option, this circuit will go ON 8 ms after the modem receives an RTS. CTS ON enables the TD pin.

Transmitted Data (2): Serial digital data in the form of marks and spaces are transmitted on this line from the DTE to the DCE. The data set samples this data stream on the ON to OFF transition of the Transmitter Signal Element Timing circuit. Data should not be transmitted on this circuit unless RTS, CTS, and DSR are ON.

Transmission Signal Element Timing (15): This circuit provides the timing for the transmission of data from the DTE to the modem. Data are sent to the modem on the OFF-to-ON transition of this line and are sampled by the modem on the ON-to-OFF transition of this line. The signal on this line is generated by an internal clock (4800 Hz) in the modem.

Transmit Signal Element Timing (24): If external (to the data set) timing is desired, the DTE can send timing information on this circuit to the modem and the internal modem clock will phase lock onto it. If the modem is designed for a 4800-bps data rate, the signal on this line should be at that frequency. The output onto TD will still be controlled by the signal from the modem on circuit 15.

Receiver Signal Element Timing (17): Clocking information on this line provides the receive DTE with timing information concerning the rate at which data are being sent from the DCE to the DTE on RD. The ON-TO-OFF transition marks the center of the data bit on the RD line.

Received Data (3): Received demodulated data are sent on this line from the DCE to the DTE. If no data are being received (RLSD OFF), a mark generator internal to the modem will keep a constant mark on the RD line.

Signal Quality Detector (21): This circuit provides a fast response indication of the presence or absence of a data carrier signal from a distant terminal. This line goes ON if a carrier is received for 4 ms or longer and goes OFF if the data carrier is absent for 2 ms or longer.

Equalizer Mode (11) (non-EIA): This circuit is used by the data set to notify the DTE when the data set begins self-adjusting because error performance is poor. When RLSD is ON and this circuit is OFF, the data set is retraining and the probability of error in the data on RD is high. When RLSD is ON and this circuit is ON, the modem is trained and the probability of error on RD is small.

New Synch (14) (non-EIA): The use of this circuit is optional and is intended for use with a data set at the primary of a multipoint dedicated network. When the primary is in a polling operation, rapid resynchronization of the receiver to many remote transmitters is required. The receiver clock normally maintains the timing information of the previous message for some interval after the message has ended. This may interfere with resynchronization on the receipt of the next message. An ON condition should be applied to this circuit by the DTE for at least 1 ms but no longer than the intermessage interval to squelch the existing timing information in the modem after the message has been received.

rate and in the type of modulation that they perform. The RLSD ON and OFF delays may differ considerably as well as the RTS/CTS delays.


**EXAMPLE: TIMING DIAGRAM OF THE SIGNALING SEQUENCE ON A RS 232C INTERFACE** (Figure 6–2)

* RTS/CTS delays
    —40 ms:   Switched carrier option.
    —20 ms:   Continuous carrier with switched RTS option.
* RLSD
    —Turn ON 10 ms after the reception of the analog carrier. For simplicity, this time will include the SQ time.
    —Turn OFF 0 ms after the loss of the analog carrier. This time will also include the SQ time.
* The primary is operating with continuous carrier, switched RTS option.
* The secondary is operating with switched carrier.
* The primary will send a 150-ms message to the secondary.
* Propagation delay (time for a signal to go from the primary modem to the secondary modem, or vice versa) = 30 ms.
* Secondary turnaround time = 100 ms. The turnaround time is the time from which a message is received until the response is transmitted. This time can be subdivided into:
    —Terminal turnaround time: the time a terminal receives a message, checks it for errors, decides what the response should be, and composes the reply.
    —RTS/CTS delay
    In this example, these two components will be handled individually.
* The secondary responds with a 20-ms message.
* Primary and secondary data sets are turned ON at $t$ = 0 ms.
* Primary turns RTS ON at $t$ = 20 ms.

On the timing diagram (Figure 6–2), the transmitted carrier (TxCar) and the received carrier (RxCar) are not signals on the RS 232C interface. The carrier is transmitted on the telephone lines from modem to modem.

$t$ = 0        Since the primary data set was turned ON at this time, and it is operating with continuous carrier option, it starts transmitting the carrier immediately.

$t$ = 30       Since the propagation delay is 30 ms, the secondary will start receiving the carrier at this time.

$t$ = 40       Ten milliseconds later, the secondary's RLSD turns ON. It took this amount of time for the secondary modem to detect the carrier and train on it.

$t$ = 20       The primary terminal turns RTS ON.

$t$ = 40       Twenty milliseconds later, the primary terminal receives a CTS ON from its modem. It has allowed time for the secondary modem to detect and train onto the carrier. When the primary receives CTS ON, it starts transmitting data. The HIGH on the TD line is not to be misinterpreted as an ON condition. The HIGH on this line represents the time duration during which

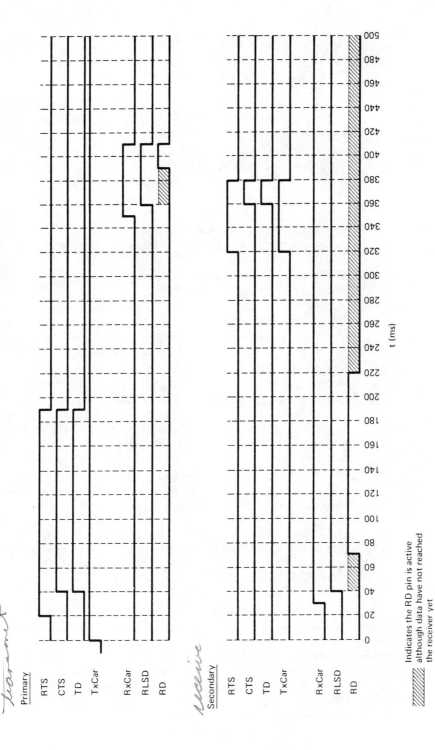

FIGURE 6–2 Timing diagram of a signal sequence on an RS 232C interface.

Indicates the RD pin is active
although data have not reached
the receiver yet

Primary

*transmit*

RTS
CTS
TD
TxCar
RxCar
RLSD
RD

Secondary

*receive*

RTS
CTS
TD
TxCar
RxCar
RLSD
RD

t (ms)

0  20  40  60  80  100  120  140  160  180  200  220  240  260  280  300  320  340  360  380  400  420  440  460  480  500

77

|          | the data are sent. During this time, the actual signals on this line will be marks and spaces. |
| $t = 70$ | Because of the propagation delay, data are received at the secondary 30 ms after they were transmitted by the primary. Comments made previously concerning the HIGH on the TD line also apply to the RD line. When there are no data on the RD line, the signal on this line is held at a constant mark condition (mark hold) by the modem. To prevent confusion, this is not shown on the timing program. |
| $t = 190$ | The primary completes the transmission of the 150-ms message. On completion of transmission, the primary returns RTS to OFF. |
| $t = 220$ | Thirty milliseconds after the primary transmitted the last piece of data, the secondary receives the last piece of data. |
| $t = 320$ | One hundred milliseconds turnaround time for the secondary. At this time, the secondary turns RTS ON and immediately starts transmitting its carrier back to the primary. |
| $t = 350$ | The primary modem receives the secondary's carrier after the propagation delay. |
| $t = 360$ | After 10 ms to detect and train onto the carrier, the primary modem turns RLSD ON. Forty milliseconds after the secondary modem received an RTS ON from the secondary terminal, it returns CTS ON to the terminal. The terminal, on receiving CTS ON, starts transmitting data. |
| $t = 390$ | Thirty milliseconds after the secondary started to transmit data, the primary starts to receive the data. |
| $t = 380$ | The secondary completes data transmission and places RTS OFF. The secondary modem, sensing this change on RTS, places an OFF on CTS and, at the same time, stops transmitting its carrier. Note that, since the primary was operating with continuous carrier option, its carrier is still being transmitted. |
| $t = 410$ | Thirty milliseconds after the secondary stopped transmitting data, the primary stops receiving data. Thirty milliseconds after the secondary stopped transmitting the carrier, the primary stopped receiving the carrier. Since there is no RLSD OFF delay, the primary modem immediately places an OFF on RLSD. Protocols such as Bisynch terminate each transmission with a trailing pad of FFH. This trailing pad allows the receive modem sufficient time to demodulate the last piece of data and send it out on RD before RLSD goes OFF. |

This example presumes a multipoint, four-wire, dedicated system. Each line in Figure 6–3 represents a two-wire pair. Since the primary is operating with continuous carrier option, all secondary station modems should be continuously trained onto it. Only one secondary station can transmit to the primary at any given time. The carrier frequency of each station may be different. Therefore, the necessity that each operate with a switched carrier option. The primary modem must be able to synchronize rapidly onto the carrier of the station that is transmitting. Hence the need of a New Sync signal from the primary terminal to the modem.

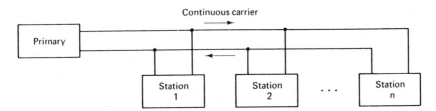

**FIGURE 6–3    Primary: continuous carrier; secondaries: switched carrier.**

## RS 422A; RS 423A; RS 449

RS 232C was intended for interfacing the computer/data terminal to the modem, where the separating distance was 50 ft or less. Realizing that this standard was unsuitable where greater distances were required between the terminal and the modem, EIA has generated three new standards: RS 422A (*Electrical Characteristics of Balanced Voltage Digital Interface Circuits*), RS 423A (*Electrical Characteristics of Unbalanced Voltage Digital Interface Circuits*), and RS 449 (*General Purpose 37-Position and 9-Position Interface for Data Terminal Equipment and Data Circuit-Terminating Equipment Employing Serial Binary Data Interchange*).

From the Appendix to RS 422A (not actually a formal part of the EIA Recommended Standard):

> The maximum permissible length of cable separating the generator and load is a function of data signaling rate and is influenced by the tolerable signal distortion, the amount of longitudinally coupled noise and ground potential difference introduced between the generator and the load circuit grounds as well as by cable balance. Increasing the physical separation and interconnecting cable length between the generator and the load interface points increases exposure to common mode noise, signal distortion, and the effects of cable imbalance. Accordingly, users are advised to restrict cable length to a minimum, consistent with the generator-load physical separation requirements [see Figure 6–4].
>
> At a 10 Mbit/s signaling rate, the cable length is limited to approximately 15 m. As the data signaling rate is reduced below 90 kbit/s, the cable length has been limited at 1200 meters (4000 feet) by the assumed maximum allowable 6 dBV signal loss.
>
> For the RS 422A, terminal A of the generator shall be negative with respect to the B terminal for a binary 1 (MARK or OFF) state. The A terminal of the generator shall be positive with respect to the B terminal for a binary 0 (SPACE or ON) state.

From the Appendix to RS 423A (not a formal part of the EIA Recommended Standard): for a 3-μs linear rise time, the maximum cable length is 90 m and the maximum signaling rate 100 kbps. For approximately a 34-μs linear rise time, the maximum cable length is 600 m and the maximum signaling rate is 60 kbps. An unbalanced digital interface circuit is shown in Figure 6–5.

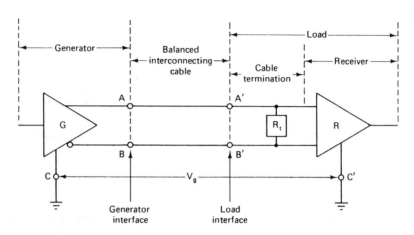

R$_t$   = optional cable termination resistance
V$_g$   = ground potential difference
A, B = generator interface points
A', B' = load interface points
C    = generator circuit ground
C'   = load circuit ground

**FIGURE 6–4   Balanced digital interface circuit. (EIA: RS 422A)**

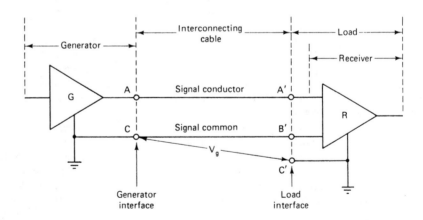

A, C = generator interface
A', B' = load interface
V$_g$   = ground potential difference
C    = generator circuit ground
C'   = load circuit ground

**FIGURE 6–5   Unbalanced digital interface circuit. (EIA: RS 423A)**

For the RS 423A, the A terminal of the generator shall be negative with respect to the C terminal for a binary 1 (MARK or OFF) state. The A terminal of the generator shall be positive with respect to the C terminal for a binary 0 (SPACE or ON) state.

A balanced interface, such as the RS 422A, transfers information to a *balanced transmission line*. With a balanced transmission line, both conductors carry current and the current in each wire is 180° out of phase with the current in the other wire. With a bidirectional *unbalanced* line, one wire is at ground potential while the other wire carries signal currents in both directions. The two signal currents are equal in magnitude with respect to electrical ground but travel in opposite directions. Currents that flow in opposite directions in a balanced wire pair are called *metallic circuit* currents. Currents that flow in the same direction are called *longitudinal* currents. A balanced pair has the advantage that most noise interference is induced equally in both wires, producing longitudinal currents that cancel in the load. Figure 6–6 shows the results of metallic and longitudinal currents on a balanced transmission line. It can be seen that longitudinal currents (generally produced by static interference) cancel in the load. Balanced transmission lines can be connected to unbalanced loads and vice versa, with special transformers called *baluns* (*bal*anced to *un*balanced).

RS 449 is an updated version of the RS 232C. It has two cables, a 37-pin connector and a 9-pin connector. The 37-pin connector is used for interfacing signals, while the 9-pin connector is used for diagnostics. With the RS 449, the two-letter pin designators much more accurately define the pin functions when compared with the RS 232C.

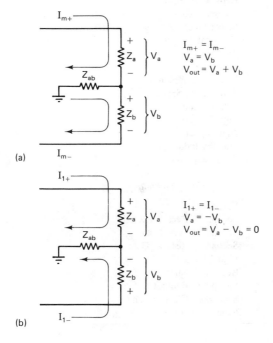

$$I_{m+} = I_{m-}$$
$$V_a = V_b$$
$$V_{out} = V_a + V_b$$

$$I_{1+} = I_{1-}$$
$$V_a = -V_b$$
$$V_{out} = V_a - V_b = 0$$

(a)

(b)

**FIGURE 6–6  Results of metallic and longitudinal currents on a balanced transmission line: (a) metallic currents due to signal voltages; (b) longitudinal currents due to noise voltages.**

RS 449 divides interchange circuits into two categories (Table 6–4): category I (two data, three timing, and five other circuits) and category II (all others). Below 20 kbps, category I circuits can be implemented with RS 422A or RS 423A drivers and receivers; above 20 kbps, RS 422 circuitry must be used. Category II circuits (generally status and maintenance circuits) always use RS 423A.

**TABLE 6–4   RS 449 Interface**

| Pin number | Mnemonic | Circuit name | Circuit category |
|---|---|---|---|
| | | *37-Pin Cable Primary* | |
| 1 | | Shield | |
| 2 | SI | Signaling Rate Indicator | II |
| 3, 21 | | Spare | |
| 4, 22 | SD | Send Data | I |
| 5, 23 | ST | Send Timing | I |
| 6, 24 | RD | Receive Data | I |
| 7, 25 | RS | Request to Send | I |
| 8, 26 | RT | Receive Timing | I |
| 9, 27 | CS | Clear to Send | I |
| 10 | LL | Local Loopback | II |
| 11, 29 | DM | Data Mode | I |
| 12, 30 | TR | Terminal Ready | I |
| 13, 31 | RR | Receiver Ready | I |
| 14 | RL | Remote Loopback | II |
| 15 | IC | Incoming Call | II |
| 16 | SF/SR | Select Frequency/Signaling Rate Select | II |
| 17, 35 | TT | Terminal Timing | I |
| 18 | TM | Test Mode | II |
| 19 | SG | Signal Ground | II |
| 20 | RC | Receive Common | |
| 28 | IS | Terminal in Service | II |
| 32 | SS | Select Standby | II |
| 33 | SQ | Signal Quality | II |
| 34 | NS | New Signal | II |
| 36 | SB | Standby Indicator | II |
| 37 | SC | Send Common | II |
| | | *9-Pin Cable Secondary* | |
| 1 | | Shield | |
| 2 | SRR | Secondary Receiver Ready | II |
| 3 | SSD | Secondary Send Data | II |
| 4 | SRD | Secondary Receive Data | II |
| 5 | SG | Signal Ground | II |
| 6 | RC | Receive Common | II |
| 7 | SRS | Secondary Request to Send | II |
| 8 | SCS | Secondary Clear to Send | II |
| 9 | SC | Send Common | II |

Much of the field equipment is presently RS 232C compatible. As the RS 449 replaces the RS 232C, problems will arise when the modem connectors are made for the RS 449 while the terminal equipment satisfies RS 232C requirements. Adapters are available through which an RS 232C terminal can hook onto an RS 449 interface.

As EIA sets standards in the United States, CCITT (Comité Consultatif International Téléphonique et Télégraphique) sets standards in the European countries. V.28 (*Electrical Characteristics for Unbalanced Double-Current Interchange Circuits*) by CCITT parallels the voltage specifications of RS 232C, and V.24 (*List of Definitions for Interchange Circuits between Data Terminal Equipment and Data Circuit Terminating Equipment*) parallels the pin designations of the RS 232C. X.20 concerns itself with the *Interface between Data Terminal Equipment for Start-Stop Transmission Services on Public Data Networks*. European standards are readily recognizable in American literature since the more common standards are identified as either X.xx or V.xx, where xx is a two-digit number.

## 20-mA LOOP

The 20-mA loop is often called the teletype (TTY) interface. Originally, it was used to interface mechanical teleprinters to the telephone line. A teleprinter is a combination of printer and keyboard. Video displays (CRTs) are not included. Interface arrangements that transfer signals with current flow rather than voltage levels (such as the 20-mA loop) are generally used to control switch closures in mechanical devices. With the 20-mA loop, a logic 1 or mark condition is indicated by the presence of the dc current, and a logic 0 or space is indicated by the absence of this current. Continuity is monitored during idle line conditions by simply testing for current on the line. This is believed to be why data communication circuits transmit idle line 1's in the absence of data.

Today, the 20-mA loop is used occasionally to interface microprocessors and other digital equipment to external line printers. Due to the mechanical nature of the printers and teleprinters, they generate noise spikes that could interfere with the data or possibly damage delicate equipment that may be attached to the other end of the interface. Very often, optical couplers are used to isolate the equipment and suppress the noise spikes.

Teleprinters are rapidly becoming extinct; they are being replaced by modern keyboard displays and high-speed printers. Consequently, the 20-mA loop is seldom used today and the RS 232C has become the standard interface.

## REFERENCES

EIA Standard RS 232C, *Interface between Data Terminal Equipment and Data Communication Equipment Employing Serial Binary Data Interchange*, August 1969.

EIA Standard RS 422A, *Electrical Characteristics of Balanced Voltage Digital Interface Circuits*, December 1978.

EIA Standard RS 423A, *Electrical Characteristics of Unbalanced Voltage Digital Interface Circuits*, December 1978.

EIA Standard RS 449, *General Purpose 37-Position and 9-Position Interface of Data Terminal Equipment and Data Circuit-Terminating Equipment Employing Serial Binary Data Interchange*, November 1977.

The references above may be obtained from: Electronic Industries Association, 2001 Eye Street, NW, Washington, DC 20006.

# QUESTIONS

1. What is the nominal maximum physical length of the RS 232C interface cable?

2. What characteristics of the RS 232C interface cable limit its physical length?

3. The Clear to Send (CTS) signal is a handshake from the DCE to the DTE in response to the _____ signal.

4. What is the purpose of the RTS/CTS delay?

5. What signal on the RS 232C interface must be active before the RTS can go active?

6. What signal on the RS 232C interface indicates the presence of a received carrier?

7. The receive data (RD) pin of the RS 232C interface is enabled by an active condition of the _____ signal.

8. + 10 V dc on a data pin represents a logic _____ .

9. − 12 V dc on a control pin represents a(an) _____ condition.

10. A driver transfers signals from the DTE to the DCE and never from the DCE to the DTE. (T, F)

11. What is the maximum terminated voltage that should be present on the RS 232C interface?

12. Analog signals are present (*between the DTE and the DCE, between the LCU and the DTE, between the transmit DCE and the receive DCE*).

13. The RS 232C interface specifies a _____ pin cable.

14. The RS 232C interface is between the (*LCU and the DTE, DTE and the DCE, LCU and the UART, transmit and receive modems*).

15. With switched carrier operation, the transmit modem will begin transmitting an analog carrier when _____ goes active.

16. RTS/CTS will be optioned (*on, off*) for switched carrier operation.

17. The RS 232C is gradually being replaced by the _____ .

18. When transmitting a message, first the (*CTS goes active, RTS goes active, DSR must be active, RLSD goes active*).

19. The RS 449 interface is comprised of a _____ wire cable for interface signaling and a _____ -wire cable for diagnostics.

20. The RS 422A and RS 423A standards specify the physical limitations for the RS 449 standard. (T, F)

21. The RS 422A standard is a (*balanced, unbalanced*) cable.

22. The maximum possible physical length of the RS 449 interface is (*less than*, *equal to*, *greater than*) that of the RS 232C interface.

23. The maximum kbps rate for the RS 423A standard remains constant regardless of the physical length. (T, F)

24. What is the European counterpart of the EIA?

25. Explain the difference between balanced and unbalanced transmission lines.

26. What is a longitudinal current?

27. What is a metallic current?

28. What is the special transformer called that matches a balanced line to an unbalanced load?

# PROBLEMS

1. For the following modem and circuit specifications, complete the timing diagram in Figure P6–1 for the RS 232C signaling sequence.

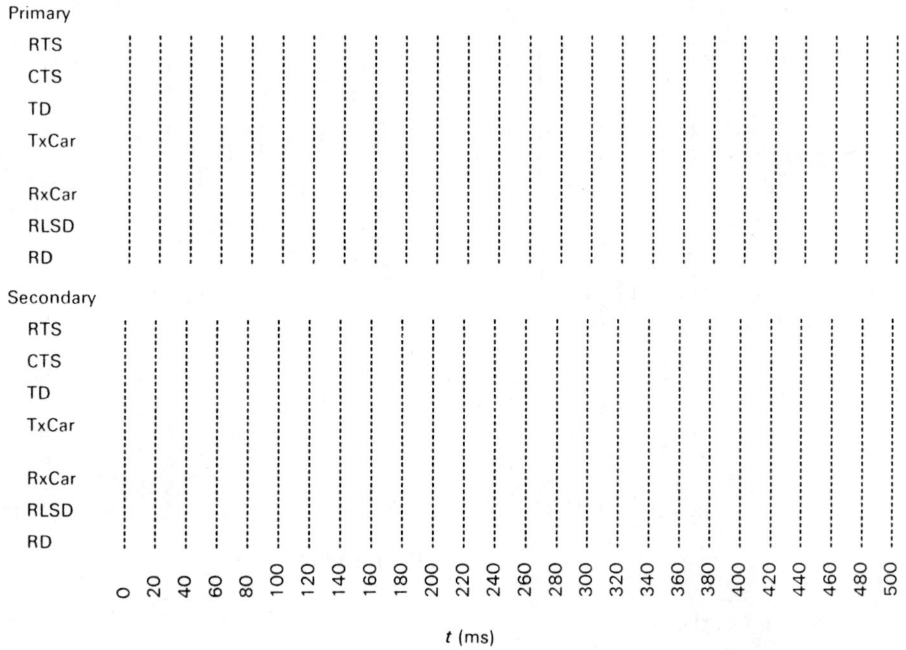

FIGURE P6–1

- RTS/CTS delay:
  —40 ms:  switched carrier option
  —20 ms:  continuous carrier option with switched RTS

- RLSD delay:
  —20 ms:  turn on
  —10 ms:  turn off

- Primary optioned for switched carrier, switched RTS
- Secondary optioned for switched carrier, switched RTS
- Propagation time:  20 mis in both directions
- Secondary turn around time:  40 ms
- Primary message length:  80 ms
- Secondary message length:  40 ms
- Primary begins sequence at time $t = 0$ ms (RTS goes active)

2. For the following modem and circuit specifications, complete the timing diagram in Figure P6–2 for the RS 232C signaling sequence.

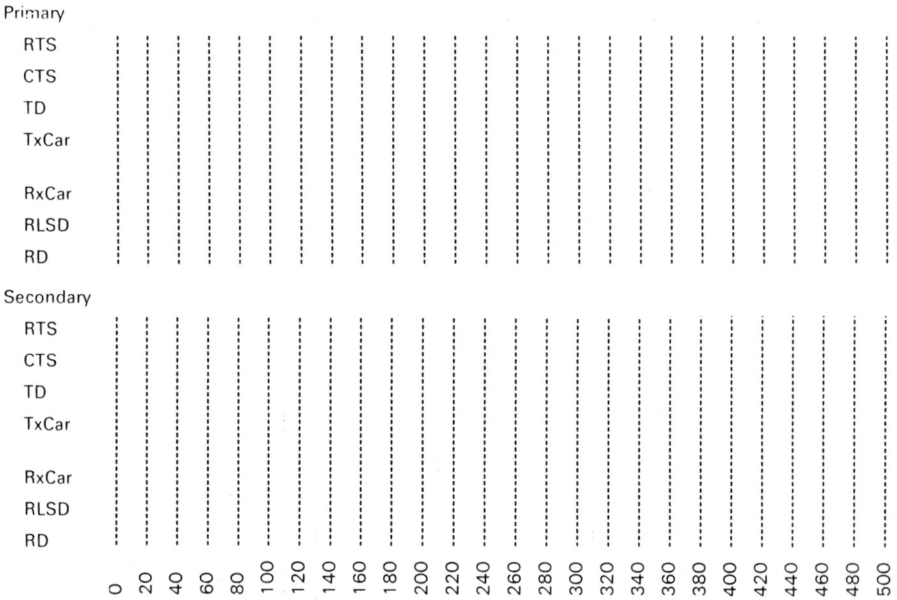

Primary
  RTS
  CTS
  TD
  TxCar

  RxCar
  RLSD
  RD

Secondary
  RTS
  CTS
  TD
  TxCar

  RxCar
  RLSD
  RD

0  20  40  60  80  100  120  140  160  180  200  220  240  260  280  300  320  340  360  380  400  420  440  460  480  500

**FIGURE P6–2**

- RTS/CTS delay:
  —20 ms:  switched carrier option
  —10 ms:  continuous carrier option with switched RTS

- RLSD delay:
  —10 ms:  turn on
  — 0 ms:  turn off

- Primary optioned for continuous carrier, switched RTS
- Secondary optioned for switched carrier, switched RTS .
- Propagation time:   40 ms. in both directions
- Secondary turnaround time:   20 ms
- Primary message length:   60 ms
- Secondary message length:   20 ms
- Primary and secondary data sets both turned on at time $t = 0$ ms
- Primary begins sequence at time $t = 20$ ms (RTS goes active)

# SEVEN

## *DATA TRANSMISSION WITH ANALOG CARRIERS*

Data transmission involves the transmittal of digital information from one DTE to another. Rather than construct special lines for this purpose, it is more convenient to use the public telephone network (PTN). Unfortunately, the vast majority of the PTN was installed long before the advent of large-scale data communications and was designed for conveying information that is analog in nature. In addition, the analog information is restricted to the frequency range from 300 to 3000 Hz. The primary function of the data set (modem) at the transmit end is to convert digital pulses to analog signals suitable for transmission on the PTN and, at the receive end, to convert the analog signals back to digital information.

Modems are categorized as being either synchronous or asynchronous. If a communications data link uses synchronous modems, the internal clocks of the transmit and the receive modems must be synchronized. These clocks are used to control the rate at which digital data flows. The type of modulation used with synchronous modems is such that the receive modem can extract clocking information from the received signal and use it to maintain synchronization with the transmit modem. In addition, carrier frequency synchronization between the transmit and receive modems must be established and maintained in order to demodulate the signal correctly at the receive end. Asynchronous modems do not have to abide by these restrictions. Clocking information is never transmitted between asynchronous modems.

Three properties of an analog signal may be varied: amplitude, frequency, and phase. Varying only the amplitude of the analog signal at a digital rate is called *amplitude shift keying* (ASK). ASK has a number of inherent disadvantages, with poor noise immunity being the most predominant; consequently, it is seldom used in practice. *Frequency shift keying* (FSK) varies the frequency at a digital rate and is used for low-speed data transmission. Modems using FSK are generally asynchronous and can operate at a maximum transmission rate of 1800 bps. Medium-speed modems are generally synchronous, operate at speeds up to 4800 bps, and use *phase shift keying* (PSK). High-speed modems (9600 bps) use *quadrature ampli-tude modulation* (QAM), in which both the amplitude and phase of the analog carrier are varied by the digital modulating signal. These facts can be ascertained from Table 7–1, which shows the characteristics of commonly used Bell modems.

## FREQUENCY SHIFT KEYING

FSK is the process of shifting (modulating) an analog carrier frequency at a digital rate. FSK closely resembles standard frequency modulation. With FSK the modulating signal is not a sinusoidal signal but a series of dc pulses that vary between two discrete voltage levels. The output of an FSK modulator can be considered a step function in the frequency domain.

### 202T Modem

The Bell System model 202T modem is an asynchronous transceiver utilizing FSK modulation. It employs a 1700-Hz carrier that can be shifted at a maximum rate of 1200 times a second. When a logic 1 (mark) is applied to the modulator, the carrier is shifted down 500 Hz to 1200 Hz. When a logic 0 (space) is applied, the carrier is shifted up 500 Hz to 2200 Hz. Consequently, as the digital signal alternates between 1 and 0, the carrier is shifted back and forth between 1200 and 2200 Hz. This process is analogous to a conventional FM system. The difference between the mark and space frequencies (1200 to 2200 Hz) relates to the peak-to-peak frequency deviation, and the digital rate of change relates to the frequency of the modulating input signal. Therefore, we can say the 1700-Hz carrier is frequency-modulated by a square-wave input signal. A figure of merit used to express the degree of modulation achieved in an FSK modulator is the $h$ factor, which is defined as

$$h = \frac{|f_m - f_s|}{\text{bps}}$$

where  $f_m$ = mark frequency
  $f_s$ = space frequency
  bps = input bit rate (bps)

For the 202T modem,

$$h = \frac{|1200 - 2200|}{1200} = \frac{1000}{1200} = 0.83$$

**TABLE 7–1  Commonly Used Bell Modems**

| Bell designation | Line facility | Operating mode | Synchronization | Type of modulation | Maximum data rate (bps) | Function |
|---|---|---|---|---|---|---|
| 103A2, A3, E | Dial-up | FDX | Asyn. | FSK | 300 | Originate/answer |
| 103F | Leased (2-wire) | FDX | Asyn. | FSK | 300 | Originate/answer |
| 113A | Dial-up | FDX | Asyn. | FSK | 300 | Originate only |
| 113B | Dial-up | FDX | Asyn. | FSK | 300 | Answer only |
| 201B | Dial-up | HDX/ FDX | Syn. | 4PSK | 2400 | |
| 201C | Leased (4-wire) | HDX/ FDX | Syn. | 4PSK | 2400 | |
| 202S | Dial-up | HDX | Asyn. | FSK | 1200 | |
| 202T | Leased (4-wire) | HDX/ FDX | Asyn. | FSK | 1200 (Basic, 3002 channel) 1800 (C1 Conditioning) | |
| 208A | Leased (4-wire) | HDX/ FDX | Syn. | 8PSK | 4800 | |
| 208B | Dial-up | HDX | Syn. | 8PSK | 4800 | |
| 209A | Leased (4-wire) | FDX | Syn. | QAM | 9600 (D1 conditioning) | |

As a general rule and for best performance, the $h$ factor is limited to a value less than 1. Note the similarity of the expression for $h$ and the expression for the modulation index (MI) in conventional FM.

$$MI = \frac{\Delta f}{f_{mod}}$$

where $\Delta f$ = peak frequency deviation
$f_{mod}$ = modulating signal frequency

The modulation index for the 202T modem is

$$\Delta f = \frac{|f_m - f_s|}{2} = \frac{|1200 - 2200|}{2} = 500 \text{ Hz}$$

This converts the peak-to-peak frequency swing of the modem to a peak frequency deviation:

$$f_{mod} = \frac{f_b}{2} = \frac{1200}{2} = 600 \text{ Hz}$$

where $f_b$ is the input bit rate (bps).

$$MI = \frac{500}{600} = 0.83$$

which is the same as the $h$ factor.

Why does $f_{mod} = f_b/2$? The fastest transitions by a 1200-bps digital signal would produce a square wave of 600 Hz. The time of each alteration would be 1/1200 s (Figure 7–1).

Fourier analysis demonstrates that any periodic wave is made up of the sum of a series of harmonically related sine waves. A square wave is a periodic wave made up of the fundamental frequency and an infinite number of odd harmonics. The fundamental frequency is equal to $1/2t_b$ ($t_b$ = time of 1 bit).

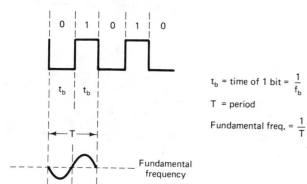

$t_b$ = time of 1 bit = $\frac{1}{f_b}$

$T$ = period

Fundamental freq. = $\frac{1}{T}$

**FIGURE 7–1   Fundamental frequency of a square wave.**

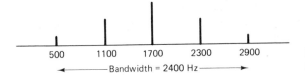

| 500 | 1100 | 1700 | 2300 | 2900 |
|-----|------|------|------|------|

← Bandwidth = 2400 Hz →

**FIGURE 7–2  Frequency spectrum of a 202T modem.**

With FM the number of side frequencies generated is directly related to the modulation index. The separation between adjacent side frequencies is equal to the frequency of the modulating signal. In the case of the 202T modem, the frequency spectrum of Figure 7–2 would result. Any additional side frequencies beyond the second set would extend beyond the passband of a basic voice band telephone channel and would be lost in transmission. Consequently, it is desirable to concentrate as much energy as possible within the 300- to 3000-Hz passband. An abbreviated Bessel function chart (Table 7–2) illustrates the significant side frequencies and their relative amplitudes for a given modulation index. Note that if the modulation index is less than 1, no significant side frequencies beyond the second set are produced. Therefore, no energy is wasted in frequency components beyond the passband of the telephone channel.

In the previous analysis, the output frequency spectrum of a 202T modem was derived for an input bit rate of 1200 bps. What if the input bit rate were less than 1200 bps?

**EXAMPLE**

Input bit rate = 500 bps.

$$MI = \frac{\Delta f}{f_{mod}} = \frac{500}{250} = 2$$

The peak frequency deviation ($\Delta f$) is still 500 Hz, because this is a function of the amplitude of the input signal and this has remained unchanged. $f_{mod}$ is a sine wave equal to one-half of the bit rate, as explained previously. From the Bessel function chart, a modulation index of 2.0 creates four pairs of significant side frequencies. These side frequencies are separated by 250 Hz (the modulating signal frequency). This frequency spectrum is shown in Figure 7–3.

**TABLE 7–2  Bessel Function Chart**

| MI | $J_0$ | $J_1$ | $J_2$ | $J_3$ | $J_4$ |
|-----|------|------|------|------|------|
| 0.0 | 1.00 | | | | |
| 0.25 | 0.98 | 0.12 | | | |
| 0.5 | 0.94 | 0.24 | 0.03 | | |
| 1.0 | 0.77 | 0.44 | 0.11 | 0.02 | |
| 1.5 | 0.51 | 0.56 | 0.23 | 0.06 | 0.01 |
| 2.0 | 0.22 | 0.58 | 0.35 | 0.13 | 0.03 |

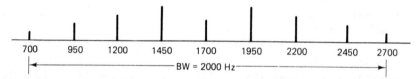

**FIGURE 7-3** Frequency spectrum for a 202T modem with an input bit rate of 500 bps.

From this we can conclude that *as the modulating frequency decreases, the bandwidth decreases.* There are more frequencies present in the modulated signal, but they are spaced closer together. The worst case (widest bandwidth) occurs with the highest modulating frequency ($f_{mod}$ = 600 Hz, $f_b$ = 1200 bps). This worst-case modulation index is called the deviation ratio and is numerically equal to

$$\text{deviation ratio} = \frac{\Delta f(\text{max})}{f_{mod}(\text{max})}$$

What about the sine-wave components present in the input square wave that are the odd harmonics of the fundamental frequency? Fourier analysis of a square wave indicates that the amplitudes of the $n$th harmonics decrease rapidly as $n$ increases. Consequently, any frequency deviation caused by them is insignificant and need not be considered. Most of the side frequencies generated by these high harmonics produce frequency components that extend beyond the passband of the telephone channel and are lost anyway.

The 202T modem is a four-wire, asynchronous modem designed to be used with four-wire, dedicated, private-line circuits. Because the transmit and receive analog signal paths are physically separated and isolated from each other, full-duplex operation is possible with no interference between the two signals. A block diagram of the 202T transmitter is shown in Figure 7-4.

The transmitting DTE initiates the message transfer by applying an ON signal to the Request to Send (pin 4) of the RS 232C interface. At this time, the VCO begins transmitting the mark frequency (1200 Hz). After a predetermined time delay,

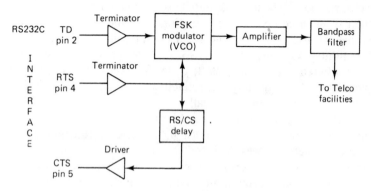

**FIGURE 7-4** Low index FSK transmitter.

the modem responds by applying an ON signal to Clear to Send (pin 5) of the interface. At this time, Transmit Data (pin 2) is enabled, and data from the DTE are allowed to FSK modulate the VCO. The amplifier and the bandpass filter provide sufficient amplification and bandlimiting to ensure that maximum allowable signal power (0 dBm) is applied to the Telco line within the 300- to 3000-Hz bandwidth.

Receiver designs vary with the manufacturer. There are a number of acceptable methods currently being used to produce satisfactory FSK demodulation. These techniques include the use of phase-locked loops (PLL), frequency-to-voltage converters, and tone detectors.

In the receive section (Figure 7–5), the bandpass filter limits the input FSK modulated signal to the 300- to 3000-Hz passband. The slope and delay equalizers provide post-equalization to compensate for long-term variations in the Telco transmission-line parameters. The equalizers may be manually or automatically adjusted. The limiter is an overdriven amplifier that provides an output square wave at either the mark or space frequency. The differentiator, full-wave rectifier, and monopulser (one-shot multivibrator) convert the square wave into a rectangular wave at twice the mark and space frequencies. The low-pass filter converts the rectangular pulses into a dc voltage and applies this voltage to the slicer, where the logic condition (1, 0) is determined. The carrier detect circuit indicates to the receive DTE when the modem is receiving an analog carrier on RLSD (pin 8) of the interface. This circuit also enables the RD driver (pin 3) and allows received data to be transferred to the DTE.

## 202S Modem

All data communication systems do not require the conveniences offered by dedicated lines. The data through-put requirements of a particular user may not justify the expense of a private line, or the type of data being transferred may not require the reliability. An affordable alternative to private-line subscription is the use of the already established direct distance dialing (DDD) network. An advantage of a dial-up system is that the subscribers pay only for the time they are actually using the Telco facilities. The user can initiate a call when there is a message to send, terminate the call upon completion, and pay only for the time of the call. The disadvantage is that the user must contend for a circuit with the rest of the network (and there are millions of other users). Also, each telephone call established could be made over different Telco facilities that do not necessarily have the same performance parameters. Consequently, the reliability (error performance) of a dial-up connection is not consistent from call to call.

A major restriction that must be considered when designing modems for dial-up applications is that they must be capable of functioning over two-wire lines. The DDD network was designed for voice communications that requires only two wires for satisfactory performance. If duplex operation is necessary or desired, both transmission and reception must be accomplished on a single pair of wires.

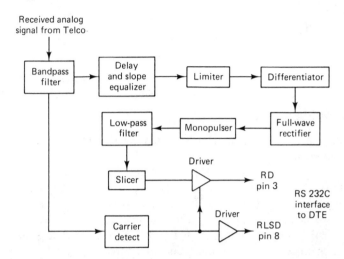

**FIGURE 7–5  202T receiver block diagram (FSK) (limited half-cycle detection).**

The Bell System 202S modem was designed for two-wire DDD applications. The mark/space frequency assignment, information rate (bps), and modulation scheme (FSK) are identical to that of the 202T (four-wire version). A mark is represented by a 1200-Hz tone and a space by a 2200-Hz tone (see Figure 7–6). The maximum bit rate is 1200 bps. Half-duplex operation is accomplished by disabling the receiver while transmitting and disabling the transmitter while receiving. Full-duplex operation is possible only through the use of a secondary reverse or backward channel. For the secondary channel not to interfere with the primary channel, a different carrier frequency must be used. The secondary amplitude shift keys a 387-Hz carrier at a very low rate (5 bps). *ASK* is simply turning the carrier on and off at a digital rate. AM sidebands are generated due to the AM nature of the modulation process, but because of the extremely low bit rate, the bandwidth required for transmission is very narrow. This process of sharing a limited bandwidth on a single facility is called *multiplexing*. Because the frequency domain of the existing channel is shared, the process is called *frequency-division multiplexing* (FDM). The 300- to 3000-Hz bandwidth of the telephone channel is divided between the primary and the secondary channels. The primary channel uses most of the bandwidth for message transfer, while the secondary channel occupies only a narrow band at the low end of the spectrum for transmission of data link control signals.

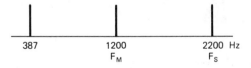

387  1200  2200 Hz
$F_M$  $F_S$

**FIGURE 7–6  Output frequency spectrum of a 202S modem.**

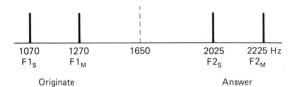

| 1070 | 1270 | 1650 | 2025 | 2225 Hz |
| $F1_S$ | $F1_M$ | | $F2_S$ | $F2_M$ |

Originate                   Answer

**FIGURE 7–7   Output frequency spectrum of the 103 modem.**

## 103 Modem

A modem more popular than the 202S for dial-up applications is the Bell System 103 modem. This is a 300-bps, asynchronous modem, designed for full-duplex operation. It also uses FDM to achieve full duplex. In this case, the 300- to 3000-Hz bandwidth of the telephone channel is divided in half; the upper band extends from 1650 to 3000 Hz and the lower band extends from 300 to 1650 Hz. Because each channel is bandlimited to one-half of the original bandwidth, the information rates possible with the 103 modem are less than that of the 202S. Each channel has its own mark and space frequencies. The mark/space frequencies associated with the lower channel are called the *originate frequencies*, while those associated with the upper channel are called the *answer frequencies*. The modem that initiates the call transmits the originate frequencies and receives the answer frequencies. The modem that terminates or answers the call transmits the answer frequencies and receives the originate frequencies. When the modem is optioned or identified to operate in the originate or answer mode, the transmit frequencies of that modem are also identified. The originate and answer frequencies are referred to as the *F1* and *F2 frequencies*, respectively (Figure 7–7).

Spectral analysis of the 103 modem:

$$h = \frac{|F1_m - F1_s|}{\text{bps}} = \frac{|1270 - 1070|}{300} = 0.67$$

$$h = \frac{|F2_m - F2_s|}{\text{bps}} = \frac{|2225 - 2025|}{300} = 0.67$$

$$\text{deviation ratio} = \frac{\Delta f(\text{max})}{f_{\text{mod}}(\text{max})} = \frac{100}{150} = 0.67$$

$$f_{\text{mod}} = \frac{f_b(\text{bps})}{2}$$

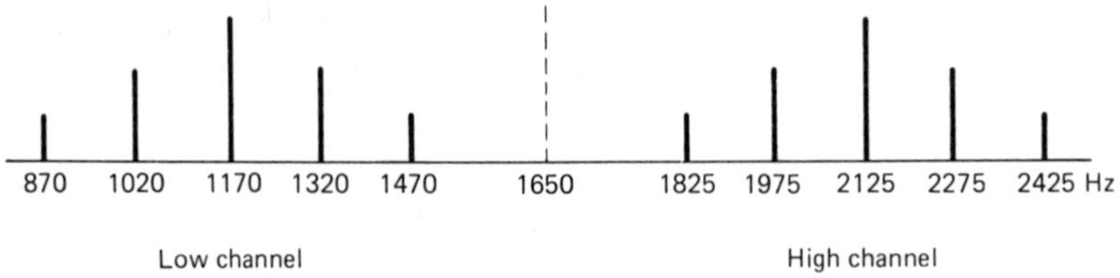

| 870 | 1020 | 1170 | 1320 | 1470 | 1650 | 1825 | 1975 | 2125 | 2275 | 2425 Hz |

Low channel                                     High channel

**FIGURE 7–8   Frequency spectrum of the high and low channels of a 103 modem with an input bit rate of 300 bps.**

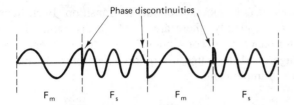

**FIGURE 7–9  Noncontinuous FSK waveform.**

A modulation index of 0.67 yields two sets of significant side frequencies centered around the low channel carrier of 1170 Hz and the high channel carrier of 2125 Hz (see Figure 7–8). On a two-wire facility, if full-duplex operation was used at a bit rate of 300 bps or less, no interference between the two channels would be encountered.

## Minimum Shift-Keying FSK

*Minimum shift-keying FSK* (MSK) is a form of *continuous-phase* frequency shift keying (CPFSK). Essentially, MSK is binary FSK except that the mark and space frequencies are *synchronized* with the input binary bit rate. Synchronous simply means that there is a precise time relationship between the two; it does not mean they are equal. With MSK, the mark and space frequencies are selected such that they are separated from the center frequency by an exact odd multiple of one-half of the bit rate [$F_m$ and $F_s = n(F_b/2)$, where $n$ = any odd integer]. This ensures that there is a smooth phase transition in the analog output signal when it changes from a mark to a space frequency, or vice versa. Figure 7–9 shows a *noncontinuous* FSK waveform. It can be seen that when the input changes from a logic 1 to a logic 0, and vice versa, there is an abrupt phase discontinuity in the analog output signal. When this occurs, the demodulator has trouble following the frequency shift; consequently, an error may occur.

Figure 7–10 shows a continuous-phase MSK waveform. Notice that when

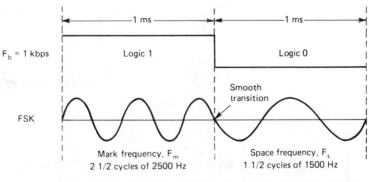

$$F_m = 5 F_b/2 = 5 (1000/2) = 2500 \text{ Hz} \quad F_s = 3 F_b/2 = 3 (1000/2) = 1500 \text{ Hz}$$

**FIGURE 7–10  Continuous-phase MSK waveform.**

the output frequency changes, it is a smooth, continuous transition. Each transition occurs at a zero crossing; consequently, there are no phase discontinuities. MSK has a better bit-error performance than conventional FSK for a given signal-to-noise ratio. The disadvantage of MSK is that it requires synchronizing circuits and is therefore more expensive to implement.

## PHASE SHIFT KEYING

### Binary Phase Shift Keying

*Phase shift keying* (PSK) is a form of modulation by a modem where the phase of the carrier is changed by the binary data. Although the data input to the modem conforms to RS 232C electrical standards, the internal circuitry of the modem can convert these voltage values to levels necessary to perform the modulation. In BPSK (binary phase shift keying), the modulator output has only two phases: it is either in phase or 180° out of phase, with the carrier oscillator depending on whether the input is a logic 1 or 0, respectively. Since a balanced modulator produces a product of its input signals, if a logic 1 is assigned a positive voltage and a logic 0 a negative voltage, the output of the modulator circuit in Figure 7–11 is either cos $\omega_c t$ or $-\cos \omega_c t$. In the receiver demodulator circuit of Figure 7–12, the carrier is detected from the incoming signal and then mixed with the incoming signal to produce either

$$\cos^2\omega_c t = \tfrac{1}{2}(1 + \cos 2\omega_c t)$$

or

$$-\cos^2\omega_c t = -\tfrac{1}{2}(1 + \cos 2\omega_c t)$$

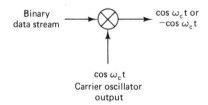

**FIGURE 7–11   BPSK modulator.**

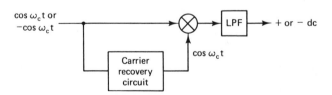

**FIGURE 7–12   BPSK demodulator.**

The low-pass filter at the output eliminates the double-frequency component, leaving only a dc value which is of the same polarity as the original modulating signal (data bit). The signal can be amplified to the level necessary to be coupled to the DTE by way of the RS 232C interface.

**Ring modulator.** One method of achieving BPSK modulation is with a balanced modulator. The ring modulator in Figure 7–13 is one form of balanced modulator. The major characteristic of a balanced modulator is that neither of the input frequencies are present at the output—only their product. For the ring modulator to function properly, the voltage applied to the X-Y input must be greater than the voltage applied at the R-S input. The voltage applied to the X-Y input always determines which diodes are conducting. The voltage at the R-S input cannot turn off a diode turned on by the X-Y voltage, nor can the R-S voltage turn on a diode held off by the X-Y voltage. If only the X-Y voltage is applied with the polarity shown in Figure 7–14, only diodes A and D will be conducting. Since this will produce equal and opposite currents in each half of the primary of the output transformer, no voltage will be coupled to the output. If the X-Y voltage is reversed as

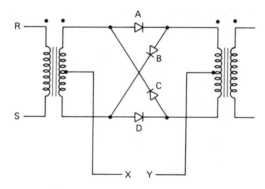

FIGURE 7–13   Ring modulator.

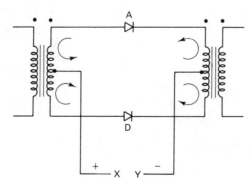

FIGURE 7–14   X-Y input only to the ring modulator.

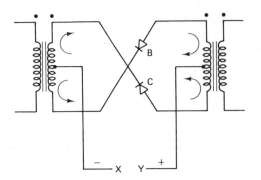

**FIGURE 7–15  X-Y input only to the ring modulator.**

shown in Figure 7–15, diodes B and C will now conduct, but again, there will be no output voltage. If only a voltage is applied to the R-S input, either diodes A and B or diodes C and D will effectively short the input from the output. This is shown in Figure 7–16. This circuit is used to produce double-sideband suppressed carrier (DSBSC) in AM with the carrier applied to the X-Y input and the modulating signal applied to the R-S input.

For this circuit to be used as a BPSK modulator, the oscillator output is applied to the R-S input and the input data stream is applied to the X-Y input.

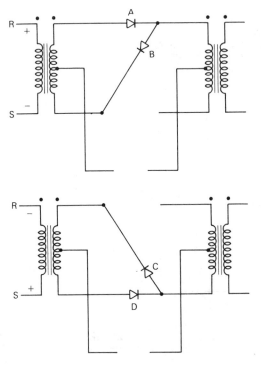

**FIGURE 7–16  R-S input only to the ring modulator.**

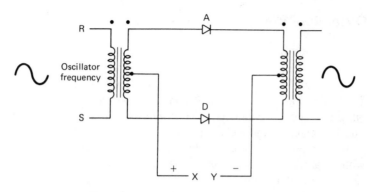

**FIGURE 7–17  In-phase output signal—BPSK.**

With the data polarity as shown in Figure 7–17, diodes A and D will be on. During the positive polarity of the oscillator voltage, diode A will conduct more than diode D. This produces a greater current in the upper half of the output primary coupling a positive voltage to the secondary. During the negative alternation of the oscillator voltage, diode D will conduct more heavily coupling a negative voltage to the output. With the data voltage as shown in Figure 7–17, the output frequency is in phase with the oscillator frequency.

If the data polarity is reversed as shown in Figure 7–18, diodes B and C will conduct. During the positive alternation of the oscillator voltage, diode B conducts more heavily, producing a greater current in the upper half of the output primary and couples a negative voltage to the output. During the negative alternation of the oscillator voltage, diode C conducts more heavily producing a greater current through the lower half of the output primary and couples a positive voltage to the output. With the data voltage as shown in Figure 7–18, the output voltage will be 180° out of phase with the oscillator voltage. Therefore, by varying the polarity of the voltage at the X-Y input, two phases, 180° apart, are generated as required by the BPSK modulator.

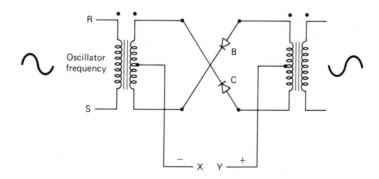

**FIGURE 7–18  Out-of-phase output signal—BPSK.**

## Quadrature PSK

With quadrature PSK (QPSK or 4-PSK) the output signal may have one of four possible phases. The output phase is determined by considering the status of two bits, called *dibits*. A block diagram of a QPSK modulator is shown in Figure 7–19. The four phases are 90° apart. The bit splitter alternately divides the input bit stream into two channels, the I (*in-phase*) channel and the Q (*quadrature*) channel. The bits entering the I channel are mixed with a carrier frequency that is in phase with the reference carrier oscillator, and the bits entering the Q channel are mixed with a carrier frequency that is 90° out of phase (in quadrature) with the reference carrier oscillator. The bit rate of the I and Q channels is equal to one-half of the input bit stream. Since bits $a$ and $b$ can be either a logic 1 or 0, the output of the top balanced modulator can be either $+ \sin \omega_c t$ or $- \sin \omega_c t$ and the output of the bottom balanced modulator can be either $+ \cos \omega_c t$ or $- \cos \omega_c t$. The four possible output phases are shown graphically in Figure 7–20.

For the input bits $a$ and $b$ shown in Figure 7–19, the output is $+ \cos \omega_c t - \sin \omega_c t$. This output will be carried through the demodulator to recover the two original bits. Figure 7–21 shows a block diagram of a QPSK demodulator. The carrier recovery circuit recovers the original transmit carrier frequency. This carrier is mixed directly with the received signal in the top channel. The output of the balanced modulator is

$$(\sin \omega_c t)(\cos \omega_c t) - \sin^2 \omega_c t = \tfrac{1}{2}\sin 2\omega_c t + \tfrac{1}{2}\sin (\omega_c t \overset{0}{-} \omega_c t) - \tfrac{1}{2}(1 - \cos 2\omega_c t)$$

The double-frequency components are filtered out by the LPF. The remaining negative voltage, which represents a logic 0, is sent to the bit combiner. The output of the bottom balanced modulator is

$$\cos^2 \omega_c t - (\sin \omega_c t)(\cos \omega_c t) = \tfrac{1}{2}(1 + \cos 2\omega_c t) - \tfrac{1}{2}\sin 2\omega_c t - \tfrac{1}{2}\sin (\omega_c t \overset{0}{-} \omega_c t)$$

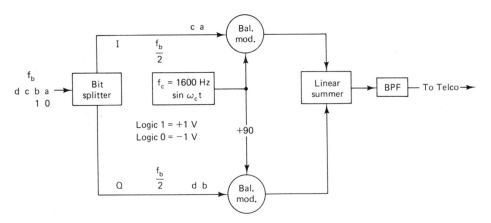

**FIGURE 7–19   QPSK modulator.**

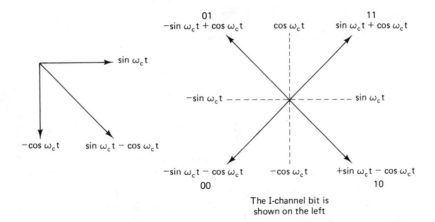

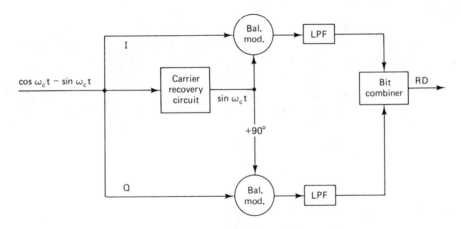

FIGURE 7–20   Output phasors for QPSK.

FIGURE 7–21   QPSK demodulator.

Again, only a positive voltage remains at the output of the LPF after the double-frequency component is removed. The recovered positive voltage represents a logic 1. The bit combiner will sequence the input bits in their proper order and output the recovered I/0 onto the RD line.

A common transmission rate for QPSK is 2400 bps ($f_b$). Therefore, in the QPSK modulator, the I and Q channels have a bit rate of 1200 bps. The fastest change in this bit stream occurs with an alternating 1/0 pattern. This would represent a square wave with a frequency of 600 Hz. A square wave is made up of a fundamental frequency and an infinite number of odd harmonics. The balanced modulator must

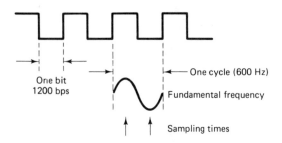

One bit
1200 bps

One cycle (600 Hz)

Fundamental frequency

Sampling times

**FIGURE 7–22   Fundamental frequency of a square wave.**

modulate the carrier with all of these frequencies, producing an upper and lower side frequency for each modulating component. If only the fundamental frequency is recovered, by sampling at the correct time, the two original bits can be recovered (Figure 7–22). The side frequencies produced by the fundamental frequency in the modulation process are 2200 Hz ($f_c + f_m$) and 1000 ($f_c - f_m$). The minimum bandwidth required to pass these frequencies is 1200 Hz (2200 − 1000). The purpose of the bandpass filter at the output of the linear summer is to exclude all other side frequencies from transmission.

**Bandwidth efficiency.**   In general, the higher the data transmission rate, the greater the bandwidth required. Data transmission rates of 2400 bps and 4800 bps have fundamental frequencies of 1200 Hz and 2400 Hz, respectively. If these frequencies amplitude-modulate the same carrier in a balanced modulator, the 1200-Hz signal would require a 2400-Hz handwidth and the 2400-Hz signal would require a 4800-Hz bandwidth (Figure 7–23). This is called the *double-sided Nyquist* or IF bandwidth.

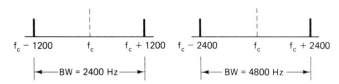

$f_c - 1200$    $f_c$    $f_c + 1200$    $f_c - 2400$    $f_c$    $f_c + 2400$

|← BW = 2400 Hz →|   |← BW = 4800 Hz →|

**FIGURE 7–23   BW requirements for AM signals.**

If the modem output is transmitted over a basic 3002 channel, the frequencies transmitted must be restricted to those allowed for voice (300 to 3000 Hz). Unless more complex modulation schemes are used, bit rates above 2400 bps quickly exceed these limits. *Bandwidth* (BW) *efficiency* is a measure of how effectively a given bandwidth is used for data transmission.

$$\text{BW efficiency (bps/Hz)} = \frac{\text{data transmission rate (bps)}}{\text{BW required (Hz)}}$$

For BPSK, since the bandwith required is equal to the transmission rate, its BW efficiency is 1 bps/Hz. With QPSK, only a 1200-Hz BW is required for a transmission rate of 2400 bps.

$$\text{BW efficiency} = \frac{2400 \text{ bps}}{1200 \text{ Hz}} = 2 \text{ bps/Hz}$$

Because of the modulation scheme used with QPSK, the data transmission rate can be doubled with no increase in bandwidth requirements as compared to BPSK.

**Baud.** Baud is a term originally used with telegraphy. It is the reciprocal of the time of the fastest signaling element that is applied *to the telephone line*. Since baud is a reciprocal of time, it is a rate. With BPSK, the times of all signaling elements are the same and are equal to the time that a constant phase is outputted by the modem as a result of modulation by a single bit. Therefore, with BPSK, the output baud is equal to the input data transmission rate. Transmission of 1200 bps to a BPSK modem is equivalent to the modem outputting a 1200-baud signal onto the telephone line. With QPSK, the time of a signaling element is the time that one of four possible phases is outputted by the modem. Since 2 bits are required to determine this phase, the output baud is one-half of the input bit rate (Figure 7–24).

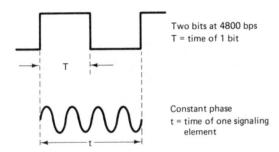

Two bits at 4800 bps
T = time of 1 bit

Constant phase
t = time of one signaling element

**FIGURE 7–24   QPSK: bit time versus signal element time.**

The output baud may be equal to or less than the input bit rate, but it can never exceed the bit rate. The latter would infer that the signaling elements, which are determined by the incoming bits, can be generated faster than the bits are received.

**Nyquist filter.** Nyquist theorems on minimum bandwidth transmission show that $f_s$ independent symbols can be passed through a low-pass filter which has a cutoff frequency of $f_s/2$. For a low-pass filter, the cutoff frequency is also equal to the bandwidth of the filter. These Nyquist filters are ideal and are referred to as *brick-wall filters* because of their sharp cutoff characteristics (Figure 7–25). For a bandpass filter, $f_N = f_s$ (symbol rate).

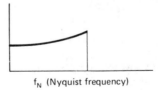

$f_N$ (Nyquist frequency)

**FIGURE 7–25   Nyquist low-pass brick-wall filter.**

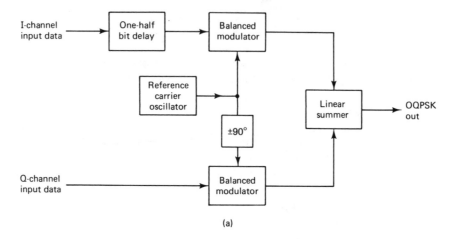

(a)

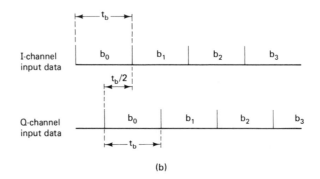

(b)

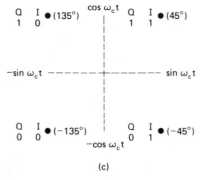

(c)

**FIGURE 7–26   Offset keyed PSK (OQPSK): (a) block diagram; (b) bit alignment; (c) constellation diagram.**

**EXAMPLE**

QPSK—4800-bps data rate. A symbol could be represented by one of four possible phases. Since 2 bits are required to determine one symbol:

$$f_s = \frac{4800 \text{ bps}}{2} = 2400 \text{ symbols/second}$$

$$f_N \text{ for BPF} = f_s = 2400 \text{ Hz}$$

## Offset QPSK

*Offset QPSK* (OQPSK) is a modified form of QPSK where the bit waveforms on the I and Q channels are offset or shifted in phase from each other by one-half of a bit time.

Figure 7–26 shows a simplified block diagram, the bit sequence alignment, and the constellation diagram for a OQPSK modulator. Because changes in the I channel occur at the midpoints of the Q-channel bits, and vice versa, there is never more than a single bit change in the dibit code, and therefore there is never more than a 90° shift in the output phase. In conventional QPSK, a change in the input dibit from 00 to 11 or 01 to 10 causes a corresponding 180° shift in the output phase. Therefore, an advantage of OQPSK is the limited phase shift that must be imparted during modulation. A disadvantage of OQPSK is that changes in the output phase occur at twice the data rate in either the I or Q channels. Consequently, with OQPSK the baud and minimum bandwidth are twice that of conventional QPSK for a given transmission bit rate. OQPSK is sometimes called OKQPSK (*offset-keyed PSK*).

## 8-PSK

With 8-PSK, the modem output signal may be one of eight possible phases. The output phase is determined by considering the status of 3 bits, called *tribits*. A block diagram of an 8-PSK modulator is shown in Figure 7–27. The bit splitter

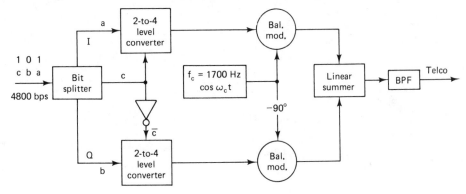

**FIGURE 7–27  8-PSK modulator.**

divides the incoming bit stream from the DTE into three channels. Bits *a* and *b* determine the polarity of the voltage at the output of the 2-to-4 level converters while bit *c* determines the dc magnitude. Two magnitudes are used, 0.34 V and 0.821 V. The reason for these values will become obvious later. When *a* and *b* are logic 1's, the outputs of the 2-to-4 level converters are positive voltages; when *a* and *b* are logic 0's, the converter outputs are negative voltages. The magnitude of the output voltages from the two converters will always be different. Whichever converter receives a logic 1 from *c* or $\bar{c}$ will have an output magnitude of 0.821 V; the other converter will have an output magnitude of 0.34 V. Since *c* and $\bar{c}$ are complements, they both cannot have the same magnitude at the same time. Since the 3 bits *a*, *b*, and *c* are independent of one another, $\pm$ 0.821 V and $\pm$ 0.34 V are the four possible magnitudes at the output of both converters. The output of the I-channel converter modulates the carrier directly while the output of the Q channel modulates a quadrature component of the oscillator output. The possible outputs of the I-channel balanced modulator are $\pm$ 0.34 cos $\omega_c t$ and $\pm$ 0.821 cos $\omega_c t$. The possible outputs of the Q-channel balanced modulator are $\pm$ 0.34 sin $\omega_c t$ and $\pm$ 0.821 sin $\omega_c t$. Again, the only restriction is that the two channels will always have different magnitudes. The linear summer combines the I- and Q-channel signals. The various possible phases that may be produced are shown in Figure 7–28.

$$\text{angle } A = \tan^{-1} \frac{0.34}{0.821} = 22.5°$$

This makes all phasors equally spaced and 45° apart. The numbers 0.34 and 0.821 are not sacred and could be any values as long as their ratio equaled tan 22.5°. Their amplitudes will, of course, determine signal strength.

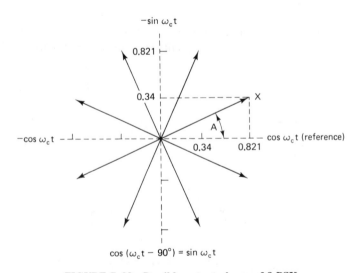

**FIGURE 7–28  Possible output phases of 8-PSK.**

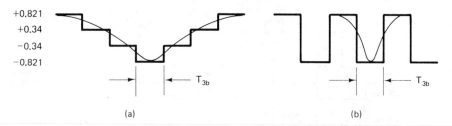

+0.821
+0.34
−0.34
−0.821

$T_{3b}$

$T_{3b}$

(a)                                                                          (b)

**FIGURE 7–29    2- to 4-level converter outputs.**

**EXAMPLE**

For the input bit combinations shown in Figure 7–27:

| | |
|---|---|
| I-channel converter output: | +0.821 |
| Q-channel converter output: | −0.34 |
| I-channel modulator output: | +0.821 cos $\omega_c t$ |
| Q-channel modulator output: | −0.34 sin $\omega_c t$ |
| Output of linear summer. | 0.821 cos $\omega_c t$ − 0.34 sin $\omega_c t$ |

The phasor produced by this sequence of bits is identified by X in Figure 7–28.

Depending on the input bit stream to the 2- to 4-level converter, the output levels could be either of the waveforms shown in Figure 7–29. $T_{3b}$ is the time of three bits. The levels change at one-third the input bit rate. If each level is considered a symbol, the symbol rate at the converter output is 1600 symbols/second. To ensure information recovery, the highest fundamental frequency contained in the converter output must be represented in the transmitted signal. The highest fundamental frequency is contained in the waveform of Figure 7–29(b) and is equal to one-half of the symbol rate or 800 Hz. If all of these frequencies amplitude-modulated the carrier, the bandwidth required for transmission would range from $f_c$ − 800 Hz (900 Hz) to $f_c$ + 800 Hz (2500 Hz) and would be equal to 1600 Hz. The purpose of the bandpass filter is to limit the transmitted signal to this range. With 8-PSK, since one signaling element is equal to the time of 3 input bits, the output baud is equal to one-third the input bit rate or 1600 baud. Since 4800-bps transmission rate is accomplished using a bandwidth of 1600 Hz, the bandwidth efficiency is 4800 bps/1600 Hz or 3 bps/Hz. Note that the bandwidth range, 900 to 2500 Hz, is well within the 300- to 3000-Hz limitation imposed by the Telco facility. Three input bits are required to produce one output phase. If each phasor is considered a symbol, the required bandwidth of the BPF $(f_N) = f_s$, where $f_s$ = 4800 bps/3 = 1600 Hz. This simply confirms a previous conclusion.

## Quadrature Amplitude Modulation

In quadrature amplitude modulation (QAM), the amplitude and the phase of the transmitted signal are varied. 8-QAM can be accomplished by the modulator in Figure 7–30. In this modulator, $a$ and $b$ still determine the polarity of the converter output; $c$, however, is fed directly to both converters. If $c = 1$, both converters

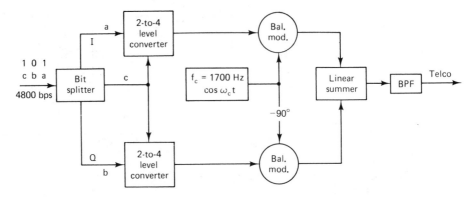

**FIGURE 7–30**   **8-QAM modulator.**

will output the higher magnitude; if $c = 0$, both converters will output the lower magnitude. With 8-PSK the output magnitudes of the two converters are never the same, while with 8-QAM the output magnitudes must always be the same. The *constellation diagram* of Figure 7–31 shows the possible output amplitudes and phases.

    With 8-QAM, two amplitudes and four phases are possible. When performance is viewed in terms of vulnerability to noise, 16-QAM outperforms 16-PSK. For an average energy per bit-to-noise density ratio of 16 dB, 16-QAM has a probability of error of $10^{-8}$, while 16-PSK has a probability of error of $10^{-4}$. Therefore, when greater bandwidth efficiency is desired, QAM modems are used instead of PSK modems. A 16-QAM modulator is shown in Figure 7–32. The bit splitter divides the incoming bit stream into two channels. The bit rate in each channel is 4800 bps. The 2- to 4-level converters require 2 bits before they produce one of four possible output levels. Bits $a$ and $b$ determine the polarity of the output, while bits $c$ and $d$ determine the magnitude.

$a, b = 0$:   negative output      $c, d = 0$:   magnitude $= 0.22$ V
$a, b = 1$:   positive output       $c, d = 1$:   magnitude $= 0.821$ V

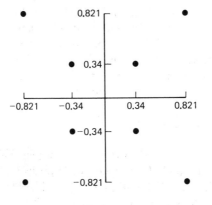

**FIGURE 7–31**   **Constellation diagram of 8-QAM.**

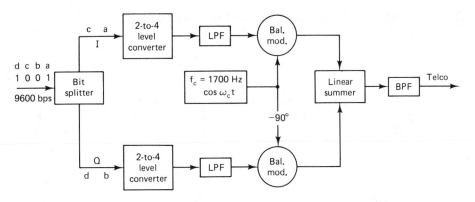

**FIGURE 7–32   16-QAM modulator.**

Since the inputs to the two converters are totally independent of each other, each converter may output $\pm 0.22$ V or $\pm 0.821$ V. The purpose of the low-pass filter (LPF) is to eliminate all frequencies above the highest fundamental frequency at the converter output. If each level at the converter output is considered a symbol, 2 input bits to the converter will be required to produce one symbol at the output.

$$\frac{f_b}{2} = f_s$$

$$\frac{4800 \text{ bps}}{2} = 2400 \text{ symbols/second}$$

$$f_N(\text{LPF}) = \frac{f_s}{2} = \frac{2400}{2} = 1200 \text{ Hz}$$

This cutoff frequency for the LPF ($f_N$) is determined by observing the fastest rate of change at the converter output. This is shown in Figure 7–33. Since 2 input bits are required to produce one output level, 4 bits are required to produce one output square wave. The fundamental frequency of the square wave is one-fourth of the input bit rate or 4800/4 = 1200 Hz. When all of the frequencies at the output of the LPF amplitude modulate the carrier, they produce frequencies ranging from 500 Hz ($f_c - f_N = 1700 - 1200 = 500$ Hz) to 2900 Hz ($f_c + f_N = 1700 + 1200$

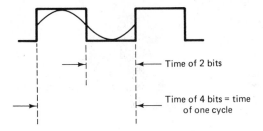

Time of 2 bits

Time of 4 bits = time of one cycle

**FIGURE 7–33   2- to 4-level converter output—QAM.**

= 2900 Hz) or a bandwidth of 2400 Hz. Since a 9600-bps transmission rate requires 2400 Hz of BW,

$$\text{BW efficiency} = \frac{9600 \text{ bps}}{2400 \text{ Hz}} = 4 \text{ bps/Hz}$$

The time of one signaling element is equal to the time of 4 bits. If 9600 bps is the input bit rate to the QAM modulator, the output baud is 2400. The phasor diagram for 16-QAM is shown in Figure 7–34. In this figure

$$\text{angle } A = \tan^{-1} \frac{0.22}{0.821} = 15°$$

Therefore, the 16 possible outputs of this modulator include three different amplitudes and 12 different phases which are equally spaced 30° apart.

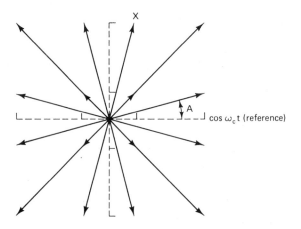

**FIGURE 7–34** **Phasor diagram for 16-QAM.**

**EXAMPLE**

For the input bit combinations shown in Figure 7–32:

| | |
|---|---|
| Output of the I-channel converter: | +0.22 V |
| Output of the Q-channel converter: | −0.821 V |
| Output of the I-channel modulator: | +0.22 cos $\omega_c t$ |
| Output of the Q-channel modulator: | −0.821 sin $\omega_c t$ |
| Output of the BPF: | +0.22 cos $\omega_c t$ − 0.821 sin $\omega_c t$ |

This output phase is identified by X in Figure 7–34.

**M-ary.**   The term *M-ary* is often used in conjunction with PSK and QAM to identify the number of bits required to generate one output symbol.

$$n = \log_2 M$$

where   $n$ = number of bits
         $M$ = number of combinations possible with $n$ bits

*M*-ary can be used to make general statements concerning PSK or QAM. *M*-ary is derived from bi-nary, implying two different levels or conditions.

**EXAMPLE**
PSK *M*-ary with $M = 8$ implies eight different phases are possible at the output. This has the same meaning as 8-PSK. To achieve eight possible phases, the input bit stream must be divided into tribits or groups of three.

**Carrier recovery.**   If FSK is used, it can be seen from Figure 7–5 that the carrier need not be recovered by the receive modem in order to demodulate the receive signal. This is not the case with PSK or QAM. Not only must the exact carrier frequency be recovered, but its phase, when reinserted in the receive balanced modulator, must be identical to that at the transmit balanced modulator. *Coherent detection* implies that demodulation is accomplished using a reinserted carrier that has the *same frequency and phase* as the transmit carrier.

The input to the BPSK demodulator of Figure 7–12 is either $+ \cos \omega_c t$ or $- \cos \omega_c t$. If either of these inputs is applied to a square-law device such as a FET, the square of either input, $\cos^2 \omega_c t = \frac{1}{2}(1 + \cos 2\omega_c t)$, will be generated. The double-frequency component is filtered out and passed through a frequency-divider circuit to recover the frequency and phase of the transmitted carrier. With 4-PSK, 8-PSK, and 16-QAM, carrier frequency and phase recovery are much more complex and expensive to implement.

One common method of achieving carrier recovery for BPSK is the *squaring loop*. Figure 7–35 shows the block diagram of a squaring loop. The received BPSK waveform is filtered and then squared. The filtering reduces the spectral width of the received noise. The squaring circuit removes the modulation and generates the second harmonic of the carrier frequency. This harmonic is phase tracked by the PLL. The VCO output frequency from the PLL is then divided by 2 and used as the phase reference for the product detectors.

With BPSK, only two output phases are possible: $+\sin \omega_c t$ and $-\sin \omega_c t$. Mathematically, the operation of the squaring circuit can be described as follows.

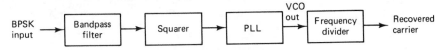

**FIGURE 7–35   Squaring loop carrier recovery circuit for a BPSK receiver.**

For a receive signal of $+\sin \omega_c t$ the output of the squaring circuit is

$$\text{output} = (+\sin \omega_c t)(+\sin \omega_c t) = +\sin^2 \omega_c t$$

$$\text{(filtered out)}$$

$$= \tfrac{1}{2}(1 - \cos 2\omega_c t) = \tfrac{1}{2} - \tfrac{1}{2}\cos 2\omega_c t$$

For a received signal of $-\sin \omega_c t$ the output of the squaring circuit is

$$\text{output} = (-\sin \omega_c t)(-\sin \omega_c t) = +\sin^2 \omega_c t$$

$$\text{(filtered out)}$$

$$= \tfrac{1}{2}(1 - \cos 2\omega_c t) = \tfrac{1}{2} - \tfrac{1}{2}\cos 2\omega_c t$$

It can be seen that in both cases the output from the squaring circuit contained a dc voltage ($\tfrac{1}{2}$ V) and a signal at twice the carrier frequency ($\cos 2\omega_c t$). The dc voltage is removed by filtering, leaving only $\cos 2\omega_c t$.

A more elaborate carrier recovery circuit is the *Costas* or *quadrature loop*, which combines carrier recovery with noise suppression and can therefore accurately recover a carrier from a poorer-quality received signal than can a conventional squaring loop.

Carrier recovery circuits for higher-than-binary encoding techniques are similar to BPSK except that circuits which raise the receive signal to the fourth, eighth, and higher powers are used.

## Differential Phase Shift Keying

As an alternative, differential phase detection is preferred. Rather than compare the phase of the received signal to a fixed reference, the phase of the present received signal is compared to the phase of the previously received signal. Differential phase

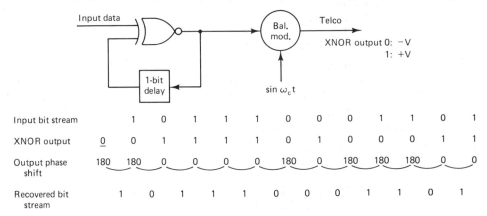

**FIGURE 7–36  DBPSK modulator.**

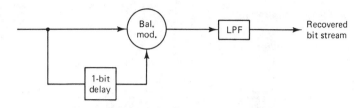

**FIGURE 7–37   DBPSK demodulator.**

detection necessitates differential phase encoding. Although differential phase detection is easier to implement, it has associated disadvantages. The signal-to-noise ratio must be 1 to 3 dB greater using differential detection to achieve the same error rate as coherent detection. A DBPSK modulator is shown in Figure 7–36. The underlined bit in the XNOR output is a reference bit. Note that the bit that modulates the carrier depends not only on the value of the current bit but also on the value of the bit immediately before it. The received signal will be $+\sin \omega_c t$ or $-\sin \omega_c t$. Since the balanced modulator in Figure 7–37 produces the product of two successive inputs, its output will be $\sin^2 \omega_c t = \frac{1}{2}(1 - \cos 2\omega_c t)$ when two consecutive phases are the same, or $-\sin^2 \omega_c t = -\frac{1}{2}(1 - \cos 2\omega_c t)$ when the two consecutive phases are different. The LPF eliminates the double-frequency component and yields either a positive or a negative dc level which represents a logic 1 or 0, respectively.

## MODEM SYNCHRONIZATION

During the RTS/CTS delay, the transmit modem outputs a special, internally generated bit pattern called the *training sequence* (Chapter 6). This bit pattern is used to synchronize (train) the receive modem. Depending on the type of modulation, transmission bit rate, and the complexity of the modem, the training sequence accomplishes one or more of the following functions in the receive modem:

1. Verify continuity (activate RLSD).
2. Initialize the descrambler circuits. (These circuits are used for clock recovery—explained later in this chapter.)
3. Initialize the automatic equalizer. (These circuits compensate for telephone line impairments—explained later in this chapter.)
4. Synchronize the transmitter and receiver carrier oscillators.
5. Synchronize the transmitter and receiver clock oscillators.
6. Disable any echo suppressors in the circuit.
7. Establish the gain of any AGC amplifiers in the circuit.

## Low-Speed Modems

Since these modems are generally asynchronous (Table 7–1) and use noncoherent FSK, the transmit carrier and clock frequencies need not be recovered by the receive modem. Therefore, scrambler and descrambler circuits are unnecessary. The pre- and post-equalization circuits, if used, are generally manual and do not require initialization. The special bit pattern transmitted during the RTS/CTS delay is usually a constant string of 1's (idle line 1's) and is used to verify continuity, set the gain of the AGC amplifiers, and disable any echo suppressors in dial-up applications.

## Medium- and High-Speed Modems

These modems are used where transmission rates of 2400 bps or more are required. In order to transmit at these higher bit rates, PSK or QAM modulation is used which requires the receive carrier oscillators to be at least frequency coherent (and possibly phase coherent). Since these modems are synchronous (Table 7–1), clock timing recovery by the receive modem must be achieved. These modems contain *scrambler* and *descrambler circuits* and *adaptive (automatic) equalizers.*

**Training.** The type of modulation and encoding technique used determines the number of bits required and therefore the duration of the training sequence. The Bell System 208 modem is a synchronous, 4800-bps modem which uses 8-DPSK. The training sequence for this modem is shown in Figure 7–38. Each symbol represents 3 bits (1 tribit) and is 0.625 ms in duration. The four-phase idle code sequences through four of the eight possible phase shifts. This allows the receiver to recover the carrier and the clock timing information rapidly. The four-phase test word allows the adaptive equalizer in the receive modem to adjust to its final setting. The eight-phase initialization period prepares the descrambler circuits for eight-phase operation. The entire training sequence (234 bits) requires 48.75 ms for transmission.

**Clock recovery.** Although timing (clock) synchronization is first established during the training sequence, it must be maintained for the duration of the transmission. The clocking information can be extracted from either the I or the Q channel, or

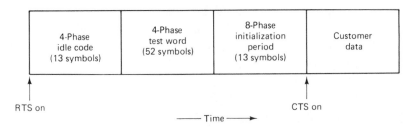

**FIGURE 7–38   Training sequence for a 208 modem.**

from the output of the bit combiner (Figure 7–21). If an alternating 1/0 pattern is assumed at the output of the LPF (Figure 7–39), a clock frequency at the bit rate of the I (or Q) channel can be recovered. The waveforms associated with Figure 7–39 are shown in Figure 7–40.

This clocking information is used to phase-lock loop the receive clock oscillator onto the transmitter clock frequency. To recover clocking information by this method successfully, there must be sufficient transitions in the received data stream. That these transitions will automatically occur cannot be assumed. In a QPSK system, an alternating I/0 pattern applied to the transmit modulator produces a sequence of all 1's in the I or Q channel, and a sequence of all 0's in the opposite channel. A prolonged sequence of all 1's or all 0's applied to the transmit modulator would not provide any transitions in either the I, Q, or the composite received data stream. Restrictions could be placed on the customer's protocol and message format to prevent an undesirable bit sequence from occurring, but this is a poor solution to the problem.

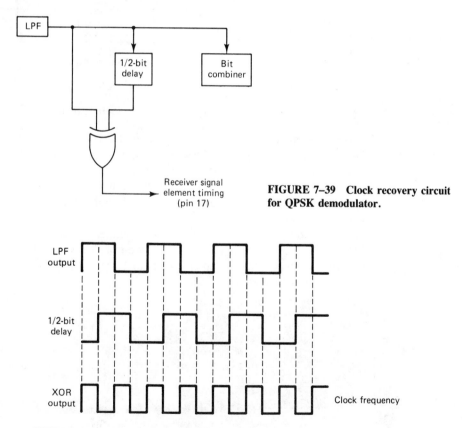

**FIGURE 7–39  Clock recovery circuit for QPSK demodulator.**

**FIGURE 7–40  Clock recovery from I (or Q) channel of a QPSK demodulator.**

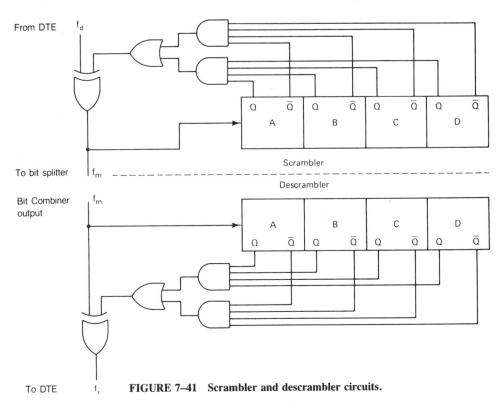

**FIGURE 7–41  Scrambler and descrambler circuits.**

**Scramblers and descramblers.**    A better method is to scramble the customer's data before it modulates the carrier. The receiver circuitry must contain the corresponding descrambling algorithm to recover the original bit sequence before data are sent to the DTE. The purpose of a scrambler is not simply to randomize the transmitted bit sequence, but to detect the occurrence of an undesirable bit sequence and convert it to a more acceptable pattern.

A block diagram of a scrambler and descrambler circuit is shown in Figure 7–41. These circuits are incomplete since an additional gate would be required to detect a varying sequence that would create an all 1 or all 0 sequence in a modulator channel after the bits were split.

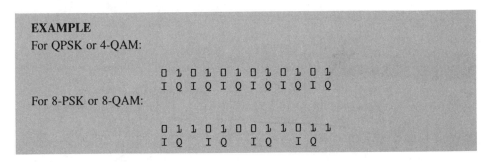

The scrambler circuit is inserted prior to the bit splitter of the QPSK modulator of Figure 7–19, and the descrambler is inserted after the bit combiner of the QPSK demodulator in Figure 7–21. In general, the output of the scrambler or descrambler OR gate is A B C D + A' B' C' D'.

$$f_m = f_d \oplus (\text{A B C D} + \text{A}' \text{B}' \text{C}' \text{D}') \quad \text{top XOR gate}$$

$$f_r = f_m \oplus (\text{A B C D} + \text{A}' \text{B}' \text{C}' \text{D}') \quad \text{bottom XOR gate}$$

Substituting for $f_m$ in the second equation, we have

$$f_r = f_d \oplus (\text{A B C D} + \text{A}' \text{B}' \text{C}' \text{D}') \oplus (\text{A B C D} + \text{A}' \text{B}' \text{C}' \text{D}')$$

Since any identity XORed with itself yields 0,

$$f_r = f_d \oplus 0$$

$$f_r = f_d$$

This simply shows that the original transmitted data ($f_d$) will be fully recovered by the receiver.

The output of either OR gate will be a 1 if the 4-bit register contains either all 1's or all 0's. Neither of these is a desirable sequence. If the OR gate output is a 1, $f_m$ will be the complement (opposite) of $f_d$, or $f_r$ will be complement of $f_m$. The intent is to create transitions in a prolonged bit stream of either all 1's or all 0's. If the output of the OR gate is a 0, neither of these undesired conditions exists and $f_m = f_d$ or $f_r = f_m$: the data pass through the XOR gate unchanged. If the other logic gates (AND, OR, NAND, NOR) were used either alone or in combination in place of the XOR gates, the necessary transitions could be created in the scrambler circuit, but the original data could not be recovered in the descrambler circuit. If a long string of all 1's or all 0's is applied to the scrambler circuit, this circuit will introduce transitions. However, there may be times when the scrambler creates an undesired sequence. The XOR output is always either a 1 or a 0. No matter what the output of the OR gate, a value of $f_d$ may be found to produce a 1 or a 0 at the XOR output. If either value for $f_d$ was equiprobable, the scrambler circuit would be unnecessary. If the 4-bit register contains all 1's, if $f_d = 1$, we would like to see it inverted. However, if $f_d = 0$, we'd prefer to pass it through the XOR gate unchanged. The scrambler circuit for this situation inverts the 0 and extends the output string of 1's. It is beyond the intended scope of this book to delve deeply into all parameters involved in scrambler design. Let it be enough to say that scramblers will cure more problems than they create.

**Equalizers.** *Equalization* is the compensation for the phase delay distortion and amplitude distortion of a telephone line. One form of equalization is C-type conditioning (Chapter 3). Additional equalization may be performed by the modems. *Compromise equalizers* are contained in the transmit section of the modem and they provide *pre-equalization*. They shape the transmitted signal by altering its delay and gain characteristics before it reaches the telephone line. It is an attempt to compensate for impairments anticipated in the bandwidth parameters of the line. When a modem is installed, the compromise equalizer is manually adjusted to provide the best *bit error rate* (BER). Typically, compromise equalizer settings affect:

1. Amplitude only
2. Delay only
3. Amplitude and delay
4. Neither amplitude nor delay

The setting above may be applied to either the high or low voice band frequencies or symmetrically to both at the same time. Once a compromise equalizer setting has been selected, it can only be changed manually. The setting that achieves the best BER is dependent on the electrical length of the circuit and the type of facilities that make it up. *Adaptive equalizers* are located in the receiver section of the modem and provide *post-equalization* to the received analog signal. Adaptive equalizers automatically adjust their gain and delay characteristics to compensate for telephone-line impairments. An adaptive equalizer may determine the quality of the received signal within its own circuitry or it may acquire this information from the demodulator or descrambler circuits. Whichever the case, the adaptive equalizer may continuously vary its settings to achieve the best overall bandwidth characteristics for the circuit.

## Probability of Error and Bit Error Rate

*Probability of error P(e)* and *bit error rate* (BER) are often used interchangeably, although they do have slightly different meanings. $P(e)$ is a theoretical (mathematical) expectation of the error rate for a given system. BER is an empirical (historical)

**TABLE 7–3**
**Performance Comparison
of Various Digital
Modulation Schemes
(BER = $10^{-6}$)**

| Modulation technique | $C/N$ ratio (dB) |
|---|---|
| BPSK | 13.6 |
| QPSK | 13.6 |
| 8QAM | 13.6 |
| 8PSK | 18.8 |
| 16PSK | 24.3 |
| 16QAM | 20.5 |
| 32QAM | 24.4 |
| 64QAM | 26.6 |

record of a system's actual error performance. For example, if a system has a $P(e)$ of $10^{-5}$, this means that mathematically, you can expect one bit error in every 100,000 bits transmitted ($1/10^{-5} = 1/100,000$). If a system has a BER of $10^{-5}$, this means that in the past there was one bit error for every 100,000 bits transmitted.

Probability of error is a function of the receiver *carrier-to-noise ratio*. Depending on the $M$-ary used and the desired $P(e)$, the minimum carrier-to-noise ratio varies. In general, the minimum carrier-to-noise ratio required for a QAM system is less than that required for a comparable PSK system (see Table 7–3). Also, the higher the level of encoding used, the higher the minimum carrier-to-noise ratio.

## SIGNAL QUALITY

A common method of determining the signal quality is to sample the power within discrete passbands in the receive signal and compare the power distribution to a known reference. The spectral distribution of the reference is based on the spectral content of a typical transmitted signal at the output of the modulator. Then, if the received distribution does not resemble a typical transmitted signal, the equalizer automatically adjusts its passband characteristics until the best received signal quality is achieved.

The quality of the received signal can also be determined by monitoring the received data entering the descrambler circuit. The receive modem cannot determine what the bit sequence should be, but it can recognize bit sequences that the transmitter scrambler circuit prohibits from occurring under normal circumstances. Typically, prolonged bit sequences of consecutive 1's and 0's are not allowed to modulate the carrier. If one of these prohibited bit sequences is detected, the equalizer concludes that transmission errors must be occurring. The equalizer then automatically adjusts its passband characteristics in an attempt to correct the suspected problem. Very often, the front panel of a modem has an LED indicator to identify when the adaptive equalizer has sequenced through all of its tab settings and is still receiving marginal data. Highspeed (synchronous) modems are usually equipped with automatic retrain circuitry. The first attempt by an adaptive equalizer to improve the quality of the received signal is accomplished with customer data or idle line 1's. If unsuccessful, the receive modem initiates a request for training by sending out its own training sequence. Whenever a modem receives a training sequence while it is transmitting, it will, in turn, suspend data transmission and send its training sequence. This is a destructive process and interrupts data transmission. Only modems capable of transmission rates of 9600 bps are equipped with automatic retraining circuitry, and this circuitry is used only if the modems were optioned for continuous carrier operation. This necessarily implies a two-point, four-wire, FDX system.

In summary, a block diagram of a synchronous PSK modem is shown in Figure 7–42. A brief functional description of each block follows.

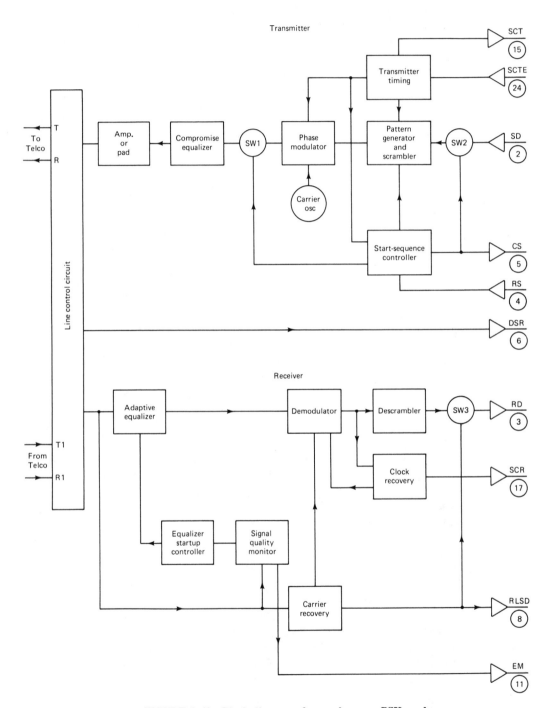

**FIGURE 7–42    Block diagram of a synchronous PSK modem.**

## Transmitter

*Transmitter timing.* This circuit generates the various clock signals required by the modem transmitter. The master clock is either internally generated or derived from the signals on SCTE from the DTE.

*Start sequence controller.* When RTS goes high, this circuit enables the transmission of the analog carrier through switch 1. The carrier is still unmodulated by terminal data because switch 2 is disabled. When RTS goes high, this circuit also signals the pattern generator/scrambler to initiate the transmission of the training sequence. This controller provides the RTS/CTS delay, activating switch 2 and placing a high on CTS after this time lapse. When RTS goes low, the start sequence controller deactivates switch 1, removing the carrier from the telephone line.

*Pattern generator and scrambler.* The pattern generator provides the special bit pattern transmitted during the training sequence, and the idle line 1's for continuous carrier applications while there is no transmission of data. The scrambler randomizes the input data to provide the transitions necessary for clock recovery.

*Phase modulator.* This circuit converts binary input data from the DTE to corresponding phase changes in the analog carrier. The phase modulator uses two binary channels to achieve differential PSK at its output.

*Compromise equalizer.* This circuit is a manually adjustable filter which provides pre-equalization to the voice band signals. This equalizer compensates for anticipated impairments in the gain and delay characteristics introduced by the telephone circuit.

*Amplifier or pad.* This is a variable-gain amplifier capable of inserting loss or gain into the transmitted signal path. It is used to set the transmit signal power during initial installation.

*T, R, T1, R1.* Transmit and receive tip and ring.

The input and output amplifiers are the drivers and terminators of the RS 232C interface.

## Receiver

*Adaptive equalizer.* This circuit provides post-equalization of the line's gain and delay characteristics.

*Signal quality monitor.* This circuit monitors the received analog signal and determines whether the adaptive equalizer is properly set to compensate for the telephone line impairment. If not, the equalizer startup controller is signaled to retrain locally the adaptive equalizer. During the retraining time, the received data may be invalid.

*Equalizer startup controller.* This circuit adjusts the adaptive equalizer when this modem is receiving a training sequence or when the signal quality of the received signal becomes substandard.

*Carrier recovery.* This circuit detects the presence of a received carrier and controls the on/off condition of the RLSD line of the RS 232C interface. This circuit also recovers the carrier and phase-lock loops the local oscillator to this frequency to produce a frequency coherent signal for demodulation.

*Demodulator.* This circuit converts the phase variations of the received carrier to the corresponding bit sequence.

*Clock recovery.* This circuit recovers the transmitted clock frequency and generates the various clocking signals required by the receiver.

*Descrambler.* After the clock has been recovered, this circuit converts the randomized data back to the original transmitted data.

## DIAL-UP MODEMS

Dial-up modems must satisfy different requirements than those designed for private-line usage. Access into and out of the DDD network must be made in a manner similar to that of a normal telephone connection. The subscriber must make the modem look and act like a telephone set to the network. There are two general methods of coupling the modem into and out of the DDD network, *hardwire* and *acoustic coupling*. Hardwire coupling is accomplished through a data coupler, while acoustic coupling permits the convenience of using the telephone set and the network as it stands.

### Message Transfer through an Acoustical Coupler

The operator at the originate end of the circuit initiates the call by picking up the telephone handset. Generally, the telephone is a standard 500-type telset [Figure 7–43(a)]. When the telset is picked up, an off-hook condition is applied to the telephone line. The Telco network responds with an audible dial tone, indicating to the operator that a successful access to the DDD network has been made. The

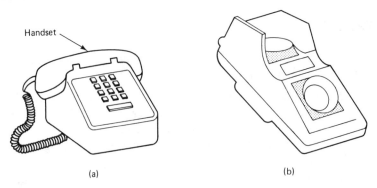

(a)                                                (b)

**FIGURE 7–43   (a) Telephone; (b) modem.**

operator then dials the telephone number associated with the answer modem. At the answer end, when an incoming ring is detected, the ring indicator (pin 22) of the RS 232C interface goes active. If the associated DTE equipment (CPU) is prepared and capable of receiving a message, it allows the data terminal ready pin (pin 20) of the RS 232C interface to go active, signaling the modem to answer the call and transmit the F2 mark frequency. On hearing this tone, the operator at the originate end places the handset in the cradle of the acoustically coupled modem [Figure 7–43(b)]. The acoustic input (F2 mark frequency) is converted to electrical energy by the modem microphone (see Figure 7–44).

The microphone allows this signal to pass through A2 to the carrier detect circuit. A2 is activated by either a mechanical switch or a photocell in the telset cradle. A2 prevents any electrical signal from entering the receive section of the modem unless the handset is in the cradle. The bandpass filter allows only the F2 mark/space frequencies to pass. When the originate modem detects the F2 mark frequency, it signals the DTE by making RLSD (pin 8) of the RS 232C interface active. After the RTS/CTS delay, F1 mark is transmitted by the originate modem. Switch A1 assures that the F1 mark is not transmitted without a carrier detect signal (without having access to the DDD network). The originate modem is now in the data mode and data transfer begins and continues until either of the DTEs breaks the carrier and terminates the call. Signaling between the modems and their corresponding DTEs is accomplished either through the RS 232C interface, a 20-mA loop, or some other serial interface.

In the transmission of data, digital pulses are sent from the line control unit (LCU) to the FSK modulator through the RS 232C interface. These pulses modulate the VCO and produce analog changes at its output. The transmitter bandpass filter allows only the F1 mark/space frequencies to pass. The speaker converts these analog signals to audible tones which enter the microphone in the handset and are transmitted over the telephone lines in the same manner as voice.

## Echo-Plex

Echo-plex is often used with low-speed acoustically coupled modems. It is a mode of transmission that achieves less than full duplex but more than half-duplex. Inexpensive, less sophisticated terminal equipment that is commonly used with acoustically coupled modems generally do not provide an electrical connection from the keyboard to the CRT. When a key is depressed on the keyboard, the corresponding character is not sent directly to the CRT for display. In Figure 7–44, with S1 in the HDX position, a transmitted character is looped back to the DTE receiver locally through the wave-shaping circuit. The receiver decodes and displays that character on the CRT. The modem is limited to half-duplex operation: if data were being received from the FSK demodulator at the same time the DTE was transmitting, the two signals would become garbled in the wave-shaping circuit. By placing S1 in the FDX position, the transmitted data cannot reach the wave-shaping circuit directly. This prevents the transmitted character from being displayed on the CRT. Echo-

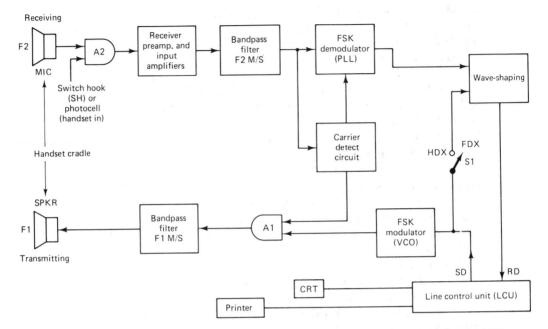

**FIGURE 7–44** Block diagram of a 103-compatible acoustic coupled modem. The AND gates here do not necessarily perform a logic AND function on the inputs. They simply indicate that both inputs must be present before an output can be obtained.

plex is achieved by having the answer DTE retransmit (echo) the received message back to the originate DTE for decoding and display. Although transmission occurs in both directions simultaneously as in FDX, only one terminal can transmit its message at any given time as in HDX. The advantage of echo-plex is that the operator at the originate DTE, when he or she sees a character displayed on his CRT, knows that it has been received by the answer terminal.

## Message Transfer through Hardwire Attachment

The modem can access the DDD network through a hardwire coupler called a *data access arrangement* (*DAA*). An *automatic data access arrangement* (*ADAA*) enables the terminal equipment to originate a call automatically (give off-hook supervision), dial into the network (provide pulses or touch-tone signals), answer a call (respond to an incoming ring), and terminate a call (give on-hook supervision). The Bell System provides two such couplers, CBT and CBS. *CBS* is designed to conform to RS 232C electrical standards, while *CBT* is designed to provide contact closures for less complex electromechanical communications equipment.

The major functions of the CBS and CBT data couplers are:

1. To provide a transmission path from the customer-provided data modems to the DDD networks

2. To protect Telco personnel and equipment from hazardous voltages which may be accidently supplied by the customer-provided modems

3. To limit the signal power from the data modem to a specified value in order to prevent interference with other telephone services

4. To protect the telephone line from longitudinal imbalance (producing currents in the same direction on a two-wire pair)

5. To provide the following functions associated with network control signaling:
    (a) To provide a loop-holding path for dc supervision
    (b) To detect ringing and alert the customer's terminal to an incoming call
    (c) To originate on-hook and off-hook signaling and to generate dial pulses in response to signals received from the customer through the interface control leads
    (d) To provide a delay of 1 to 3 s after an incoming call is answered in order to prevent data signals from interfering with automatic message accounting equipment

6. To provide the capability of remote testing the data coupler.

7. To provide an indication of the status of the switch hook of the associated telephone

A block diagram of a CBS data coupler is shown in Figure 7–45. The tip and ring (T, R) pins correspond to the tip and ring of a two-wire telephone facility.

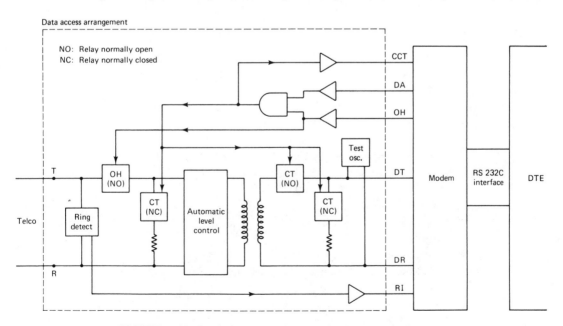

**FIGURE 7–45 Simplified block diagram of a data access arrangement.**

The incoming 20-Hz ring signal is rectified and integrated by the ring detector until there is sufficient buildup of charge to ensure the DAA that the ring is legitimate. This prevents normal line transients, produced by switching and dialing, from supplying false ring indications. The coupler supplies a dc voltage on the RI pin to notify the modem of an incoming ring. The modem responds with an ON signal on the OH (off-hook) pin. This signal actuates the OH relay, allowing a dc current path for the two-wire telephone loop. The Telco dial switch recognizes this as an off-hook condition and stops the ringing signal. The modem now applies an ON signal to the DA pin, which actuates the CT relay. Actuating this relay removes the resistive load from the automatic level control in the modem and allows the T and R leads to be coupled (cut through—CT) to the DT and DR (data tip and data ring) leads. Activation of the CT relay also produces an ON signal on the CCT pin, indicating to the modem that the Telco facilities are coupled to the modem. The modem now waits for the carrier to be sent. A call is terminated when the modem supplies a low to the OH pin which disables the OH relay and opens the dc path to the dial switch. A call is originated in much the same manner. The modem applies an ON signal to the OH pin to complete a dc path to the dial switch. The OH signal is now pulsed to represent dial pulses to the telephone switching equipment. After the dial pulses have been completed, the DA pin is made high to activate the CT relay and cut the Telco facilities through to the modem. The modem, on receipt of an ON signal on CCT, will begin message transmission. After the message has been sent, the modem allows OH to go low, terminating the Telco Connection.

# QUESTIONS

1. A modem is a form of analog-to-digital/digital-to-analog converter.   (T, F)
2. If, when you made a phone call, your voice frequencies were in analog form end to end, would modems be required for transmission or reception?
3. What is meant by a synchronous modem?
4. Asynchronous modems do not require a clock input.   (T, F)
5. What is the fundamental frequency of a 1200-Hz square wave?
6. What is the maximum data transmission rate of modems using:
   (a) FSK?    (b) PSK?    (c) QAM?
7. FSK can be considered a form of FM.   (T, F)
8. A backward or reverse channel is normally provided to modems designed to operate on a four-wire system.   (T, F)
9. The reverse channel uses FSK modulation.   (T, F)
10. The reverse channel is used mainly for data link control signals.   (T, F)
11. Modems with originate/answer options are normally used in full-duplex systems.   (T, F)
12. With originate/answer modems, F2 refers to the originate mark/space frequencies.   (T, F)

13. As the transmission rate increases:
    (a) The *h* factor increases.  (T, F)
    (b) The number of significant frequencies present in the transmitted signal increases. (T, F)
    (c) The separation between the frequencies contained in the transmitted signal increases. (T, F)
    (d) The bandwidth required to transmit all of the significant frequencies increases (T, F)

14. The value of $J_0$ on the Bessel function chart indicates the relative amplitude of the first pair of side frequencies contained in the transmitted waveform.   (T, F)

15. If dial-up lines are used for FDX operation, some form of multiplexing must be accomplished.   (T, F)

16. With FSK, bandwidth limitations restrict the maximum bit transmission rate.   (T, F)

17. The carrier frequency is present at the output of a balanced modulator.   (T, F)

18. In BPSK, what is the function of the low-pass filter (Figure 7–12)?

19. By increasing the bandwidth efficiency, the data transmission rate can be increased without an accompanying increase in bandwidth requirements.   (T, F)

20. Baud is always synonymous with bits per second.   (T, F)

21. The output baud can be greater than the input bit rate.   (T, F)

22. The output baud can be less than the input bit rate.   (T, F)

23. Tribits are associated with (*BPSK, QPSK, 8-PSK,* QAM).

24. In an *M*-ary system with $M = 16$:
    (a) How many different output conditions are possible?
    (b) How many bits are required to determine each of these conditions?

25. DPSK uses coherent phase detection.   (T, F)

26. Identify one advantage of DPSK over PSK.

27. Identify one disadvantage of DPSK over PSK.

28. A 4800-bps data stream is applied to a:
    (a) QPSK modem     (b) 8-PSK modem     (c) QAM modem
    Answer the following questions for each of the three modems.
    (1) How many bits are considered before the output phase and amplitude are determined?
    (2) How many different output phases are possible?
    (3) Ideally, what is the degree separation between phasors?
    (4) How many different amplitudes can the output have?
    (5) What is the output baud?

29. If 8-DPSK was used and the following sequence of phases was detected:

$$\ldots, \ +45, \ +90, \ +135, \ +45$$
$$\qquad\quad A \qquad\qquad\qquad\quad B$$

    Phasor B would be (*leading, lagging, in phase with, 180° out of phase with*) phasor A. By how many degrees?

30. What is meant by ''coherent frequency detection'' and ''coherent phase detection''?

**31.** The bit pattern in the training sequence for high-speed modems is a continuous sequence of 1's. (T, F)

**32.** Identify all possible locations where clocking information can be extracted in a PSK or a QAM demodulator?

**33.** What is the purpose of a scrambler circuit?

**34.** What does equalization provide?

**35.** What is the difference between a compromise equalizer and an adaptive equalizer?

**36.** In reply to a transmitted message, a modem receives a training sequence. This high-speed modem will (*retransmit the message, accept the training sequence as a positive acknowledgment and continue with message transmission, suspect internal malfunctions and train onto the training sequence, send out its own training sequence*).

**37.** Only high-speed modems are equipped with automatic retraining circuitry. (T, F)

**38.** With dial-up modems, how does the operator at the originate end know when the circuit has been completed to the receive modem?

**39.** A dial-up modem cannot transmit unless it is detecting a receive carrier signal. (T, F)

**40.** How does FDX in echo-plex operation differ from normal FDX?

**41.** Describe minimum shift-keying FSK.

**42.** What is the difference between a continuous and a noncontinuous FSK waveform?

**43.** What is an advantage of MSK over conventional FSK?

**44.** What is a disadvantage of MSK?

**45.** Describe offset QPSK.

**46.** What is an advantage of OQPSK over conventional QPSK?

**47.** What is a disadvantage of OQPSK?

**48.** Describe the operation of a squaring loop.

**49.** What is the difference between probability of error and bit error rate?

# PROBLEMS

**1.** $f_c = \sin \omega_c t$ at 1500 Hz; input bit rate = 2000 bps; $b, a = 0, 0$.

**2.** $f_c = \cos \omega_c t$ at 1550 Hz; input bit rate = 1800 bps; $b, a = 1, 0$.

With reference to Figure P7–1 and for the data given in Problems 1 and 2, answer the following questions.

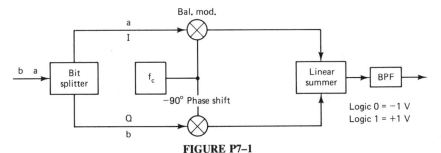

**FIGURE P7–1**

(a) What is the bit rate in the I channel?

(b) What is the fundamental frequency of the highest-frequency square wave applied to the balanced modulator?

(c) What is the trigonometric expression for the output of the BPF for the input bits given ($\omega_c t \pm 90°$ to be eliminated from the output expression)?

(d) What is the bandwidth at the output of the BPF?

(e) The bandwidth of part (d) extends from _____ to _____ Hz.

(f) What is the output baud?

(g) Identify the values of input bits $a$ and $b$ that will provide the output phasors shown in Figure P7–1g. Use the phase of $f_c$ as reference.

**FIGURE P7–1g**

(h) For the corresponding demodulator circuit (Figure 7–21), show trigonometrically how the original data bits are recovered.

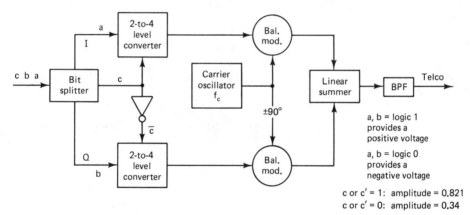

**FIGURE P7–3**

**3.** $f_c = \sin \omega_c t$ at 1600 Hz; phase shift = $-90°$; $c, b, a = 1, 1, 0$; input bit rate = 4500 bps.

**4.** $f_c = \cos \omega_c t$ at 1500 Hz; phase shift = $+90°$; $c, b, a = 0, 1, 1$; input bit rate = 4200 bps. With reference to Figure P7–3 and for the data given in Problems 3 and 4, answer the following questions.

(a) What is the bit rate of the I channel?

(b) What is the fundamental frequency of the highest-frequency square wave applied to the balanced modulator?

(c) What is the trigonometric expression for the output of the BPF for the input bits given ($\omega_c t \pm 90°$ to be eliminated from the output expression)?

(d) What is the bandwidth at the output of the BPF?

(e) The bandwidth of part (d) extends from _____ to _____ .

(f) What is the output baud?

(g) What is the bandwidth efficiency?

**5.** $f_c = \sin \omega_c t$ at 1500 Hz; input bit rate = 9200 bps; $d, c, b, a = 1, 1, 0, 0$; phase shift = $+ 90°$.

**6.** $f_c = \cos \omega_c t$ at 1600 Hz; input bit rate = 8800 bps; $d, c, b, a = 0, 1, 1, 0$; phase shift = $- 90°$.

In Figure P7–5, the first bit into the 2- to 4-level converter determines the sign: $0 = -$; $1 = +$. The second bit into the 2- to 4-level converter determines the amplitude: for Problem 5, $0 = 0.6$, $1 = 0.9$; for Problem 6, $0 = 5$, $1 = 8.66$. Using this information, together with the data given for Problems 5 and 6, answer the following questions.

**(a)** What is the bit rate into the 2- to 4-level converter?

**(b)** How many input bits into the 2- to 4-level converter are required to obtain one symbol (one of the four levels) out?

**(c)** What is the symbol rate into the LPF?

**(d)** Each symbol can have four possible values. What are they?

**(e)** What is the $f_N$ for the LPF?

**(f)** What is the double-sided $f_N$ for the BPF?

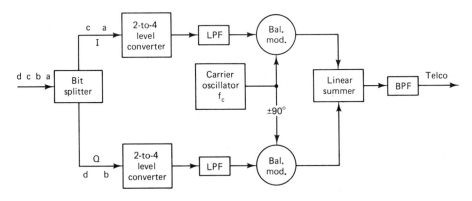

**FIGURE P7–5**

**(g)** The bandwidth of the BPF extends from _____ to _____ .

**(h)** What is the output baud?

**(i)** What is the BW efficiency?

**(j)** For the given input bits, what is the trigonometric expression for the output of the BPF?

**(k)** What are the three possible amplitude values at the BPF output?

**(l)** The degrees separating two successive phasors is either _____ or _____ .

**7.** For the DBPSK modulator in Figure 7–36, for the given input bit stream, identify the binary output sequence of the XNOR gate with the provided reference bit. The order of the arrow is to the left.

| | *Reference Bit* | *Input Bit Sequence* |
|---|---|---|
| **(a)** | 1 | 1  1  0  0  1  0  1  1  0 |
| **(b)** | 0 | 1  1  0  0  1  0  1  1  0 |

**8.** For the DBPSK demodulator in Figure 7–37, identify the output bit sequence for the phase inputs identified and the reference phase given.

|     | *Reference Phase* | *Phase Inputs* | | | | | | | | |
| --- | --- | --- | --- | --- | --- | --- | --- | --- | --- | --- |
| **(a)** | 0° | 0° | 180° | 180° | 180° | 0° | 180° | 0° | 180° | 0° |
| **(b)** | 0° | 180° | 180° | 0° | 180° | 0° | 180° | 180° | 180° | 180° |

# EIGHT

## *VOICE OR DATA TRANSMISSION WITH ANALOG CARRIERS*

### FREQUENCY-DIVISION MULTIPLEXING

*Multiplexing* is the transmission of information from more than one source on the same media. In *frequency-division multiplexing* (FDM), many information channels are transmitted simultaneously, with each channel occupying a different frequency band (Figure 8–1). If each information channel originally occupied the same frequency range, the frequencies must be translated to different areas of the frequency spectrum before they are combined. To achieve frequency separation, each channel amplitude-modulates a different carrier frequency. If a carrier is amplitude-modulated with a single frequency, the resultant waveform is mathematically described as

$$A \sin \omega_c t \; + \; \frac{mA}{2} \cos (\omega_c - \omega_m)t \; - \; \frac{mA}{2} \cos (\omega_c + \omega_m)t$$

$$\text{(a)} \qquad\qquad \text{(b)} \qquad\qquad\qquad \text{(c)}$$

where  $A$ = peak carrier amplitude
         $m$ = modulation coefficient
         $\omega_c = 2\pi f_c$
         $\omega_m = 2\pi f_m$
         $f_c$ = carrier frequency
         $f_m$ = modulating frequency

Expression (a) is the original carrier frequency, (b) the lower side or difference frequency, and (c) the upper side or sum frequency.

134

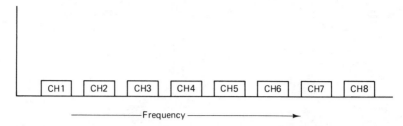

**FIGURE 8–1   Frequency-division multiplexing.**

If a carrier is amplitude-modulated by a band of frequencies, an upper and a lower sideband are produced. The *upper sideband* (USB) is made up of the sum of the carrier frequency and the individual frequencies present in the modulating signal: the *lower sideband* (LSB) is made up of the difference between the carrier frequency and the individual frequencies present in the modulating signal (Figure 8–2).

In amplitude modulation the carrier contains no intelligence; therefore, it is suppressed through some form of balanced modulator (see ''Ring modulator'' in Chapter 7). Since the upper and lower sidebands contain identical information, the transmission of only a single sideband is necessary to convey the information. With FDM, a single sideband is transmitted without the carrier. This signal is described as single-sideband suppressed carrier (SSBSC).

An A-type (analog) channel bank performs frequency-division multiplexing of twelve voice band channels. Each voice band channel can carry either voice information or digital information from a modem. Each channel amplitude-modulates a different carrier frequency. The lower sideband of each modulation process is extracted and combined with the lower sidebands from the eleven other channels to form a *group* (Figure 8–3). A group has a bandwidth of 48 kHz (12 × 4 kHz) and occupies the frequency band from 60 to 108 kHz. Although each voice channel is allocated a frequency range of 0 to 4 kHz, signal information is normally limited to a 300- to 3000-Hz passband. Consequently, a group has a natural guard band of 1.3 kHz (Figure 8–4) between adjacent channel signals.

If further multiplexing is desired, five groups may be similarly combined to produce a *supergroup* (SG). The bandwidth of an SG, which results from combining

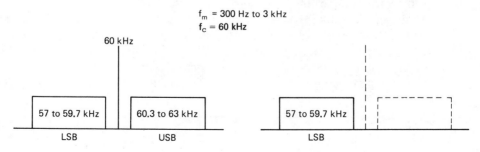

**FIGURE 8–2   Frequency spectrum of an AM signal and a SSB-SC signal.**

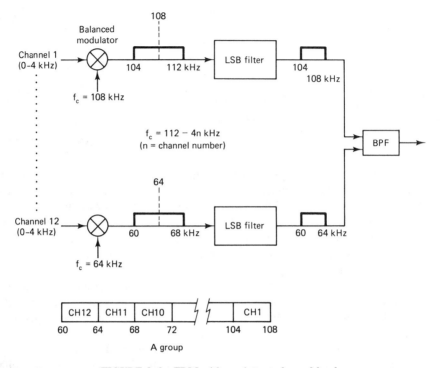

**FIGURE 8–3   FDM with an A-type channel bank.**

the LSBs of each modulation process, is 240 kHz and extends from 312 to 552 kHz (Figure 8–5).

Supergroups may be combined to form a *mastergroup*. A mastergroup is made up of ten supergroups and contains information from 600 voice band channels. A scheme that forms the U600 mastergroup is shown in Figure 8–6. The guard band between the adjacent supergroups in Figure 8–6 is 8 kHz and the guard band between adjacent mastergroups in Figure 8–7 is 80 kHz. This guard band makes it easier to separate adjacent supergroups and mastergroups with filters in the FDM receiver. An L600 mastergroup combines supergroups 1 through 10 in a slightly different method and occupies the bandwidth from 60 to 2788 kHz. The L600 mastergroups are transmitted either directly on coaxial cable or are modulated further and then transmitted on microwave radio. U600 mastergroups may be transmitted as is or used as stepping stones for further multiplexing. In Figure 8–7, three mastergroups are multiplexed to form one *microwave radio channel*.

A supergroup may be added to the three multiplexed mastergroups (1860 VB

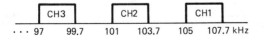

**FIGURE 8–4   Guard band between signals in FDM.**

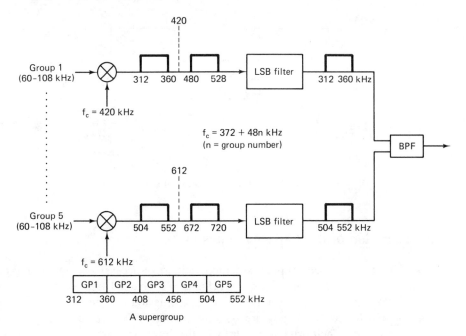

**FIGURE 8–5**   **Formation of a supergroup.**

channels) and transmitted on either coaxial cable or microwave radio. *Jumbogroups* (3600 VB channels) combine six mastergroups and are transmitted on coaxial cable. Three jumbogroups may be further multiplexed (10,800 VB channels) and transmitted on coaxial cable.

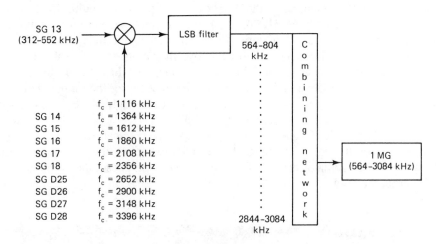

**FIGURE 8–6**   **U600 mastergroup.**

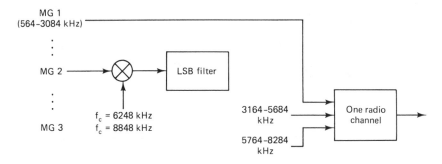

**FIGURE 8–7   Formation of a radio channel.**

## Generation of the Carrier Frequency

An FDM receiver accomplishes demodulation by successively mixing the baseband signal down in frequency until the original VB channel frequencies are recovered. In FDM systems, it is essential for the transmit and the receive carrier frequencies to be synchronized. If they are not, the recovered VB signals would be offset in frequency from their original spectrum by the difference in the two carrier frequencies. For the purpose of power conservation, balanced modulators, which suppress the carrier, are used in FDM transmitters. Therefore, the transmit carrier cannot be recovered at the receiver directly from the composite baseband signal. Note that all of the carrier frequencies in Figures 8–3 through 8–6 are multiples of 4 kHz. In a communications system, one station is designated as the master station. The 4-kHz master oscillator, to which all stations in the system are synchronized, is located at this station. Generally, the 4-kHz base frequency is multiplied to a higher pilot frequency (typically 64,312, or 552 kHz) and combined with the composite baseband spectrum. Each secondary station demodulates the pilot, then recovers and regenerates the 4-kHz base frequency. Consequently, all stations in a system derive their carrier frequencies from coherent 4-kHz carrier supplies. In a large system, such as the Bell System or General Telephone, it is impractical for the master station to transmit the pilot directly to each slave station. Instead, many slave stations also serve as repeaters to other slave stations for the pilot.

The carrier frequencies used in FDM are obtained by passing the 4-kHz base frequency through a nonlinear device, then extracting the appropriate harmonics. If the 4-kHz oscillator drifts in frequency or shifts in phase, the harmonics will change proportionally.

**EXAMPLE**
What is the radio channel output frequency range of a voice band channel belonging to channel 10, group 4, supergroup 17, mastergroup 3?

| | Ch 10 at GP | GP 4 at SG | SG 17 at MG | MG 3 at radio channel |
|---|---|---|---|---|
| $f_c$ | 72 kHz | 564 kHz | 2108 kHz | 8848 kHz |
| LSB | 68–72 kHz | 492–496 kHz | 1612–1616 kHz | 7232–7236 kHz |

From this example it can be seen that as a voice band channel is elevated in frequency, it still occupies the original 4-kHz bandwidth.

**EXAMPLE**

If the 4-kHz oscillator drifts 10 Hz, by how much would a 1-kHz tone on channel 3, group 2, supergroup 17, mastergroup 2 differ from what it should be?

Assuming that all carrier frequencies that are amplitude-modulated are derived as harmonics of the 4-kHz signal:

For Channel 3
    Ideal $f_c$:     100 kHz
    Actual $f_c$:    4.01 kHz $\times$ 25 = 100.25 kHz
    Ideal LSF:    100 kHz − 1 kHz = 99 kHz
    Actual LSF:   100.25 kHz − 1 kHz = 99.25 kHz

For Group 2
    Ideal $f_c$:     468 kHz
    Actual $f_c$:    4.01 kHz $\times$ 117 = 469.17 kHz
    Ideal LSF:    468 kHz − 99 kHz = 369 kHz
    Actual LSF:   469.17 kHz − 99.25 kHz = 369.92 kHz

For SuperGroup 17
    Ideal $f_c$:     2108 kHz
    Actual $f_c$:    4.01 kHz $\times$ 527 = 2113.27 kHz
    Ideal LSF:    2108 kHz − 369 kHz = 1739 kHz
    Actual LSF:   2113.27 kHz − 369.92 kHz = 1743.35 kHz

For MasterGroup 2
    Ideal $f_c$:     6248 kHz
    Actual $f_c$:    4.01 kHz $\times$ 1562 = 6263.62 kHz
    Ideal LSF:    6248 kHz − 1739 kHz = 4509 kHz
    Actual LSF:   6263.62 kHz − 1743.35 kHz = 4520.27 kHz

A 10-Hz variation caused the tone to be 11 kHz off at the radio channel output. Since each channel occupies a 4-kHz bandwidth, this drift offsets the tone approximately three channels in the composite baseband spectrum.

In Figure 8–6, supergroups 25 through 28 include a prefix D in their supergroup number. This indicates that the carrier frequencies which these supergroups modulate are *derived* and not produced by simply extracting harmonics of the 4-kHz base frequency. The carriers for supergroups 15 through 18 are mixed with another lower

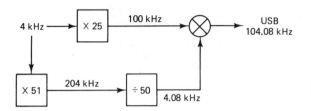

**FIGURE 8–8   Derivation of the 104.08-kHz pilot frequency.**

harmonic (1040 kHz), and the upper side frequency is filtered for use as the carrier frequency for the D supergroups. This is done to reduce the amount of phase jitter present in the higher supergroup carriers. Any phase jitter present in the 4-kHz base oscillator is multiplied, together with the frequency, when the carrier frequencies are generated. By deriving the higher supergroup carrier frequencies through a combination of multiplying and mixing, the total phase jitter is reduced. The phase jitter present in two signals that are mixed in a nonlinear device is not algebraically additive in the resultant. For example, if a 10-kHz signal with 5° of phase jitter is mixed with a 5-kHz signal with 5° of phase jitter, the upper frequency produced is 15 kHz, but it does not necessarily have 10° of phase jitter. The carrier frequencies for mastergroups and higher are also derived in a similar fashion.

### Pilot Carrier

In the process of transmission, variations in signal amplitude may occur. The nature of these variations must first be determined before compensation can be made. A pilot frequency of 104.08 kHz is inserted at the group level. The derivation of this pilot frequency from the 4-kHz master oscillator signal is shown in Figure 8–8. Since five groups modulate five different pilot frequencies in the formation of supergroups, each supergroup contains five different pilot frequencies (Figure 8–9).

As each group continues to be raised in frequency through modulation, the associated pilot frequency is also raised. At the receiver, it is assumed that each group experienced the same amplitude variations as its pilot frequency. Since the pilot is inserted at a known signal level, the variations that this signal sustained

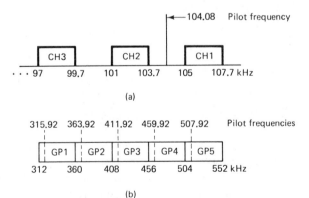

**FIGURE 8–9   Pilot frequencies in (a) groups and (b) supergroups.**

can readily be determined. As a result, these pilot frequencies are used for automatic gain control at all levels.

## Amplitude Regulation

Figure 8–10a shows the gain characteristics for an ideal transmission medium. For the ideal situation (Figure 8–10a), the gain for all baseband frequencies is the same. In a more practical situation, the gain is not the same for all frequencies (Figure

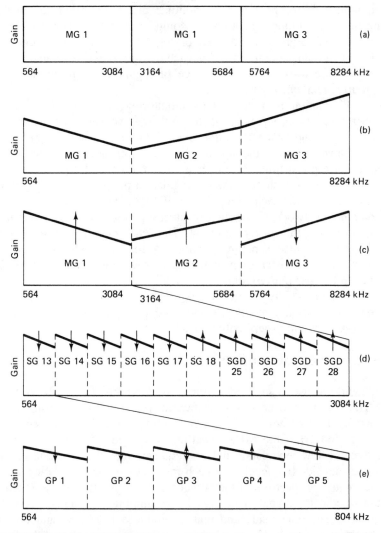

**FIGURE 8–10    Gain characteristics: (a) ideal gain versus frequency characteristics; (b) amplitude distortion; (c) mastergroup regulation; (d) supergroup regulation; (e) group regulation.**

8–10b). Therefore, the demultiplexed voice band channels do not have the same amplitude characteristics as the original voice band signals. This is called *amplitude distortion*. To reduce amplitude distortion, filters with the opposite characteristics as those introduced in the transmission medium can be added to the system, thus canceling the distortion. This is impractical because every transmission system has different characteristics and a special filter would have to be designed and built for each system.

Automatic gain devices (regulators) are used in the receiver demultiplexing equipment to compensate for amplitude distortion introduced in the transmission medium. Amplitude regulation is accomplished in several stages. First, the amplitude of each mastergroup is adjusted or regulated (mastergroup regulation; Figure 8–10c), then each supergroup within each mastergroup is regulated (supergroup regulation; Figure 8–10d). The last stage of regulation is performed at a group level (group regulation; Figure 8–10e).

Regulation is performed by monitoring the power level of a mastergroup, supergroup, or group *pilot*, then regulating the entire frequency band associated with it, depending on the pilot level. Pilots are monitored rather than the actual signal levels because the signal levels vary depending on how many channels are in use at a given time. A pilot is a continuous signal with a constant power level.

Figure 8–11 shows how the regulation pilots are nested within the composite baseband signal. Each group has a 104.08-kHz pilot added to it in the channel combining network. Consequently, each supergroup has five group pilots. The group 1 pilot is also the supergroup pilot. Thus each mastergroup has 50 group pilots, of which 10 are also supergroup pilots. A separate 2840-kHz mastergroup pilot is added to each mastergroup in the supergroup combining network, making a total of 51 pilots per mastergroup.

Figure 8–12 is a partial block diagram for an FDM demultiplexer that shows how the pilots are monitored and used to separately regulate the mastergroups, supergroups, and groups automatically.

## Microwave Transmission

A radio channel comprised of three mastergroups (564 kHz to approximately 8.3 MHz), in order to be sent as a microwave (> 1 GHz) transmission, must still be raised further in frequency. An IF carrier frequency of 70 MHz is frequency modulated by the 564 kHz—8.3 MHz signal at a low modulation index (approximately 0.4). At such a low modulation index, frequency modulation produces only one pair of significant side frequencies for each frequency in the modulating signal. As a result, the frequency spectrum of the output resembles the frequency spectrum of an amplitude-modulated signal. This signal now amplitude-modulates a 6-GHz carrier frequency. The upper sideband that is produced is filtered and transmitted. This is shown in Figure 8–13. The bandwidth required to transmit 1800 VB channels is 16.6 MHz. Since FCC allows a 29-MHz bandwidth, the transmitted bandwidth is well within limits.

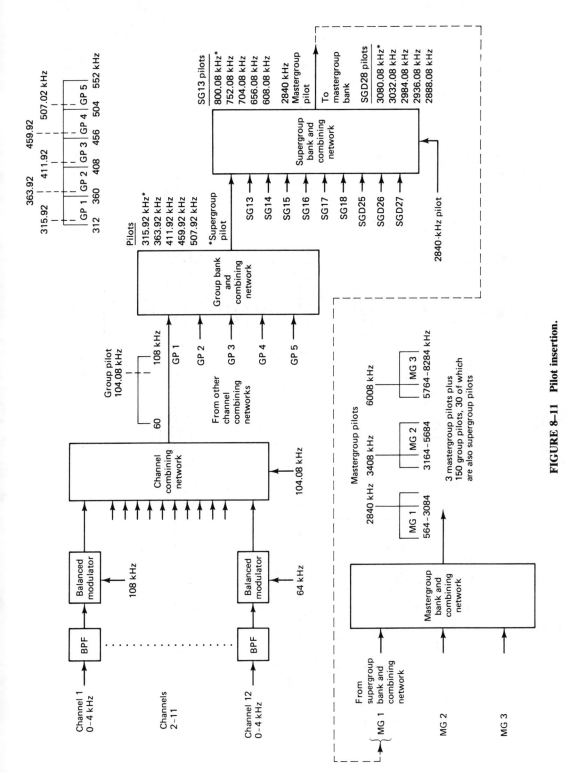

**FIGURE 8–11  Pilot insertion.**

143

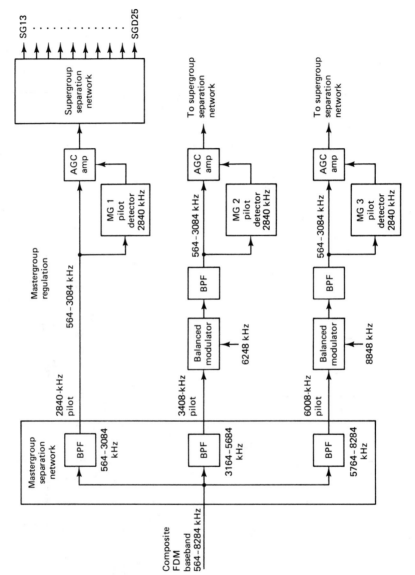

**FIGURE 8-11** (Continued)

144

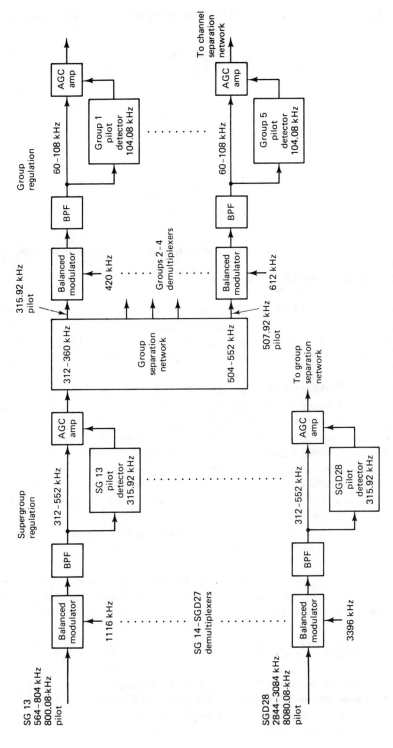

**FIGURE 8-12** Amplitude regulation.

145

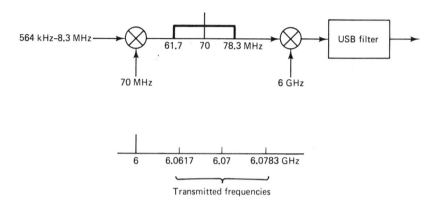

**FIGURE 8–13    Raising frequencies to the gigahertz range.**

## Baseband Signal

The composite FDM spectrum, whether it is a group, supergroup, mastergroup, or jumbogroup, is called the baseband signal. The *baseband signal* is the signal which modulates the signal that becomes the final carrier. In Figure 8–13, since the final carrier is 6.07 GHz, the frequencies 564 KHz to 8.3 MHz make up the baseband signal.

In FDM transmission, transmitted channels that leave a particular locale need not all be going to the same destination. Groups, supergroups, mastergroups, and so on, may be broken down at various points, channels extracted, additional channels inserted, and the various groupings entirely reformed. If there are 1800 VB channels on a microwave circuit, only about 450 are anticipated to be used at a given instant: only 450 channels will have information going the same way at the same time. If there are 1800 ongoing conversations, only half (900) are expected to be speaking in the direction of transmission. When people talk, there are many pauses in their speech. Of the 900 people speaking in the direction of transmission, half (450) are expected to be actually talking at any given instant. Should all 1800 people actually speak in the same direction at the same time, the system would suffer a severe overload.

FDM is used extensively by telephone companies, utilities, municipalities, the government, and the military to reduce the number of carrier facilities required in communication systems.

## SINGLE-CHANNEL TERMINALS

When the bandwidth of the signals to be transmitted is such that after digital conversion it occupies the entire capacity of a digital transmission line, a single-channel terminal is provided. Examples of such single-channel terminals are picturephone, mastergroup, and commercial television terminals.

## Mastergroup and Commercial Television Terminals

Figure 8–14 shows the block diagram of a mastergroup and commercial television terminal. The mastergroup terminal receives voice band channels that have already been frequency-division multiplexed without requiring that each voice band channel be demultiplexed to voice frequencies. The signal processor provides frequency shifting for the mastergroup signal (shifts it from a 564- to 3084-kHz bandwidth to a 0- to 2520-kHz bandwidth) and dc restoration for the television signal. By shifting the mastergroup band, it is possible to sample at a 5.1-MHz rate. Sampling of the commercial television signal is at twice that rate or 10.2 MHz.

To meet the transmission requirements, a 9-bit PCM code is used to digitize each sample of the mastergroup or television signal. The digital output from the terminal is therefore approximately 46 Mbps for the mastergroup and twice that much (92 Mbps) for the television signal.

The digital terminal shown in Figure 8–14 has three specific functions: it converts the parallel data from the output of the encoder to serial data, it inserts frame synchronizing bits, and it converts the serial binary signal to a form more suitable for transmission. In addition, for the commercial television terminal, the 92-Mbps digital signal must be split into two 46-Mbps digital signals because there is no 92-Mbps line speed in the digital hierarchy.

## Picturephone Terminals

Essentially, *picturephone* is a low-quality video transmission for use between nondedicated subscribers. For economic reasons it is desirable to encode a picturephone

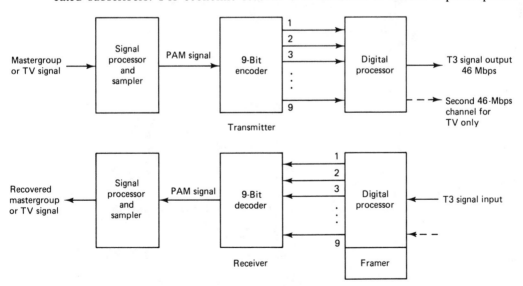

**FIGURE 8–14  Block diagram of a mastergroup or commercial television digital terminal.**

signal into the T2 capacity of 6.312 Mbps, which is substantially less than that for commercial network broadcast signals. This substantially reduces the cost and makes the service affordable. At the same time, this permits the transmission of adequate detail and contrast resolution to satisfy the average picturephone subscriber. Picturephone service is ideally suited to a differential PCM code. Differential PCM is similar to conventional PCM except that the exact magnitude of a sample is not transmitted. Instead, only the difference between that sample and the previous sample is encoded and transmitted. To encode the difference between samples requires substantially fewer bits than encoding the actual sample.

## Data Terminals

The portion of communications traffic that involves data (signals other than voice) is increasing exponentially. Also, in most cases, the data rates generated by each individual subscriber are substantially less than the data rate capacities of digital lines. Therefore, it seems only logical that terminals be designed that transmit data signals from several sources over the same digital line.

Data signals could be sampled directly; however, this would require excessively high sample rates resulting in excessively high transmission bit rates, especially for sequences of data with few or no transitions. A more efficient method is one that codes the transition times. Such a method is shown in Figure 8–15. With the coding format shown, a 3-bit code is used to identify when transitions occur in the data and whether that transition is from a 1 to a 0, or vice versa. The first bit of the code is called the address bit. When this bit is a logic 1 this indicates that no

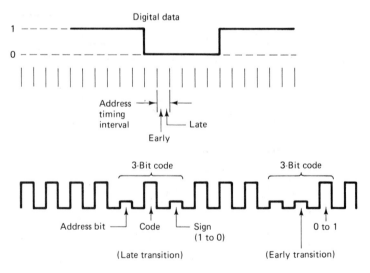

**FIGURE 8–15   Data coding format.**

transition occurred, a logic 0 indicates that a transition did occur. The second bit indicates whether the transition occurred during the first half (0) or during the second half (1) of the sample interval. The third bit indicates the sign or direction of the transition; a 1 for this bit indicates a 0-to-1 transition and a 0 indicates a 1-to-0 transition. Consequently, when there are no transitions in the data, a signal of all 1's is transmitted. Transmission of only the address bit would be sufficient; however, the sign bit provides a degree of error protection and limits error propagation (when one error leads to a second error, etc.). The efficiency of this format is approximately 33%; there are 3 code bits for each data bit. The advantage of using a coded format rather than the original data is that coded data are more efficiently substituted for voice in analog systems. To transmit a 250-kbps data signal, the same bandwidth is required to transmit 60 voice channels with analog multiplexing. With this coded format, a 50-kbps data signal displaces three 64-kbps PCM encoded channels, and a 250-kbps data stream displaces only 12 voice band channels.

## HYBRID DATA

With *hybrid* data it is possible to combine digitally encoded signals with FDM signals and transmit them as one composite baseband signal. There are four primary types of hybrid data: data under voice (DUV), data above voice (DAV), data above video (DAVID), and data in voice (DIV).

### Data under Voice

Figure 8–16 shows the block diagram of AT&T's 1.544-Mbps *data under FDM voice* system. With the L1800 FDM system explained earlier in this chapter, the 0 to 564-kHz frequency spectrum is void of baseband signals. With FM transmission, the lower baseband frequencies realize the highest signal-to-noise ratios. Consequently, the best portion of the baseband spectrum was unused. DUV is a means of utilizing this spectrum for the transmission of digitally encoded signals. A T1 carrier system can be converted to a quasi-analog signal and then frequency-division multiplexed onto the lower portion of the FDM spectrum.

In Figure 8–16a, the *elastic store* removes timing jitter from the incoming data stream. The data are then *scrambled* to suppress the discrete high-power spectral components. The advantage of scrambling is the randomized data output spectrum is continuous and has a predictable effect on the FDM radio system. In other words, the data present a load to the system equivalent to adding additional FDM voice channels. The serial seven-level *partial response* encoder (correlative coder) compresses the data bandwidth and allows a 1.544-Mbps signal to be transmitted in a bandwidth less than 400 kHz. The low-pass filter performs the final spectral shaping of the digital information and suppresses the spectral power above 386 kHz. This prevents the DUV information from interfering with the 386-kHz pilot control tone.

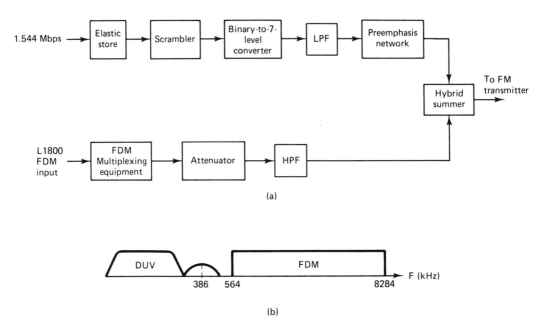

**FIGURE 8–16** Data under voice (DUV): (a) block diagram; (b) frequency spectrum.

The DUV signal is preemphasized and combined with the L1800 baseband signal. The output spectrum is shown in Figure 8–16b.

AT&T uses DUV for *digital data service* (DDS). DDS is intended to provide a communications medium for the transfer of digital data from station to station without the use of a data modem. DDS circuits are guaranteed to average 99.5% error-free seconds at 56 kbps.

## Data above Voice

Figure 8–17 shows the block diagram and frequency spectrum for a *data above voice* system. The advantage of DAV is for FDM systems that extend into the low end of the baseband spectrum; the low-frequency baseband does not have to be vacated for data transmission. With DAV, data PSK modulates a carrier which is then up-converted to a frequency above the FDM message. With DAV, up to 3.152 Mbps can be cost-effectively transmitted using existing FDM/FM microwave systems.

## Data above Video

Essentially, *data above video* is the same as DAV except that the lower baseband spectrum is a *vestigial sideband* video signal rather than a composite FDM signal. Figure 8–18 shows the frequency spectrum for a DAVID system.

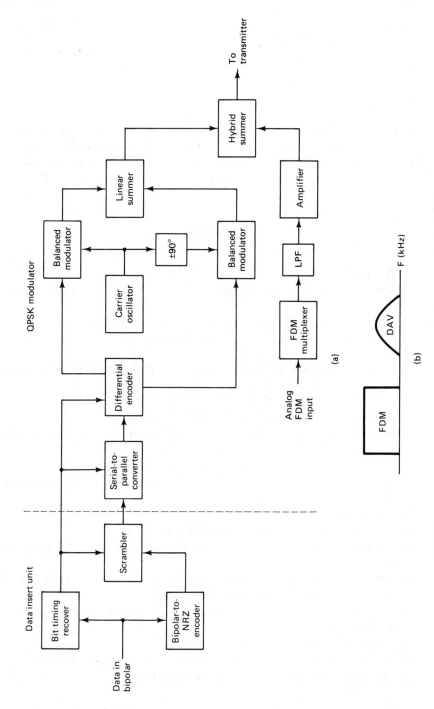

**FIGURE 8–17** Data above voice (DAV): (a) block diagram; (b) frequency spectrum.

151

FIGURE 8–18  Data above video (DA-VID).

## Data in Voice

*Data in voice* (DIV), developed by Fujitsu of Japan, uses an eight-level PAM-VSB modulation technique with steep filtering. It uses a highly compressed partial response encoding technique which gives it a high bandwidth efficiency of nearly 5 bps/Hz (1.544-Mbps data are transmitted in a 344-kHz bandwidth).

# QUESTIONS

1. Define FDM.
2. The A-type channel bank frequency-division multiplexes incoming channels to form _____ .
3. Identify the number of voice band channels in each of the following.
   (a) Group    (b) Supergroup    (c) Mastergroup
   (d) Radio channel    (e) Jumbogroup
4. Identify the frequency range and the bandwidth at the output of each of the following.
   (a) Group    (b) Supergroup    (c) U600 mastergroup
   (d) Radio channel made up of three U600 mastergroups
5. At what level is the 104.08-kHz pilot frequency inserted?
6. The 104.08-kHz pilot frequency is inserted for the purpose of supplying a carrier frequency to the receiver for demodulation purposes.   (T, F)
7. What is the difference between phase jitter and frequency drift?
8. The guard bands between supergroups in a U600 mastergroup all have the same bandwidth.   (T, F)
9. What is meant by a "radio channel"?
10. Amplitude modulation is used to generate FDM.   (T, F)
11. All carrier frequencies are generated solely as harmonics of the base frequency (4 kHz).   (T, F)
12. Why are the carrier frequencies for supergroups D25, D26, D27, and D28 derived differently than the carriers for the other supergroups?
13. Describe how a single-channel mastergroup and television terminal digitally encodes the baseband signals.
14. What is picturephone transmission?
15. Describe how a data terminal samples, encodes, and transmits the input data.
16. Describe how amplitude regulation is achieved with FDM.
17. What are hybrid data?
18. Describe how data are transmitted along with conventional FDM signals with a DUV system.

**19.** What is the purpose of an elastic store?

**20.** Describe how data above voice transmit digital data along with conventional FDM signals.

**21.** Briefly describe the advantage of data in voice over the other types of hybrid data transmission.

# PROBLEMS

**1.** Construct a table showing the lower sideband output frequency range for each of the following.
   **(a)** The 12 channels in a group
   **(b)** The 5 groups in a supergroup
   **(c)** The supergroups in a U600 mastergroup

**2.** Location:

   A = group output
   B = supergroup output
   C = mastergroup output
   D = radio channel output

   **(a)** What frequency range would a voice band channel occupy in a radio channel if it belonged to the following?
   **(1)** MG 1, SG 14, GP 2, Ch 11
   **(2)** MG 2, SG D27, GP 4, Ch 3
   **(b)** Find the frequency of a 1-kHz tone belonging to MG 2, SG 18, GP 5, Ch 7.
   **(c)** Find the frequency of a 1.5-kHz tone belonging to MG 3, SG D26, GP 4, Ch 6.

|           | A | B | C | D |
|-----------|---|---|---|---|
| **(a) (1)** | _____ | _____ | _____ | _____ |
| **(2)**   | _____ | _____ | _____ | _____ |
| **(b)**   | _____ | _____ | _____ | _____ |
| **(c)**   | _____ | _____ | _____ | _____ |

**3.** What is the pilot carrier frequency at the mastergroup output if it was derived from GP 1, SG 16, MG 2?

**4.** If a 4-kHz master oscillator drifted to 4000.1 Hz, what would be the frequency of a 1-kHz tone at the radio channel output for Ch 3, GP 4, SG 17, MG 2?

# NINE

# *VOICE OR DATA TRANSMISSION WITH DIGITAL CARRIERS*

## DIGITAL COMMUNICATIONS

The boundaries encompassed by the phrase "digital communications" are not clear cut. *Digital communications* is commonly understood to be the transmittal of information from one point to another in digital form. Signals from voiceband channels are converted to digital pulses, transmitted on metallic lines (called *T-carriers* by the Bell system), and reconverted to analog signals at the receive end. The *source signal* can be *analog* (voice) or *digital* (data). In Chapter 7 it was shown that modems convert digital pulses into signals suitable for transmission on voice-grade lines. Therefore, digital data from a DTE can be converted to an analog signal by a modem, reconverted to digital (but of a different form), and then be transmitted as such on T-carriers. Digital information may be highly multiplexed. Multiplexed digital signals can PSK modulate an analog carrier and then be transmitted as a microwave signal. Microwave transmission of digital information is called *digital radio*. These various forms of digital transmission are depicted in Figure 9–1.

## DDS

Additional digital services called DDS are provided by Telco. Unfortunately, these same initials represent two different types of services, Digital Data Service and Dataphone Digital Service. In *Digital Data Service*, the customer's digital signals are immediately placed on a T-carrier. This service was initially used to provide short-range digital communications. To extend the distance capability of this service,

154

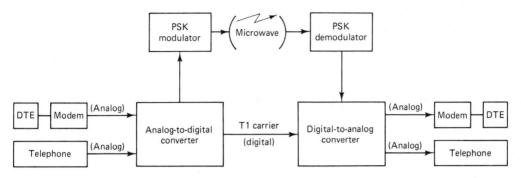

**FIGURE 9–1   Digital transmission.**

these digital signals are converted to seven-level PAM signals. The 1A-Radio Digital Terminal (1A-RDT) then converts the multilevel PAM signals to a signal that occupies a bandwidth of approximately 500 kHz and transmits it as *data under voice (DUV)*. We have seen in Chapter 8 that FDM mastergroups are transmitted at frequencies of 564 kHz and above. A DUV signal is so called because it uses the same transmitting and receiving facilities as the mastergroups (Figure 9–2). With Digital Data Service, the signal is considered in digital form from the customer's originating equipment to the customer's terminating equipment. Again, allowances have to be provided in this terminology since neither the multilevel PAM nor the DUV signals are in digital form.

With *Dataphone Digital Service*, a modem is involved. Figure 9–3 contrasts Digital Data Service with Dataphone Digital Service.

Various DTEs, transmitting at 2400, 4800, 9600 bps, or 56 kbps, have modems that convert these digital data to analog signals which are then transmitted to a Telco central office. At the central office, modems reconvert these signals to digital, multiplex them with other dataphone circuits, and transmit the resulting signal at a 1.544-Mbps rate. Dataphone Digital Service differs from a "regular" digital channel in that the transmission rate is slower and voice signals are never multiplexed with the digital information.

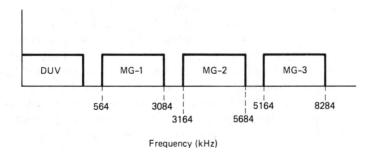

Frequency (kHz)

**FIGURE 9–2   Data under voice (DUV).**

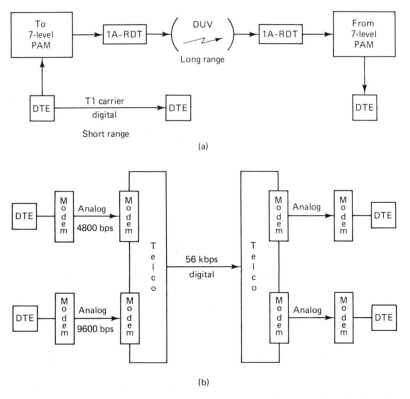

**FIGURE 9–3** Digital data service and dataphone digital service: (a) digital data service; (b) dataphone digital service.

## Advantages of Digital Transmission

The major advantage of digital transmission is noise immunity. In the transmission of analog information, undesired amplitude, frequency, and phase variations, introduced by either hardware or external noise sources, can produce errors in the received signal. With digital transmission, precise evaluation of the amplitude, frequency, and phase are not required. Although the digital pulses deteriorate as they travel down the transmission line, to extract the original information at the receive end, it is only necessary to determine whether the received signal at the time of sampling is either above or below a particular voltage threshold. *Regenerative repeaters* (not amplifiers), spaced approximately 6000 ft apart, detect the digital pattern, reconstruct the signal, and then retransmit it. The signal quality at the repeater output is equal to what it was when it was originally transmitted. The repeater distance of 6000 ft is chosen strictly from convenience. Digital pulses are transmitted along metallic facilities and these facilities must first be established. The use of existing facilities was the most convenient solution. Loading coils which are used to improve the quality of analog voice channel lines are spaced approximately 6000 ft apart. Telco manholes are available to access the lines at these points. Since these coils have to be removed to provide the BW required of T-carriers, it was logical to install repeaters in their place.

The trend toward digital communications is ever-increasing in the United States. Although not yet used for long-distance communications, there are more miles of T-carriers than miles of lines carrying analog signals. Previously, digital transmission was used for distances of 50 miles or less. Improved transmission techniques now permit digital transmission for distances of 500 miles with repeaters spaced approximately every 15,000 ft.

# PULSE CODE MODULATION

Digital transmission is the transmission of voltages which represent logic 1's and 0's. There are several ways of transforming analog information into digital information:

1. *Pulse width modulation* (PWM) or *pulse duration modulation* (PDM): The pulse width is proportional to the amplitude of the analog signal.
2. *Pulse position modulation* (PPM): Varying the position of identical pulses with respect to a particular time reference.
3. *Pulse amplitude modulation* (PAM): Converting analog amplitudes to discrete voltage levels. Although the resultant signal is made up of different discrete voltage levels, it cannot be considered digital information since these levels do not represent logic 1's and 0's. The output of the 2-to-4-level converter in the modulator circuit of Figure 7–32 is a PAM signal.

*Pulse code modulation* (PCM) is the most prevalent method of converting analog information to digital pulses. The analog voltage is sampled at regular intervals by a *sample-and-hold* (*S/H*) *circuit* and, while this sampled value of voltage is held constant, an analog-to-digital converter (ADC) converts it to a binary number. This number is proportional to the amplitude of the analog signal at the time of sampling. This is shown in Figure 9–4. The digital numbers will be explained shortly. The binary equivalent of the sampled voltage level is transmitted and reconverted at the receive end to a voltage level by a digital-to-analog converter (DAC).

## Sampling Rate

The Nyquist sampling theorem establishes the minimum rate at which a signal must be sampled for accurate information reproduction. Essentially, if a signal is sampled at a rate ($f_N$) that is equal to or greater than twice the highest frequency ($f_a$) contained in the signal, the original signal can be accurately reproduced from these samples. If $f_N$ is less than $2f_a$, distortion is introduced in the sample that will be carried through the system and appear in the demodulated output. This distortion is called

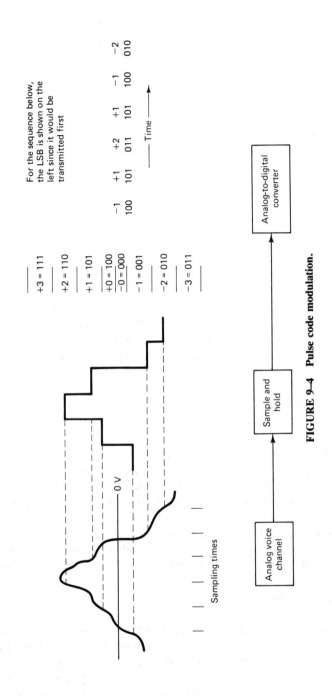

**FIGURE 9-4  Pulse code modulation.**

158

*foldover distortion* or *aliasing*. Basically, a sampling circuit is a switch which is a nonlinear device. It has two inputs, the analog signal ($f_a$) and the rectangular sample pulse ($f_N$). A nonlinear device with two or more inputs acts as an AM modulator. $f_N$ is amplitude-modulated by $f_a$. The two original frequencies ($f_N$ and $f_a$), their sum ($f_N + f_a$), and their difference ($f_N - f_a$) are present at the output of the sample circuit. Since the sample pulse is rectangular and repetitive, it is made up of the fundamental frequency ($f_N$) and all of the higher harmonics of $f_N$. Therefore, $f_N$ and all of its harmonics are amplitude-modulated by $f_a$. The output frequency spectrum is shown in Figure 9–5.

---

**EXAMPLE**

Maximum analog input frequency to a sampling circuit = 3 kHz. Sampling frequency = 6 kHz. The output will contain the original frequency range of 0 to 3 kHz and the band of frequencies 3 to 9 kHz produced by the amplitude modulation of the 6-kHz signal. These higher frequencies can be removed by filtering and the original signals will not be contaminated. If the sampling rate is reduced to 5 kHz, a frequency range from 2 to 8 kHz is produced by the modulation process. If frequencies above 3 kHz are filtered out, a band from 2 to 3 kHz, produced by modulation, still overlaps (folds over) the original frequency range. The presence of these frequencies in the original passband, after demodulation by the receiver, is foldover distortion. Since voice channels will be digitized for transmission, and a 0- to 4-kHz BW is allocated for each voice channel, a minimum sampling rate of 2 × 4 kHz = 8000 samples/second is required. A sample of the input signal must be taken every 1/800 s or every 125 μs. For the frequency range 0 to 4 kHz, since the highest frequency in this range (4 kHz) corresponds to the BW of the channel, the minimum sample rate = 2 × BW (low-pass filter).

---

## Sample and Hold

*Aperture time* is the time during which the analog input signal is being sampled. Since the analog signal may be changing during this time, if the ADC attempted to converted this varying voltage to digital, distortion would be introduced. This distortion is called *aperture distortion*. This distortion is eliminated by the use of

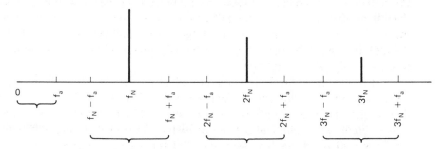

**FIGURE 9–5  Frequency spectrum produced by the amplitude modulation of $f_N$ and all of its harmonics by $f_a$.**

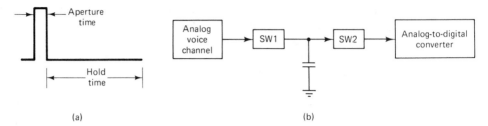

**FIGURE 9–6   Sample and hold: (a) S/H pulse; (b) S/H circuit.**

sample-and-hold (S/H) circuits. In a sample-and-hold circuit, the aperture time is made very small. During this time, the analog voltage produces a charge across a capacitor. During the hold time, the capacitor voltage is converted into a binary number by the ADC. The sampled voltage is essentially nonvarying during the conversion process. In the S/H circuit of Figure 9–6, SW 1 is first closed and SW 2 is opened. SW 1 should provide a low-impedance output and a high current drive to allow the capacitor to adjust rapidly to the sampled analog voltage. At the end of the aperture time, SW 1 is opened and SW 2 is closed, coupling the capacitor voltage to the ADC. SW 2 should have a high input impedance to keep the capacitor voltage constant during the time it is being converted to a binary number by the ADC.

## Quantization Noise

Between the maximum and minimum amplitudes of the analog signal, an infinite number of levels exist. In the encoding process, this infinite number of levels is converted to a finite combination of binary numbers dependent on the number of bits allocated for each sample. Therefore, each binary number will have to represent a range of sampled voltages. At the receive end, each binary number is decoded to a single voltage level. The error introduced in the conversion process is called *quantization error* or *quantization noise*. As can be seen in Table 9–1 by comparing the maximum difference between the input and the decoded output voltages, the maximum error is equal to one-half of the voltage represented by the step value of the ADC [one-half of the voltage of the least significant bit (LSB)]. Because of the high immunity to noise by a PCM signal, quantization noise is the major source of error in digital transmission.

If $n$ bits are used, $2^n$ possible combinations are available at the output of the ADC. The most significant bit represents the sign: 1 for $+$, 0 for $-$. (In digital theory, the opposite scheme is used for signed binary number representation.) For $n = 3$, eight 3-bit combinations are available. Two of these combinations are used for zero: 100 for a positive zero and 000 for a negative zero. This leaves seven combinations (levels) for the DAC at the receive end to convert into a single voltage

**TABLE 9–1**

| Binary | Decimal | Decoded voltage value (mV) | Input voltage range (mV) |
|--------|---------|----------------------------|--------------------------|
| 1 1 1  | +3      | 12                         | 10 to 14                 |
| 1 1 0  | +2      | 8                          | 6 to 10                  |
| 1 0 1  | +1      | 4                          | 2 to 6                   |
| 1 0 0  | +0      | 0                          | 0 to 2                   |
| 0 0 0  | −0      | 0                          | −2 to 0                  |
| 0 0 1  | −1      | −4                         | −6 to −2                 |
| 0 1 0  | −2      | −8                         | −10 to −6                |
| 0 1 1  | −3      | −12                        | −14 to −10               |

value. Since one of these bits is used as a sign bit, $n - 1$ bits remain to express the magnitude of the number. The maximum magnitude that can be obtained with $n$ bits is $2^n - 1$. If $n = 3$ and 1 bit is used for the sign, 2 bits remain to represent the magnitude. The maximum number that can be expressed with 2 bits is $2^n - 1$ = 3. If the voltage of the LSB is 4 mV, the decoded voltage values are as shown in Table 9–1.

Since a range of input voltages is capable of producing each binary number, *the maximum value of this error is equal to one-half of the value of the LSB*. At the input, a range of 4 mV was assigned to each binary number. Note that although two combinations were used to express zero, the range of values expressing ±0 is still 4 mV.

**EXAMPLE**

Given: 7-bit code (1 bit for sign, 6 bits for magnitude). LSB = 0.15 V.
(a) How many different voltage levels may be represented with this voltage scheme?
   *Ans.*: $2^7 - 1 = 127$.
(b) What is the maximum positive decoded voltage that can be represented with this coding scheme?
   *Ans.*: 1111111 = +63. $63 \times 0.15 = +9.45$ V.
(c) What is the maximum peak-to-peak decoded voltage that can be represented with this coding scheme?
   *Ans.*: $2 \times 9.45 = +18.9$ V.
(d) What is the voltage represented by the binary code of 1101010?
   *Ans.*: $42 \times 0.15 = +6.3$ V.
(e) What is the voltage represented by the binary code of 0011010?
   *Ans.*: $-(26 \times 0.15) = -3.9$ V.
(f) What is the binary code for +4.75 V?
   *Ans.*: $4.75/0.15 = 31.7$. Round off to 32. The code is 1100000.
(g) What is the binary code for −2.72 V?
   *Ans.*: $2.72/0.15 = 18.13$. Round off to 18. The code is 0010010.

**EXAMPLE**

Given: Maximum peak-to-peak input voltage of 31.5 V. Six bits (5 for magnitude, 1 for sign) will be used for encoding.

(a) How many different voltage levels may be represented?

$Ans.$: $2^6 - 1 = 63$.

(b) Voltage of LSB?

$Ans.$: $31.5/63 = 0.5$ V.

(c) What is the maximum quantization error?

$Ans.$: 0.25 V; one-half of the LSB.

(d) What is the decoded value for 101101?

$Ans.$: $13 \times 0.5 = +6.5$ V.

(e) What is the decoded value for 011001?

$Ans.$: $-(25 \times 0.5) = -12.5$ V.

(f) To what binary number will $+ 13.62$ V be encoded?

$Ans.$: $13.62/0.5 = 27.24$. Round off to 27. The binary number is 111011.

(g) To what binary number will $-9.37$ V be encoded?

$Ans.$: $9.37/0.5 = 18.74$. Round off to 19. The number is 010011.

## Resolution

The ADCs under discussion are linear—all of the voltage steps are equal. Therefore, the maximum amount of quantization error is the same for all input amplitudes. However, all amplitudes do not have the same probability of occurring in normal speech. The majority of voice conversations have relatively low signal amplitudes. The *resolution* of an ADC or a digital-to-analog converter (DAC) identifies the smallest analog voltage value that can be discerned by these converters and is equal to the voltage of the least significant bit. For a given voltage range, the resolution can be increased by increasing the number of bits in these converters and decreasing the magnitude of the LSB.

## Signal-to-Quantization Noise Ratio

The 3-bit PCM coding scheme described in the preceding section is a linear code. That is, the magnitude change between any two successive codes is uniform. Consequently, the magnitude of their quantization error is also equal. The maximum quantization noise is the voltage of the least significant bit divided by 2. Therefore, the worst possible *signal-to-quantization noise ratio* (SQR) occurs when the input signal is at its minimum amplitude (101 or 001). Mathematically, the worst-case SQR is

$$SQR = \frac{\text{minimum voltage}}{\text{quantization noise}} = \frac{V_{lsb}}{V_{lsb}/2} = 2$$

For a maximum amplitude input signal of 3 V (either 111 or 011), the maximum quantization noise is also the voltage of the least significant bit divided by 2. Therefore, the SQR for a maximum input signal condition is

$$SQR = \frac{\text{minimum voltage}}{\text{quantization noise}} = \frac{V_{\text{max}}}{V_{\text{lsb}}/2} = \frac{3}{0.5} = 6$$

From the preceding example it can be seen that even though the magnitude of error remains constant throughout the entire PCM code, the percentage of error does not; it decreases as the magnitude or the input signal increases. As a result, the SQR is not constant.

The preceding expression for SQR is for voltage and presumes the maximum quantization error and a constant-amplitude analog signal; therefore, it is of little practical use and is shown only for comparison purposes. In reality and as shown in Figure 9–4, the difference between the PAM waveform and the analog input waveform varies in magnitude. Therefore, the signal-to-quantization noise ratio is not constant. Generally, the quantization error or distortion caused by digitizing an analog sample is expressed as an average signal power-to-average noise power ratio. For linear PCM codes (all quantization intervals have equal magnitudes), the signal-to-quantizing noise ratio (also called *signal-to-distortion ratio* or *signal-to-noise ratio*) is determined as follows:

$$SQR \text{ (dB)} = 10 \log \frac{v^2/R}{(q^2/12)/R}$$

where
$R$ = resistance
$v$ = rms signal voltage
$q$ = size of one voltage step
$v^2/R$ = rms signal power
$(q^2/12)/R$ = rms quantization noise power

If the resistances are assumed to be equal,

$$SQR \text{ (dB)} = 10 \log \frac{v^2}{q^2/12}$$

$$= 10.8 + 20 \log \frac{v}{q}$$

## Linear versus Nonlinear PCM Codes

Early PCM systems used *linear codes* (i.e., the magnitude change between any two successive steps is uniform). With linear encoding, the accuracy (resolution) for the higher-amplitude analog signals is the same as for the lower-amplitude signals, and the SQR for the lower-amplitude signals is less than for the higher-amplitude signals. With voice transmission, low-amplitude signals are more likely to occur than large-amplitude signals. Therefore, if there were more codes for the lower amplitudes, it would increase the accuracy where the accuracy is needed. As a result, there would be fewer codes available for the higher amplitudes, which would increase the quantization error for the larger-amplitude signals (thus decreasing the

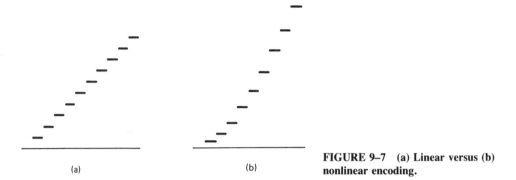

**FIGURE 9–7**   (a) Linear versus (b) nonlinear encoding.

(a)                    (b)

SQR). Such a coding technique is called *nonlinear* or *nonuniform encoding*. With nonlinear encoding, the resolution is increased for low amplitude input signals. Figure 9–7 shows the step outputs from a linear and a nonlinear ADC.

Note, with nonlinear encoding, there are more codes at the bottom of the scale than there are at the top, thus increasing the accuracy for the smaller signals. Also note that the distance between successive codes is greater for the higher-amplitude signals, thus increasing the quantization error and reducing the SQR. Also, because the ratio of $V_{max}$ to $V_{min}$ is increased with nonlinear encoding, the dynamic range is larger than with a uniform code. It is evident that nonlinear encoding is a compromise; SQR is sacrificed for the high-amplitude signals to achieve more accuracy for the low-amplitude signals and to achieve a larger dynamic range.

It is difficult to fabricate nonlinear ADCs; consequently, alternative methods of achieving the same results have been devised.

### Idle Channel Noise

During times when there is no analog input signal, the only input to the sampler is random, thermal noise. This noise is called *idle channel noise* and is converted to a PAM sample just as if it were a signal. Consequently, even input noise is quantized by the ADC. Figure 9–8 shows a way to reduce idle channel noise by a method called *midtread quantization*. With midtread quantizing, the first quantization interval is made larger in amplitude then the rest of the steps. Consequently, input noise can be quite large and still be quantized as a positive or negative zero code. As a result, the noise is suppressed during the encoding process.

In the PCM codes described thus far, the lowest-magnitude positive and negative codes have the same voltage range as all the other codes (+ or − one-half the resolution). This is called *midrise quantization*. Figure 9–8 contrasts the idle channel noise transmitted with a midrise PCM code to the idle channel noise transmitted when midtread quantization is used. The advantage of midtread quantization is less idle channel noise. The disadvantage is a larger possible magnitude for Qe in the lowest quantization interval.

Residual noise that fluctuates slightly above and below 0 V is converted to

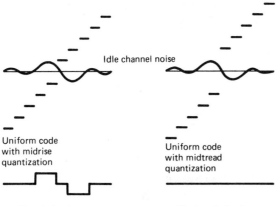

Idle channel noise

Uniform code
with midrise
quantization

Uniform code
with midtread
quantization

Decoded noise

No decoded noise

**FIGURE 9–8   Idle channel noise.**

either a + or − zero PCM code and is consequently eliminated. In systems that do not use the two 0-V assignments, the residual noise could cause the PCM encoder to alternate between the zero code and the minimum + or − code. Consequently, the decoder would reproduce the encoded noise. With a folded binary code, most of the residual noise is inherently eliminated by the encoder.

## Analog Companding

Before conversion to digital, the analog signals are passed through a nonlinear amplifier which provides greater amplification for the lower-amplitude signals. This action compresses the analog voltage range from $V_{max}$ to $V_{min}$ and increases the probability of the presence of voltage amplitudes at all levels for the ADC. At the receive end, after the digital signal has been decoded, the amplitudes must be expanded—amplified nonlinearly, with the lower amplitudes amplified the least to restore the signals to their original proportions. This process of compression and expansion is called *companding*. Figure 9–9 shows a block diagram displaying analog companding. In the United States, $\mu$-law is used for companding.

$$V_{out} = \frac{V_{max} \times \ln\left(1 + \mu V_{in}/V_{max}\right)}{\ln\left(1 + \mu\right)}$$

The amount of compression can be varied by varying $\mu$. This is shown in Figure 9–10.

---

**EXAMPLES USING $\mu$ = 255**

For $V_{in} = V_{max}$, $V_{out} = V_{max}$, providing unity gain.
For $V_{in} = 0.75\ V_{max}$, $V_{out} = 0.948\ V_{max}$.
Gain $= V_{out}/V_{in} = 0.948\ V_{max}/0.75\ V_{max} = 1.26$.
For $V_{in} = 0.5\ V_{max}$, $V_{out} = 0.876\ V_{max}$, gain $= 1.75$.
For $V_{in} = 0.25\ V_{max}$, $V_{out} = 0.752\ V_{max}$, gain $= 3$.

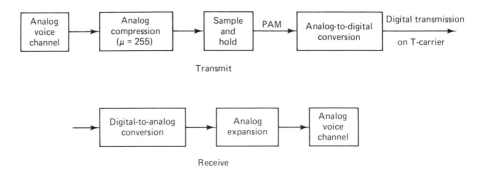

Transmit

Receive

**FIGURE 9–9    PCM with analog companding.**

From these examples, it can be seen that the lower amplitudes are amplified by a greater amount than the higher-amplitude input signals.

### Dynamic Range

The *dynamic range* is a ratio of maximum to minimum expected voltage values in a system and is normally expressed as a dB value. It is an important factor in establishing the number of bits that will be used for PCM encoding.

**EXAMPLE**
If a system has a dynamic range of 30 dB,

$$30 \text{ dB} = 20 \log \frac{V_{max}}{V_{min}}$$

$$\frac{V_{max}}{V_{min}} = 31.623$$

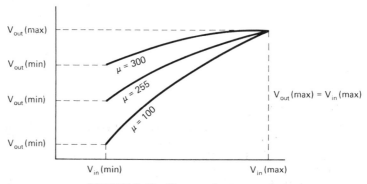

**FIGURE 9–10    Compression versus μ.**

This is the number of conditions desired. To produce this number of conditions:

$$2^n = 31.623 \qquad \text{where } n \text{ is the number of bits}$$

$$n \log 2 = \log 31.623$$

Thus $n = 1.5/.3 = 5$ bits would be required.

$V_{min}$ establishes the voltage of the least significant bit and the dynamic range affects the number of bits for each sample. There are other factors that must be considered which will further increase the number of bits per sample.

## D1 Channel Bank

When digital transmission was first implemented, Telco's D (digital) channel banks performed the conversion of analog voice samples to digital pulses. The D1 channel bank used analog companding with a $\mu = 100$ and converted each voice sample into a 7-bit code (sign plus 6 bits for magnitude). An additional bit was added for signaling. *Signaling functions* include off-hook, on-hook, dialing, and so on. As increased requirements were placed on these digital transmission systems, improved resolution was required. One method of improving resolution is simply to increase the number of steps in the ADC. This would require that each sampled voltage be converted into a greater number of bits. If the rate of information transmission was to remain the same, the transmission bit rate would have to be increased. If a single sample is converted into 12 bits instead of 8, the transmission bit rate would have to be 1.5 times greater. If the transmission bit rate is increased, the time of each pulse is shortened, necessitating the use of a greater BW. This is not desired.

## Digital Companding

D2, D3, and D4 channel banks use a $\mu = 255$ and convert each sample into 8 bits (sign plus 7 bits for magnitude). This is not totally true and the variations will be explained later.

Although D2, D3, and D4 channel banks use a $\mu = 255$, digital, not analog, companding is used to achieve this result. *Digital companding* involves compression at the transmit end after the signal has been sampled and expansion at the receive end before the received pulses are decoded. This is shown in the block diagram of Figure 9–11. Compare this figure with Figure 9–9 regarding the location where companding is accomplished.

In digital compression, each sample is first converted to a 12-bit binary number to obtain the greater resolution. However, instead of transmitting these 12 bits, they are first compressed to 8 bits and then transmitted. A compression of 12 bits to 8 bits cannot be accomplished without introducing error. Since increased resolution is desired for the lower amplitudes, this scheme minimizes the absolute error at these signal levels. In the 12-bit to 8-bit compression, the sign bit (MSB) is not

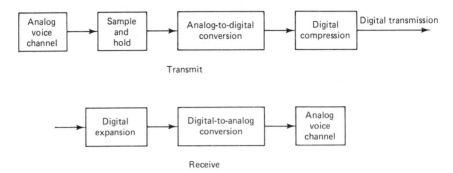

FIGURE 9–11   PCM with digital companding.

affected. The remaining 11 bits are broken up into segments. Each segment is identified by a 3-bit binary number which is determined by subtracting the number of leading 0's of the 11-bit magnitude representation from 7.

---

**EXAMPLE**

   12-bit code:    s00001101010
   Segment code:   $7 - 4 = 3 = 011$

---

The digit following the leading 0's is understood to be a 1 (otherwise, there would be an additional leading 0) and is not transmitted. As seen in Table 9–2, segment zero is an exception. The next 4 bits following the first 1 are transmitted as is and make up the last 4 bits of the compressed digital number. Any remaining bits are truncated. This is shown in Figure 9–12.

When the 8-bit code is received, it must first be expanded back to a 12-bit code before decoding.

---

**EXAMPLE**

   Received 8-bit code: s 011 1010
   $7 - 3 = 4$ leading 0's followed by a 1
   12-bit code: s 0000 1 1010 xx

---

In this case, the receiver has no information concerning the two LSBs which could range from 00 to 11. To minimize this companding error, a leading 1 followed by 0's is inserted by the receiver to replace all truncated information so that the final 12-bit expanded code would be s00001101010. The absolute error due to companding will vary depending on the segment transmitted. Table 9–2 shows the original 12-bit values, the 8-bit compressed values, and finally the expanded values of these 8 bits.

**TABLE 9–2    Twelve Bit-to-Eight Bit Compression**

| Segment | 12-bit code | 8-bit code | Expanded 12-bit code |
|---------|-------------|------------|----------------------|
| 0 | s0000000abcd | s000abcd | s0000000abcd |
| 1 | s0000001abcd | s001abcd | s0000001abcd |
| 2 | s000001abcdx | s010abcd | s000001abcd1 |
| 3 | s00001abcdxx | s011abcd | s00001abcd10 |
| 4 | s0001abcdxxx | s100abcd | s0001abcd100 |
| 5 | s001abcdxxxx | s101abcd | s001abcd1000 |
| 6 | s01abcdxxxxx | s110abcd | s01abcd10000 |
| 7 | s1abcdxxxxxx | s111abcd | s1abcd100000 |

$a$, $b$, $c$, and $d$ are transmitted as is. Bits identified by $x$ are lost in the companding process. Note that for segments 0 and 1, the original 12 bits are accurately reproduced, whereas for segment 7, only the 6 most significant bits have been recovered. Discounting the sign bit, the remaining 11 bits could produce $2^{11}$ = 2048 combinations. Segments 0 and 1 can each be any of 16 different combinations, depending on the specific values of $a$, $b$, $c$, and $d$. In segment 2, the last 5 bits, $a$, $b$, $c$, $d$, and $x$, could range from 00000 to 11111, or 32 different combinations. However, in the companding process, these 32 combinations will result in only 16 levels, represented by $a$, $b$, $c$, $d$, and 1. Thirty-two levels have been compressed into 16. Similarly, for segment 7, 1024 levels have been compressed into 16. This nonlinear digital compression is shown graphically in Figure 9–13.

The 8-bit PCM code produced by digital compression has the longer sequence of consecutive zeros for the lower-amplitude signals. If the receiver encounters a long stream of zeros, it may lose bit synchronization, and the lower-amplitude voice signals are much more likely to occur. For this reason, the magnitude bits may be passed through an inverter prior to transmission. The sign bit remains unchanged. The resultant code is called *folded binary* and is shown in Figure 9–14. The obvious question now is: Why should long sequences of 0's cause the receiver to lose synchronization and not a long stream of 1's? This question will be answered in a later section where the types of pulses used in transmitting 1's and 0's are described.

The curve shown in Figure 9–13 closely approximates the $\mu$-law curve for a $\mu$ = 255. The amount of companding error introduced is proportional to the amount of compression. The smaller amplitudes are represented by smaller binary numbers and, for these numbers, there is very little if any compression. This was the goal of digital companding—to obtain the greater resolution offered by a 12-bit code

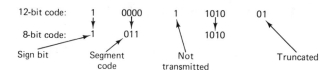

12-bit code:    1    0000    1    1010    01

8-bit code:    1    011    1010

Sign bit    Segment code    Not transmitted    Truncated

**FIGURE 9–12    Twelve bit-to-eight bit code conversion.**

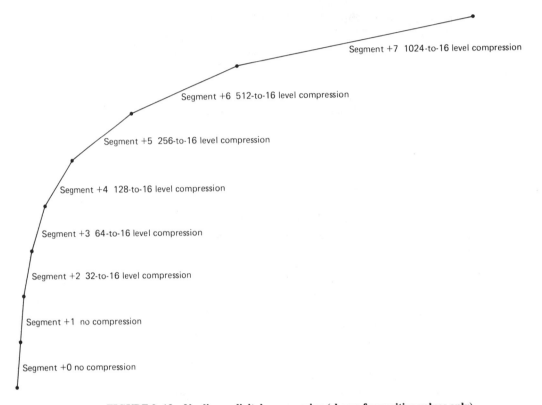

**FIGURE 9–13**   Nonlinear digital compression (shown for positive values only).

for the smaller-amplitude signals. As the signal amplitude increases, the amount of companding error increases. However, the maximum amount of percentage error is constant for all segments.

The following formula is used for the computation of percentage error:

$$\% \text{ error} = \frac{\left|\text{Tx voltage level} - \text{Rx voltage level}\right|}{\text{receiver voltage level}} \times 100$$

The maximum percentage error will occur for the smallest number within a segment.

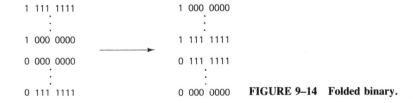

FIGURE 9–14   Folded binary.

**EXAMPLE**

For segment 3:

        Transmit:   s00001000000

        Receive:    s00001000010

$$\% \text{ error} = \frac{|64 - 66|}{66} \times 100 = 3.03\%$$

For segment 7:

        Transmit:   s10000000000

        Receive:    s10000100000

$$\% \text{ error} = \frac{|1024 - 1056|}{1056} \times 100 = 3.03\%$$

## Vocoders

The PCM coding and decoding processes described in the preceding sections were concerned primarily with reproducing waveforms as accurately as possible. The precise nature of the waveform was unimportant as long as it occupied the voice band frequency range. When digitizing speech signals only, special voice encoders/decoders called *vocoders* are often used. To achieve acceptable speech communications, the short-term power spectrum of the speech information is all that must be preserved. The human ear is relatively insensitive to the phase relationship between individual frequency components within a voice waveform. Therefore, vocoders are designed to reproduce only the short-term power spectrum, and the decoded time waveforms often only vaguely resemble the original input signal. Vocoders cannot be used in applications where analog signals other than voice are present, such as output signals from voice band data modems. Vocoders typically produce *unnatural* sounding speech and are therefore generally used for recorded information such as ''wrong number'' messages, encrypted voice for transmission over analog telephone circuits, computer output signals, and educational games.

    The purpose of a vocoder is to encode the minimum amount of speech information necessary to reproduce a perceptible message with fewer bits than those needed by a conventional encoder/decoders. Vocoders are used primarily in limited bandwidth applications. Essentially, there are three vocoding techniques available: the *channel vocoder*, the *formant vocoder*, and the *linear predictive coder*.

    **Channel vocoders.**   The first channel vocoder was developed by Homer Dudley in 1928. Dudley's vocoder compressed conventional speech waveforms into an analog signal with a total bandwidth of approximately 300 Hz. Present-day digital vocoders operate at less than 2 kbps. Digital channel vocoders use bandpass filters to separate the speech waveform into narrower *subbands*. Each subband is full-wave rectified, filtered, then digitally encoded. The encoded signal is transmitted to the destination receiver, where it is decoded. Generally speaking, the quality of the signal at the output a vocoder is quite poor. However, some of the more advanced channel vocoders operate at 2400 bps and can produce a highly intelligible, although slightly synthetic sounding speech.

**Formant vocoders.** A formant vocoder takes advantage of the fact that the short-term spectral density of typical speech signals seldom distributes uniformly across the entire voice band spectrum (300 to 3000 Hz). Instead, the spectral power of most speech energy concentrates at three or four peak frequencies called *formants*. A formant vocoder simply determines the location of these peaks and encodes and transmits only the information with the most significant short-term components. Therefore, formant vocoders can operate at lower bit rates and thus require narrower bandwidths. Formant vocoders sometimes have trouble tracking changes in the formants. However, once the formants have been identified, a formant vocoder can transfer intelligible speech at less than 1000 bps.

**Linear predictive coders.** A linear predictive coder extracts the most significant portions of speech information directly from the time waveform rather than from the frequency spectra as with the channel and formant vocoders. A linear predictive coder produces a time-varying model of the *vocal tract excitation* and transfer function directly from the speech waveform. At the receive end, a *synthesizer* reproduces the speech by passing the specified excitation through a mathematical

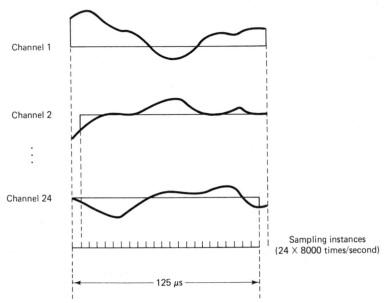

**FIGURE 9–15  PCM/TDM sampling sequence.**

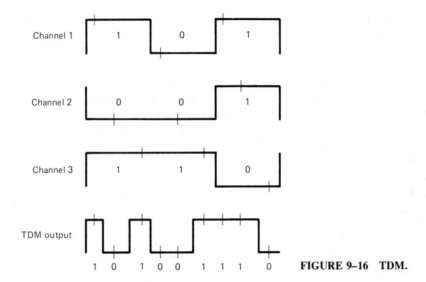

Channel 1

Channel 2

Channel 3

TDM output

1   0   1   0   0   1   1   1   0          **FIGURE 9–16   TDM.**

model of the vocal tract. Linear predictive coders provide more-natural-sounding speech than does either the channel or formant vocoder. Linear predictive coders typically encode and transmit speech at between 1.2 and 2.4 kbps.

## Time-Division Multiplexing

In TDM, the information from the various channels are interleaved and separated by time. This is shown in Figure 9–15. The various channels that are to be multiplexed are alternately sampled and transmitted. PCM easily lends itself to TDM. Telco's D channel banks accomplished the sampling (Figure 9–16), encoding, and multiplexing. Twenty-four voiceband channels are each sampled at a 8000-sample/second rate (at the same rate but not at the same time) and each sample is encoded into an 8-bit number.

$$24 \text{ samples} \times 8 \text{ bits/sample} = 192 \text{ bits}$$

For every 192 bits, an additional bit, the *frame bit*, is added for the purpose of synchronization. The resulting 193 bits make up a *frame* (Figure 9–17). Since frames are generated at the same frequency as the sampling rate, time-division multiplexing 24 channels requires a transmission rate of $193 \times 8000 = 1.544$ Mbps.

The D1 channel bank used analog companding with a $\mu = 100$ and a 7-bit code for the sample. A signaling bit was transmitted with every 7-bit code for

| S | Channel 1 | 8 bits | Channel 2 | 8 bits | | | Channel 24 | 8 bits |

**FIGURE 9–17   One TDM frame.**

every channel. The frame bits transmitted by the D1 channel bank simply alternated between 1 and 0. Signaling bits are not required that often. With the D2, D3, and D4 channel banks, digital companding is used with a $\mu = 255$ and 8 bits are used for each sample. Frames are grouped in units of 12. In the sixth and twelfth frames, the least significant PCM bit of every channel is used as a signaling bit, leaving only 7 bits for data information in these frames. The framing bits now must not only provide the necessary synchronization information but also must identify the sixth and twelfth frames. This is accomplished by using the odd-channel frame bits for synchronization and the even-channel frame bits to identify the sixth and twelfth frames. These frame bit patterns are shown in Figure 9–18. If only the framing bits are observed, the sixth frame (A-channel signaling) is identified by a 0-to-1 transition and the twelfth frame (B-channel signaling) is identified by a 1-to-0 transition.

| Frame number | 1 | 2 | 3 | 4 | 5 | 6 | 7 | 8 | 9 | 10 | 11 | 12 |
|---|---|---|---|---|---|---|---|---|---|---|---|---|
| Synch bits | 1 | | 0 | | 1 | | 0 | | 1 | | 0 | |
| Signal bits | | 0 | | 0 ——▶ 1 | | 1 | | 1 | | 1 ——▶ 0 | | |

**FIGURE 9–18   Frame bit sequence.**

The outputs from a D channel bank are placed on T-carriers. In order to be able to maintain timing, the D-channel banks are required to ensure that no sequence is transmitted that has more than 14 consecutive zeros. Each PCM encoded sample (8 bits) is monitored for the presence of at least one nonzero bit. If such a bit is not present, a 1 is substituted for the bit in bit position 7 (bit positions numbered 1 through 8, left to right).

> **WORST-CASE EXAMPLES**
> (a) 1000 0000    0000 0001
>     No substitution, 14 consecutive 0's, a 1 present in each PCM encoded sample.
> (b) 1000 0000    0000 0000
>     A 1 would be substituted for bit 7 of the second PCM encoded sample:
>     1000 0000    0000 0010 (order of arrow is to the left)

## Digital Channel Banks

The D1 channel bank sequentially samples the analog signal of 24 voiceband channels. Common equipment is used to perform analog compression ($\mu = 100$) on each sample. The resultant signal is then converted into a 7-bit code (sign plus 6 bits for magnitude). An additional bit, the signaling bit, is added to every 7-bit code of every channel. These channel banks are now obsolete. Saying that they are obsolete does not necessarily imply that they no longer exist. When opportunity permits, Telco is replacing them with D3 channel banks.

D2 and D3 channel banks are functionally similar. However, the D3 channel banks are the first to incorporate integrated circuitry which reduces the size and the power consumption required. Both D2 and D3 channel banks time-division multiplex 24 channels and use digital companding which approximates the $\mu = 255$ curve. Eight bits (sign plus 7 bits for magnitude) are used to represent each sample, with the least significant bit stolen every sixth frame for signaling.

The D4 channel bank uses customized LSI circuitry and time-division multiplexes 48 voice channels using digital companding with $\mu = 255$. Rather than use common equipment for encoding, each channel has its own encoder/decoder. The D4 channel bank multiplexes the encoded signals. The obvious advantage is that, should an encoder malfunction, only one channel is lost. The recent decreases in cost of *codecs* (coder/decoder) has made individual channel encoding/decoding cost-effective. The ultimate goal is to digitize voice transmission entirely. Rather than have an analog signal transmitted from the telephone to the digital channel bank, by incorporating a codec right into the telephone set, the voice signal is digitized at the outset. The transmission rate of the D4 channel bank is 3.152 Mbps. This bit rate is slightly higher than twice the output bit rate of a 24-channel multiplexed output. Since a larger number of channels are multiplexed, additional control bits are required.

## North American Digital Hierarchy

Although the transmission bit rate at the output of the D1, D2, D3, D4 channels banks is either 1.544 Mbps or 3.152 Mbps, these outputs are further multiplexed to achieve an even greater transmission bit rate. Table 9–3 shows the line nomenclature and the type of signal transmitted on that line.

> **EXAMPLE**
> A DS-1 signal has a transmission rate of 1.544 Mbps. This signal represents samples from 24 voice band channels time-division-multiplexed together. The DS-1 signal is transmitted on a T1 line.

## Multiplexers and Cross-Connects

Figure 9–19 shows the multiplexers used to achieve this higher transmission rate. Bell's nomenclature is used to identify the multiplexers. These multiplexers are

**TABLE 9–3  North American Digital Hierarchy**

| Line type | DS (digital signal) | Bit rate (Mbps) | Number of VB channels |
|-----------|---------------------|-----------------|-----------------------|
| T1        | DS-1                | 1.544           | 24                    |
| T1C       | DS-1C               | 3.152           | 48                    |
| T2        | DS-2                | 6.312           | 96                    |
| T3        | DS-3                | 44.736          | 672                   |
| T4M       | DS-4                | 274.176         | 4032                  |

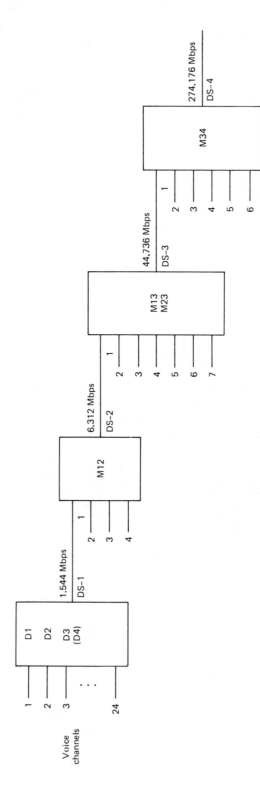

**FIGURE 9–19  North American Digital hierarchy.**

| Cross-Connect | Maximum Distance (ft) |
|---|---|
| DSX-1 | 750 |
| DSX-2 | 1000 |
| DSX-3 | 450 |
| DSX-4 | 150 |

**FIGURE 9–20  Maximum distances to cross-connects.**

sometimes called *muldems* (multiplexer/demultiplexer) since they must handle signal conversion in either direction. For simplicity, the number of inputs shown to each multiplexer reflects the number required from the multiplexer of the next lower order. This is not a restriction—M13 may receive either DS-1 or DS-2 or the proper combination of both to provide a DS-3 output.

DS-1, DS-2, DS-3, and DS-4 interface points are called the DSX-1, DSX-2, DSX-3, and DSX-4 cross-connects, respectively. The Bell system provides detailed specification for data stream requirements at these cross-connect points, namely, how the data stream appears at the end of a definite length of a particular cable. Maximum distances to cross-connect points are shown in Figure 9–20. These specifications set the maximum separation distance between the digital equipment. This is shown in Figure 9–21.

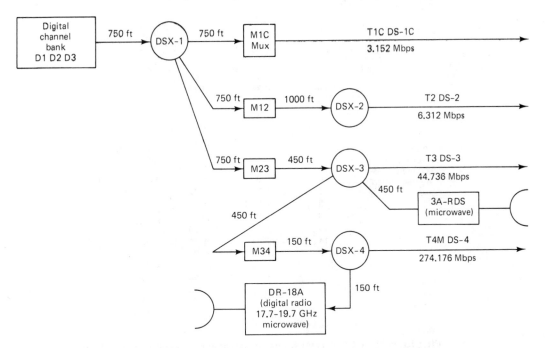

**FIGURE 9–21  Distances between equipment dictated by cross-connects.**

## CCITT Time-Division-Multiplexed Carrier System

Figure 9–22 shows the frame alignment for the CCITT (Comité Consultatif International Téléphonique et Télégraphique) European standard PCM-TDM system. With the CCITT system, an 125-μs frame is divided into 32 equal time slots. Time slot 0 is used for a frame alignment pattern and for an alarm channel. Time slot 17 is used for a common signaling channel. The signaling for all the voice band channels is accomplished on the common signaling channel. Consequently, there are 30 voice band channels time-division multiplexed into each CCITT frame.

With the CCITT standard, each time slot has 8 bits. Consequently, the total number of bits per frame is

$$\frac{8 \text{ bits}}{\text{time slot}} \times \frac{32 \text{ time slots}}{\text{frame}} = \frac{256 \text{ bits}}{\text{frame}}$$

| Time slot 0 | Time slot 1 | Time slots 2–16 | Time slot 17 | Time slots 18–30 | Time slot 31 |
|---|---|---|---|---|---|
| Framing and alarm channel | Voice channel 1 | Voice channels 2–15 | Common signaling channel | Voice channels 16–29 | Voice channel 30 |
| 8 bits | 8 bits | 112 bits | 8 bits | 112 bits | 8 bits |

(a)

Time slot 17

| | Bits | |
|---|---|---|
| Frame | 1234 | 5678 |
| 0 | 0000 | xyxx |
| 1 | ch 1 | ch 16 |
| 2 | ch 2 | ch 17 |
| 3 | ch 3 | ch 18 |
| 4 | ch 4 | ch 19 |
| 5 | ch 5 | ch 20 |
| 6 | ch 6 | ch 21 |
| 7 | ch 7 | ch 22 |
| 8 | ch 8 | ch 23 |
| 9 | ch 9 | ch 24 |
| 10 | ch 10 | ch 25 |
| 11 | ch 11 | ch 26 |
| 12 | ch 12 | ch 27 |
| 13 | ch 13 | ch 28 |
| 14 | ch 14 | ch 29 |
| 15 | ch 15 | ch 30 |

16 frames equal one multiframe; 500 multiframes are transmitted each second

x = spare
y = loss of multiframe alignment if a 1

4 bits per channel are transmitted once every 16 frames, resulting in a 500-bps signaling rate for each channel

(b)

**FIGURE 9–22   CCITT TDM frame alignment and common signaling channel alignment: (a) CCITT TDM (125 μs, 256 bits, 2.048 Mbps); (b) common signaling channel.**

and the line speed is

$$\text{line speed} = \frac{256 \text{ bits}}{\text{frame}} \times \frac{8000 \text{ frames}}{\text{second}} = 2.048 \text{ Mbps}$$

## Digital Line Signals

When information is transmitted in digital form, various factors must be considered:

1. Timing recovery
2. Bandwidth required
3. Ease of detection
4. Error detection

Some of the choices for pulse format are shown in Figure 9–23.

    **Transmission voltages.**    Unipolar transmission involves the transmission of only a single nonzero voltage level (+ V for a 1, 0 V for a 0). If the bit voltage is maintained for the entire bit time, the waveform is termed *Non-Return-to-Zero* (NRZ). In *Return-to-Zero* (RZ) transmission, the bit voltage is maintained for only a portion of the bit time. RZ transmission has an associated *duty cycle* which identifies the percentage of bit time that a nonzero polarity is maintained. Since a 50% duty cycle is used with RZ transmission, this duty cycle will be assumed in all future references to RZ. In *bipolar transmission*, two nonzero voltage levels are involved— a positive value for a 1 and a negative value for a 0. Bipolar transmission may be

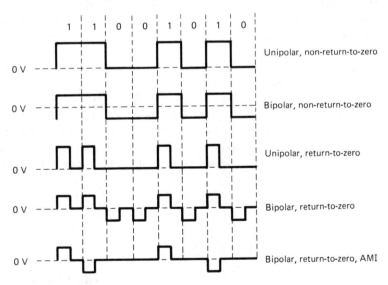

**FIGURE 9–23  Possible transmission waveforms.**

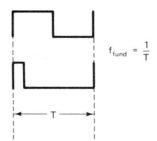

$$f_{fund} = \frac{1}{T}$$

**FIGURE 9–24   Fundamental frequency of rectangular waveforms.**

NRZ or RZ. Because of clear-cut advantages, T1 and T1C carriers use bipolar RZ with *alternate mark inversion* (AMI). In this line format, 0's are transmitted at a 0-V level and 1's alternate in polarity.

**Timing recovery.**   For timing recovery, a sufficient amount of transitions in the transmitted signal is essential. If NRZ transmission is used without any modifications, long strings of 1's and 0's would produce a constant line voltage. Since these would make timing recovery extremely difficult, unmodified NRZ is not a suitable form for digital transmission. RZ transmission forces transitions in a constant string of 1's. These transitions are absent in RZ for a constant string of 0's if a unipolar format is used.

**Bandwidth requirements.**   Line signals undergo deterioration en route to a repeater or to the final destination. To reconstruct the digital signal, the received signals are filtered to extract the fundamental frequency and then sampled. The low-pass filters pass the highest fundamental frequency of the incoming signal. The fundamental frequency of any rectangular waveform is the reciprocal of the time for one cycle no matter what the duty cycle (Figure 9–24).

Digital pulses are described as having a (*sin x*)/*x frequency spectrum*. If the factors of this ratio are considered separately, its meaning is much more readily understood. The numerator is a sine wave whose instantaneous amplitudes depend on *x*; the denominator increases with *x*. Therefore, this ratio simply represents a dampened sine wave. A square wave is made up of the fundamental frequency and an infinite number of odd harmonics whose amplitudes decrease with frequency. If these frequencies are plotted (Figure 9–25), their envelope would have a (sin *x*)/*x* curve. A dc component would be present if the square wave is not symmetric

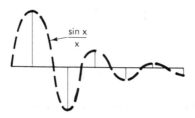

$$\frac{sin\ x}{x}$$

**FIGURE 9–25   (sin *x*)/*x* frequency spectrum.**

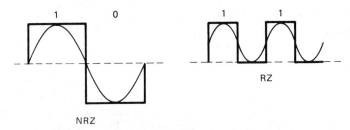

**FIGURE 9–26** Highest fundamental frequency for RZ and NRZ.

around the 0-V level. By filtering out the fundamental frequency and then sampling it at the time of maximum amplitudes, the binary 0–1 values can be extracted.

In NRZ transmission, the fastest line variations occur when alternate 1's and 0's are transmitted. From Figure 9–26 it can be seen that RZ require a bandwidth equal to the transmission bit rate. RZ introduces variations in the line signal which NRZ does not have, but it requires twice the BW. Bipolar RZ with AMI combines the best of both worlds. The fastest variations occur when consecutive 1's are transmitted. Since these 1's will be of opposite polarity, the BW, as seen in Figure 9–27, is one-half of the bit rate, and line variations are introduced when consecutive 1's are transmitted.

**Ease of detection.**    Unipolar transmission involves the transmission of only a single nonzero voltage level. To extract the digital data from the received signal, decision circuitry is required to determine whether a signal value is either above or below a positive threshold value. A long string of 1's (NRZ or RZ) may produce variations in the threshold value which may cause erroneous conversions. Variations in the threshold values may also be encountered in the bipolar transmission, again caused by long strings of 1's and 0's. With RZ AMI transmission, since all nonzero line voltages are produced by 1's and since these 1's alternate in polarity, the average dc line voltage is 0 V. Therefore, for this type of transmission, line signals have very little effect on the decision threshold levels.

**Error detection.**    By the very nature of RZ AMI transmission, received pulses should alternate in polarity. If two successive pulses of the same polarity are encountered, a noise impairment is definitely present: either a pulse was sufficiently reduced in amplitude to eliminate a 1 or noise in a 0-V region was of sufficient amplitude to be interpreted as a 1.

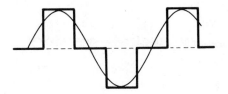

**FIGURE 9–27  Fundamental frequency of bipolar RZ with AMI.**

T2 and T3 carriers essentially use the same form of line signals as those used by the T1 and the T1C carriers. Bipolar RZ AMI transmission creates variations in long strings of consecutive 1's. However, long strings of 0's could still present a problem in trying to maintain timing. Earlier in this chapter it was seen that in a multiframe transmission, provisions are taken to prevent more than 14 consecutive 0's from being transmitted. The burden of not transmitting more than 14 consecutive 0's is placed on the source (D channel banks). Since the DS-2 signal on a T2 carrier is constructed by multiplexing four DS-1 signals and the DS-3 signal on the T3 carrier is comprised of multiplexed DS-1 and DS-2 signals, alternative measures are taken to eliminate long zero strings on these lines.

B6ZS stands for *Binary Six Zero Substitute*. Whenever six 0's are encountered, a substitution of either $0 - + 0 + -$ or $0 + - 0 - +$ is made for them. Here $+$ and $-$ represent the polarity of the voltage representing a 1. These substitutions purposely cause the transmission of consecutive 1's with the same polarity. If the pattern produced can be recognized by the receiver, the receiver can replace the substituted pattern with 0's. The substitute pattern chosen should produce a pulse of the same polarity at the second and fifth bits. This then would depend on the polarity of the last 1 preceding the string of six 0's.

---

**EXAMPLE**

$+$ and $-$ are used to indicate the polarity of the 1 bit.

(a) Original:               1 000 1 0 1 0000 1 0000000 1 $\cdots$
     With substitution:   $+$ 000 $-$ 0 $+$ 0000 $-$ 0 $-$ $+$ 0 $+$ $-$ 0 $+$ $\cdots$
(b) Original:               1 00000 1 00 1 0000000 1 00 1 $\cdots$
     With substitution:   $+$ 00000 $-$ 00 $+$ 0 $+$ $-$ 0 $-$ $+$ 0 $-$ 00 $+$ $\cdots$

---

B3ZS stand for *Binary Three Zero Substitute*. If DS-2 signals are multiplexed with DS-1 signals to form a DS-3 signal, the B6ZS must first be detected and stripped from the DS-2 signal and, before transmission on a T3 carrier, substitutions are made for every three consecutive 0's. The substitute patterns are one of the following: $00 +$, $00 -$, $-0 -$, or $+0 +$. The pattern chosen should cause a polarity repetition at the third substitute bit. The pattern chosen is dependent not only on the polarity of the 1 immediately preceding the three consecutive 0's but also on whether the number of 1's transmitted since the last substitution is odd or even. $00 -$ and $00 +$ are substituted if an odd number of 1's were transmitted since the last substitution.

---

**EXAMPLE**

Original Transmission:

                    0 1 0000 1 0 1 0000 1 00 11 0000 1 00 111 00000 1 $\cdots$

With B3ZS:

    0 $+$ 00 $+$ 0 $-$ 0 $+$ $-$ 0 $-$ 0 $+$ 00 $-$ $+$ 00 $+$ 0 $-$ 00 $+$ $-$ $+$ $-$ 0 $-$ 00 $+$ $\cdots$

---

| T1, T1C | Bipolar RZ AMI |
| T2 | B6ZS |
| T3 | B3ZS |
| T4 | Scrambled unipolar NRZ |

**FIGURE 9–28  Line signals used by various T-carriers.**

As mentioned earlier, unmodified NRZ is generally unsuitable for binary transmission. On T4 carriers, NRZ transmission is used but the data are first scrambled. Scrambling was discussed in Chapter 7. Figure 9–28 lists the line signals of the various T-carriers.

## DELTA MODULATION

Delta modulation is an alternative method of digitizing analog information. Delta modulation is related to digital encoding in a manner similar to the way DPSK is related to PSK. Rather than determine the binary code value solely from the sampled value of voltage, the binary output is dependent on whether the present sample is more or less positive than a reference signal which is determined from a history of previous sample sizes. If the present sample is more positive than this reference, a logic 1 is outputted; if the present sample is less positive than the reference, a logic 0 is outputted. The major advantage of delta modulation is the simplicity in encoding and decoding and the associated cheaper cost. However, the encoded signals cannot be integrated with other digital signals which use the D channel banks, nor do these signals lend themselves easily to being multiplexed. A block diagram of a linear delta modulator is shown in Figure 9–29.

As long as the analog input remains more positive than the reference voltage supplied by the digital-to-analog converter, the Up Enable signal for the counter will be active and the DAC output will increase in step increments. The rate at which the steps change values (the width of a step) is determined by the clock frequency; the size of one step is determined by the resolution of the DAC and its input voltages. In the idealized situation shown in Figure 9–30, the DAC output follows the input voltage. At the receive end (Figure 9–31), the received 1's and

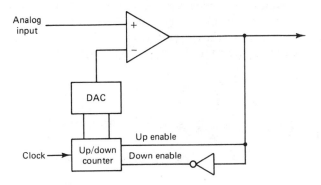

**FIGURE 9–29  Linear delta modulator.**

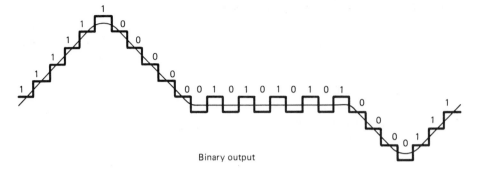

**FIGURE 9–30   Ideal action by a linear delta modulator.**

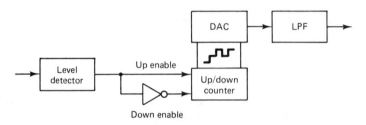

**FIGURE 9–31   Delta demodulator.**

0's can be used to reproduce the staircase waveforms and the low-pass filter would smooth out this waveform recreating the analog waveform.

Two problems arise with delta modulation—quantization or granular noise and slope overload—and each requires contrasting cures. *Granular noise* is the difference between the value of the sampled voltage and the value of the step size. In Figure 9–32, if waveform A is the original voltage and waveform B is the waveform reconstructed from the received digital data, the reproduced waveform possesses greater variations than were present in the original signal. This problem can be alleviated by increasing the resolution of the ADC (making the step sizes smaller) and increasing the clock frequency.

Granular noise is more prevalent in analog signals whose amplitudes vary slowly. *Slope overload* occurs when the analog voltage varies at a rate faster than the delta modulator can maintain. This is shown in Figure 9–33. This problem would be solved if the step size was increased so that the staircase waveform could

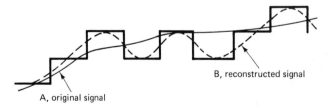

B, reconstructed signal

A, original signal

**FIGURE 9–32   Distortion due to granular noise.**

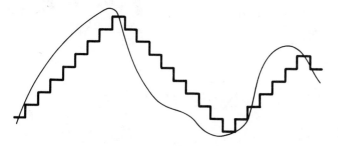

**FIGURE 9–33 Slope overload.**

catch up faster with the analog signal. To minimize granular noise, a small step size is desired; to minimize problems presented by slope overload, a large step size is desired. Obviously, for linear delta modulators, a compromise in step size must be achieved.

*Adaptive delta modulators* (ADM) automatically vary the step size based on signal requirements. *Continuous variable-slope delta modulator (CVSD)* is one form of an adaptive delta modulator. A block diagram of a CVSD is shown in Figure 9–34. Whenever an output sequence of either four 1's or four 0's is detected (an indication that the ADC output is trying to catch up with either a continuously rising or falling voltage), the step size is automatically increased. When this condition is no longer present—indicated by four nonsimilar bits in the shift register—the step size is returned to normal. Various other algorithms are available for adaptive delta modulators.

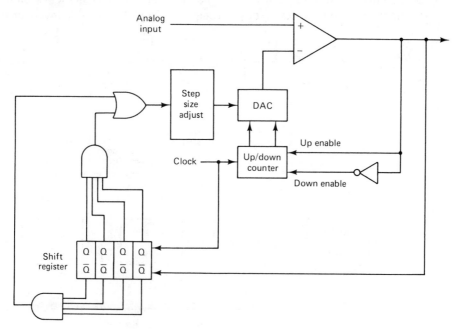

**FIGURE 9–34 Continuous slope delta modulator.**

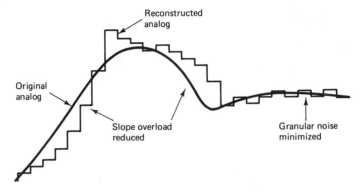

**FIGURE 9-35** **Adaptive delta modulation.**

## DIFFERENTIAL PULSE CODE MODULATION

In a typical PCM-encoded speech waveform, there are often successive samples taken in which there is little difference between the amplitudes of the two samples. This facilitates transmitting several identical PCM codes, which is redundant. Differential pulse code modulation (DPCM) is designed specifically to take advantage of the sample-to-sample redundancies in typical speech waveforms. With DPCM, the difference in the amplitude of two successive samples is transmitted rather than the actual sample. Since the range of sample differences is typically less than the range of individual samples, fewer bits are required for DPCM than conventional PCM.

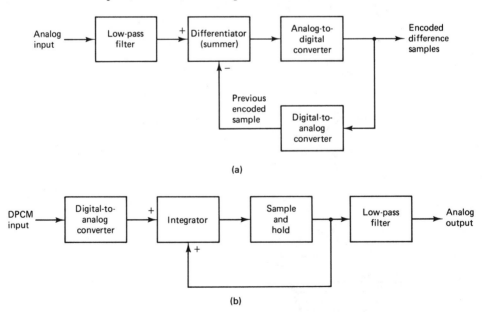

**FIGURE 9-36** **DPCM: (a) Transmitter; (b) Receiver**

Figure 9–36a shows a simplified block diagram of a DPCM transmitter. The analog input signal is bandlimited to one-half of the sample rate, then compared to the preceding DPCM signal in the differentiator. The output of the differentiator is the difference between the two signals. The difference is PCM encoded and transmitted. The A/D converter operates the same as in a conventional PCM system except that it typically uses fewer bits per sample.

Figure 9–36b shows a simplified block diagram of a DPCM receiver. Each received sample is converted back to analog, stored, and then summed with the next sample received. In the receiver shown in Figure 9–36b the integration is performed on the analog signals, although it could also be performed digitally.

## FRAME SYNCHRONIZATION

With TDM systems it is imperative that a frame is identified and that individual time slots (samples) within the frame are also identified. To acquire frame synchronization, there is a certain amount of overhead that must be added to the transmission. There are five methods commonly used to establish frame synchronization: added digit framing, robbed digit framing, added channel framing, statistical framing, and unique line signal framing.

### Added Digit Framing

T1 carriers using D1, D2, or D3 channel banks use *added digit framing*. There is a special *framing digit* (framing pulse) added to each frame. Consequently, for an 8-kHz sample rate (125-µs frame), there are 8000 digits added per second. With T1 carriers, an alternating 1/0 frame synchronizing pattern is used.

To acquire frame synchronization, the receive terminal searchs through the incoming data until it finds the alternating 1/0 sequence used for the framing bit pattern. This encompasses testing a bit, counting off 193 bits, then testing again for the opposite condition. This process continues until an alternating 1/0 sequence is found. Initial frame synchronization is dependent on the total frame time, the number of bits per frame, and the period of each bit. Searching through all possible bit positions requires $N$ tests, where $N$ is the number of bit positions in the frame. On the average, the receiving terminal dwells at a false framing position for two frame periods during a search; therefore, the maximum average synchronization time is

$$\text{synchronization time} = 2NT = 2N^2t$$

where   $T$ = frame period or $Nt$
        $N$ = number of bits per frame
        $t$ = bit time

For the T1 carrier, $N = 193$, $T = 125$ µs, and $t = 0.648$ µs; therefore, a maximum of 74,498 bits must be tested and the maximum average synchronization time is 48.25 ms.

## Robbed Digit Framing

When a short frame time is used, added digit framing is very inefficient. This occurs in single-channel PCM systems such as those used in television terminals. An alternative solution is to replace the least significant bit of every $n$th frame with a framing bit. The parameter $n$ is chosen as a compromise between reframe time and signal impairment. For $n = 10$, the SQR is impaired by only 1 dB. *Robbed digit framing* does not interrupt transmission, but instead, periodically replaces information bits with forced data errors to maintain clock synchronization.

## Added Channel Framing

Essentially, *added channel framing* is the same as added digit framing except that digits are added in groups or words instead of as individual bits. The CCITT multiplexing scheme previously discussed uses added channel framing. One of the 32 time slots in each frame is dedicated to a unique synchronizing sequence. The average frame synchronization time for added channel framing is

$$\text{synchronization time (bits)} = \frac{N^2}{2(2^L - 1)}$$

where  $N = $ number of bits per frame
    $L = $ number of bits in the frame code

For the CCITT 32-channel system, $N = 256$ and $L = 8$. Therefore, the average number of bits needed to acquire frame synchronization is 128.5. At 2.048 Mbps, the synchronization time is approximately 62.7 μs.

## Statistical Framing

With *statistical framing*, it is not necessary to either rob or add digits. With the Gray code, the second bit is a 1 in the central half of the code range and 0 at the extremes. Therefore, a signal that has a centrally peaked amplitude distribution generates a high probability of a 1 in the second digit. A mastergroup signal has such a distribution. With a mastergroup encoder, the probability that the second bit will be a 1 is 95%. For any other bit it is less than 50%. Therefore, the second bit can be used for a framing bit.

## Unique Line Code Framing

With *unique line code framing*, the framing bit is different from the information bits. It is either made higher or lower in amplitude or of a different time duration. The earliest PCM/TDM systems used unique line code framing. D1 channel banks used framing pulses that were twice the amplitude of normal data bits. With unique line code framing, added digit or added word framing can be used with it or data bits can be used to simultaneously convey information and carry synchronizing signals. The advantage of unique line code framing is synchronization is immediate and automatic. The disadvantage is the additional processing requirements required to generate and recognize the unique framing bit.

## BIT INTERLEAVING VERSUS WORD INTERLEAVING

When time-division multiplexing two or more PCM systems, it is necessary to interleave the transmissions from the various terminals in the time domain. Figure 9–37 shows two methods of interleaving PCM transmissions: *bit interleaving* and *word interleaving*.

T1 carrier systems use word interleaving; 8-bit samples from each channel are interleaved into a single 24-channel TDM frame. Higher-speed TDM systems and delta modulation systems use bit interleaving. The decision as to which type of interleaving to use is usually determined by the nature of the signals to be multiplexed.

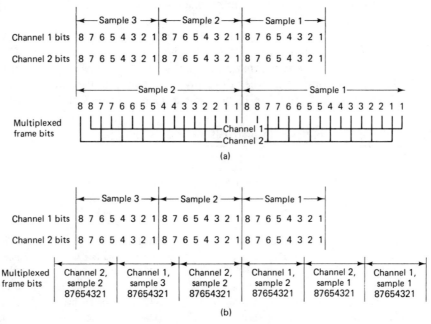

**FIGURE 9–37**  Interleaving: (a) bit; (b) word.

## CODECS

A *codec* is a large-scale-integration (LSI) chip designed for use in the telecommunications industry for *private branch exchanges* (PBXs), central office switches, digital handsets, voice store-and-forward systems, and digital echo suppressors. Essentially, the codec is applicable for any purpose that requires the digitizing of analog signals, such as in a PCM-TDM carrier system.

"Codec" is a generic term that refers to the *co*ding and *dec*oding functions performed by a device that converts analog signals to digital codes and digital codes to analog signals. Recently developed codecs are called *combo* chips because they combine codec and filter functions in the same LSI package. The input/output filter performs the following functions: bandlimiting, noise rejection, antialiasing, and reconstruction of analog audio waveforms after decoding. The codec performs the following functions: analog sampling, encoding/decoding (analog-to-digital and digital-to-analog conversions), and digital companding.

## 2913/14 COMBO CHIP

The 2913/14 is a combo chip that can provide the analog-to-digital and the digital-to-analog conversions and the transmit and receive filtering necessary to interface a full-duplex (four-wire) voice telephone circuit to the PCM highway of a TDM carrier system. Essentially, the 2913/14 combo chip replaces the older 2910A/11A codec and 2912A filter chip. The 2913 (20-pin package)/2914 (24-pin package) combo chip is manufactured with HMOS technology. There is a newer CHMOS version (29C13/14) that is functionally identical to the HMOS version except that it comes in a 28-pin package and has three low-power modes of operation. In addition, the 2916/17 and 29C16/17 are 16-pin limited-feature versions of the 2913/14 and 29C13/

**TABLE 9–4    Features of Several Codec/Filter Combo Chips**

| 2916 (16-pin) | 2917 (16-pin) | 2913 (20-pin) | 2914 (24-pin) |
|---|---|---|---|
| μ-law companding only | A-law companding only | μ/A-law companding | μ/A-law companding |
| Master clock 2.048 MHz only | Master clock 2.048 MHz only | Master clock 1.536 MHz, 1.544 MHz, or 2.048 MHz | Master clock 1.536 MHz, 1.544 MHz, or 2.048 MHz |
| Fixed data rate | Fixed data rate | Fixed data rate | Fixed data rate |
| Variable data rate 64 kbps–2.048 Mbps | Variable data rate 64 kbps–4.048 Mbps | Variable data rate 64 kbps–4.096 Mbps | Variable data rate 64 kbps–4.096 Mbps |
| 78-dB dynamic | 78-dB dynamic range | 78-dB dynamic range | 78-dB dynamic range |
| ATT D3/4 compatible | ATT D3/4 compatible | ATT D3/4 compatible | ATT D3/4 compatible |
| Single-ended input | Single-ended input | Differential input | Differential input |
| Single-ended output | Single-ended output | Differential output | Differential output |
| Gain adjust transmit only | Gain adjust transmit only | Gain adjust transmit and receive | Gain adjust transmit and receive |
| Synchronous clocks | Synchronous clocks | Synchronous clocks | Synchronous clocks |
|  |  |  | Asynchronous clocks |
| — | — | — | Analog loopback |
| — | — | — | Signaling |

14. The following discussion is limited to the 2914 combo chip, although extrapolation to the other versions is quite simple.

Table 9–4 lists several of the combo chips available and their prominent features. Table 9–5 lists the pin names for the 2914 and gives a brief description of each of their functions. Figure 9–38 shows the block diagram of a 2914 combo chip.

## General Operation

The following major functions are provided by the 2914 combo chip:

1. Bandpass filtering of the analog signals prior to encoding and after decoding
2. Encoding and decoding of voice and call progress signals
3. Encoding and decoding of signaling and supervision information
4. Digital companding

## System Reliability Features

The 2914 combo chip is powered up by pulsing the *transmit frame synchronization input* (FSX) and/or the *receive frame synchronization input* (FSR), while a TTL high (inactive condition) is applied to the *power down select pin* ($\overline{\text{PDN}}$) and all clocks and power supplies are connected. The 2914 has an internal reset on all power-ups (when VBB or VCC are applied or temporarily interrupted). This ensures the validity of the digital output and thereby maintains the integrity of the PCM highway.

On the transmit channel, PCM *data output* (DX) and *transmit timeslot strobe* ($\overline{\text{TSX}}$) are held in a high-impedance state for approximately four frames (500 µs) after power-up. After this delay DX, $\overline{\text{TSX}}$, and signaling are functional and will occur in the proper time slots. Due to the auto-zeroing circuit on the transmit channel, the analog circuit requires approximately 60 ms to reach equilibrium. Therefore, signaling information such as on/off hook detection is available almost immediately while analog input signals are not available until after the 60-ms delay.

On the receive channel, the *signaling bit output pin* SIGR is also held low (inactive) for approximately 500 µs after power-up and remains inactive until updated by reception of a signaling frame.

$\overline{\text{TSX}}$ and DX are placed in the high-impedance state and SIGR is held low for approximately 20 µs after an interruption of the *master clock* (CLKX). Such an interruption could be caused by some kind of fault condition.

## Power-Down and Standby Modes

To minimize power consumption, two power-down modes are provided in which most 2914 functions are disabled. Only the power-down, clock, and frame synchronization buffers are enabled in these modes.

**TABLE 9–5  2914 Combo Chip**

| Symbol | Name | Function |
|---|---|---|
| VBB | Power (−5 V) | Negative supply voltage. |
| PWRO+ | Receive power amplifier output | Noninverting output of the receive power amplifier. This output can drive transformer hybrids or high-impedance loads directly in either a differential or single-ended mode. |
| PWRO− | Receive power amplifier output | Inverting output of the receive power amplifier. Functionally, PWRO− is identical and complementary to PWRO+. |
| GSR | Receive gain control | Input to the gain-setting network on the receive power amplifier. Transmission level can be adjusted over a 12-dB range depending on the voltage at GSR. |
| $\overline{\text{PDN}}$ | Power-down select | When $\overline{\text{PDN}}$ is a TTL high, the 2914 is active. When $\overline{\text{PDN}}$ is low, the 2914 is powered down. |
| CLKSEL | Master clock frequency select | Input that must be pinstrapped to reflect the master clock frequency at CLKX, CLKR. <br> CLKSEL = VBB . . . . . . . . 2.048 MHz <br> CLKSEL = GRDD . . . . . . . 1.544 MHz <br> CLKSEL = VCC . . . . . . . . 1.536 MHz |
| LOOP | Analog loopback | When this pin is a TTL high, the analog output (PWRO+) is internally connected to the analog input (VFXI+), GSR is internally connected to PWRO−, and VFXI− is internally connected to GSX. |
| SIGR | Receive signaling bit output | Signaling bit output from the receiver. In the fixed data rate mode only, SIGR outputs the logic state of the eighth bit of the PCM word in the most recent signaling frame. |
| DCLKR | Receive variable data rate | Selects either the fixed or variable data rate mode of operation. When DCLKR is tied to VBB, the fixed data rate mode is selected. When DCLKR is not connected to VBB, the 2914 operates in the variable data rate mode and will accept TTL input levels from 64 kHz to 4096 MHz. |
| DR | Receive PCM highway input | PCM data are clocked in on this lead on eight consecutive negative transitions of the receive data rate clock; CLKR in the fixed data rate mode and DCLKR in the variable data rate mode. |
| FSR | Receive frame synchronization clock | 8-kHz frame synchronization clock input/time slot enable for the receive channel. Also in the fixed data rate mode, this lead designates the signaling and nonsignaling frames. In the variable data rate mode this signal must remain active high for the entire |

**TABLE 9–5   (Continued)**

| Symbol | Name | Function |
|---|---|---|
| | | length of the PCM word (8 PCM bits). The receive channel goes into the standby mode whenever this input is TTL low for 300 ms. |
| GRDD | Digital ground | Digital ground for all internal logic circuits. This pin is not internally tied to GRDA. |
| CLKR | Receive master clock | Receive master clock and data rate clock in the fixed data rate mode, master clock only in the variable data rate mode. |
| CLKX | Transmit master clock | Transmit master clock and data rate clock in the fixed data rate mode, master clock only in the variable data rate mode. |
| FSX | Transmit frame synchronization clock | 8-kHz frame synchronization clock input/time slot enable for the transmit channel. Operates independently but in an analogous manner to FSR. |
| DX | Transmit PCM output | PCM data are clocked out on this lead on eight consecutive positive transitions of the transmit data rate clock; CLKX in the fixed data rate mode and DCLKX in the variable data rate mode. |
| $\overline{\text{TSX}}$/ DCLKX | Times-slot strobe/ buffer enable Transmit variable data rate | Transmit channel timeslot strobe (output) or data rate clock (input) for the transmit channel. In the fixed data rate mode, this pin is an open drain output designed to be used as an enable signal for a three-state buffer. In variable data rate mode, this pin is the transmit data rate clock input which can operate at data rates between 64 kbps and 4096 kbps. |
| SIGX/ ASEL | Transmit signaling input μ- or A-law select | A dual-purpose pin. When connected to VBB, A-law companding is selected. When it is not connected to VBB, this pin is a TTL-level input for signaling bits. This input is substituted into the least significant bit position of the PCM word during every signaling frame. |
| GRDA | Analog ground | Analog ground return for all internal voice circuits. Not internally connected to GRDD. |
| VFXI+ | Noninverting analog input | Noninverting analog input to uncommitted transmit operational amplifier. |
| VFXI− | Inverting analog input | Inverting analog input to uncommitted transmit operational amplifier. |
| GSX | Transmit gain control | Output terminal of on-chip uncommitted operational amplifier. Internally, this is the voice signal input to the transmit BPF. |
| VCC | Power (+5 V) | Positive supply voltage. |

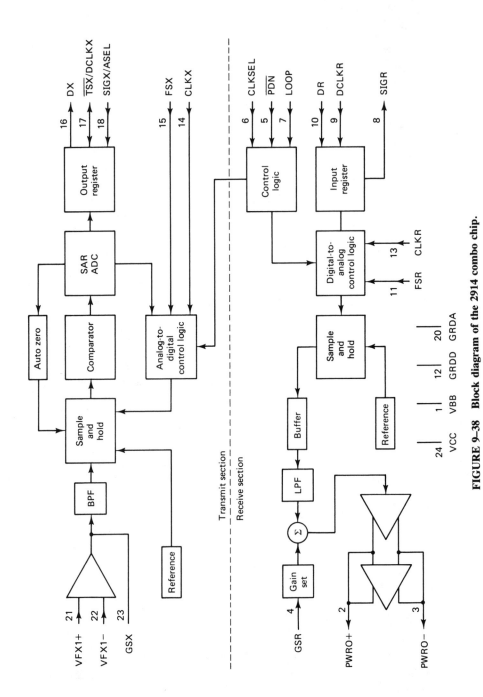

FIGURE 9-38  Block diagram of the 2914 combo chip.

194

The power-down is enabled by placing an external TTL low signal on $\overline{\text{PDN}}$. In this mode power consumption is reduced to an average of 5 mW.

The standby mode for the transmit and receive channels is separately controlled by removing FSX and/or FSR.

## Fixed-Data-Rate Mode

In the *fixed-data-rate mode*, the master *transmit* and *receive clocks* (CLKX and CLKR) perform the following functions:

1. Provide the master clock for the on-board switched capacitor filters
2. Provide the clock for the analog-to-digital and digital-to-analog converters
3. Determine the input and output data rates between the codec and the PCM highway

Therefore, in the fixed-data-rate mode, the transmit and receive data rates must be either 1.536, 1.544, or 2.048 Mbps, the same as the master clock rate.

Transmit and receive frame synchronizing pulses (FSX and FSR) are 8-kHz inputs which set the transmit and receive sampling rates and distinguish between *signaling* and *nonsignaling* frames. $\overline{\text{TSX}}$ is a *time-slot strobe buffer enable* output which is used to gate the PCM word onto the PCM highway when an external buffer is used to drive the line. $\overline{\text{TSX}}$ is also used as an external gating pulse for a time-division multiplexer (see Figure 9–39).

Data are transmitted to the PCM highway from DX on the first eight positive transitions of CLKX following the rising edge of FSX. On the receive channel, data are received from the PCM highway from DR on the first eight falling edges of CLKR after the occurrence of FSR. Therefore, the occurrence of FSX and FSR must be synchronized between codecs in a multiple-channel system to ensure that only one codec is transmitting to or receiving from the PCM highway at any given time.

Figure 9–39 shows the block diagram and timing sequence for a single-channel PCM system using the 2914 combo chip in the fixed-data-rate mode and operating with a master clock frequency of 1.536 MHz. In the fixed-data-rate mode, data are inputted and outputted in short bursts. (This mode of operation is sometimes called the *burst mode*.) With only a single channel, the PCM highway is active only $\frac{1}{24}$ of the total frame time. Additional channels can be added to the system provided that their transmissions are synchronized so that they do not occur at the same time as transmissions from any other channel.

From Figure 9–39 the following observations can be made:

1. The input and output bit rates from the codec are equal to the master clock frequency of 1.536 Mbps.
2. The codec inputs and outputs 64,000 PCM bits per second.

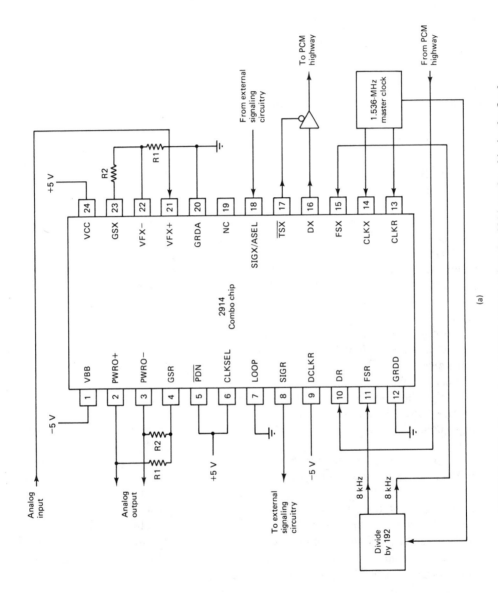

**FIGURE 9–39** Single-channel PCM system using the 2914 combo chip in the fixed-data-rate mode: (a) block diagram; (b) timing sequence.

(a)

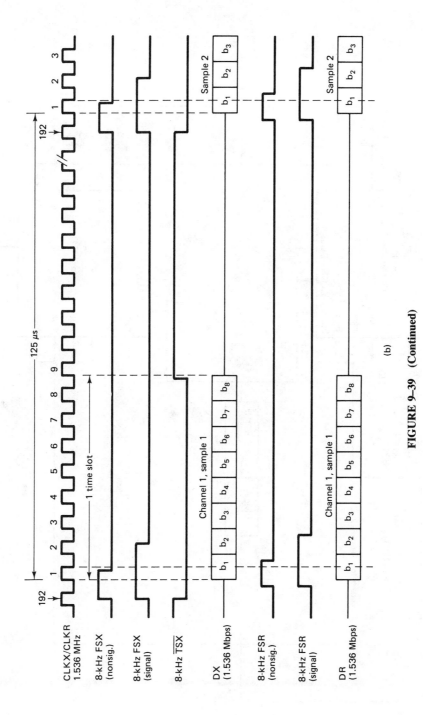

**FIGURE 9–39** (Continued)

197

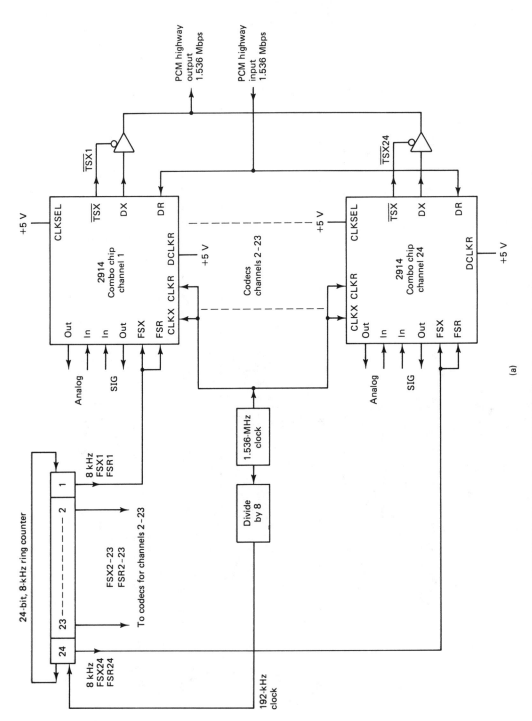

**FIGURE 9-40** **24-channel PCM-TDM system using the 2914 combo chip in the fixed-data-rate mode and operating with a master clock frequency of 1.536 MHz: (a) block diagram; (b) timing diagram.**

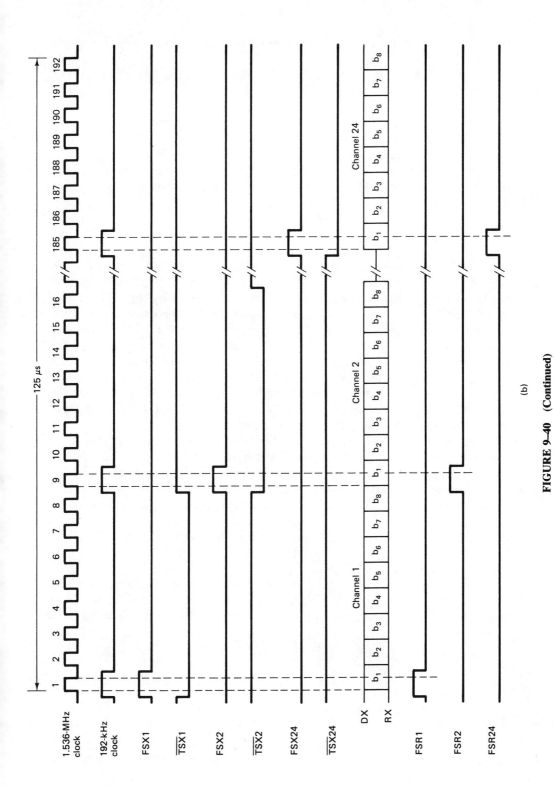

**FIGURE 9-40** (Continued)

(b)

199

3. The data output (DX) and data input (DR) are active only $\frac{1}{24}$ of the total frame time (125 μs).

To add channels to the system shown in Figure 9–39, the occurrence of the FSX, FSR, and $\overline{\text{TSX}}$ signals for each additional channel must be synchronized so that they follow a timely sequence and do not allow more than one codec to transmit or receive at the same time. Figure 9–40 shows the block diagram and timing sequence for a 24-channel PCM-TDM system operating with a master clock frequency of 1.536 MHz.

## Variable-Data-Rate Mode

The *variable-data-rate mode* allows for a flexible data input and output clock frequency. It provides the ability to vary the frequency of the transmit and receive bit clocks. In the variable data rate mode, a master clock frequency of 1.536, 1.544, or 2.048 MHz is still required for proper operation of the on-board bandpass filters and the analog-to-digital and digital-to-analog converters. However, in the variable-data-rate mode, DCLKR and DCLKX become the data clocks for the receive and transmit PCM highways, respectively. When FSX is high, data are transmitted onto the PCM highway on the next eight consecutive positive transitions of DCLKX. Similarly, while FSR is high, data from the PCM highway are clocked into the codec on the next eight consecutive negative transitions of DCLKR. This mode of operation is sometimes called the *shift register mode*.

On the transmit channel, the last transmitted PCM word is repeated in all remaining time slots in the 125-μs frame as long as DCLKX is pulsed and FSX is held active high. This feature allows the PCM word to be transmitted to the PCM highway more than once per frame. Signaling is not allowed in the variable-data-rate mode because this mode provides no means to specify a signaling frame.

Figure 9–41 shows the block diagram and timing sequence for a two-channel PCM-TDM system using the 2914 combo chip in the variable-data-rate mode with a master clock frequency of 1.536 MHz, a sample rate of 8 kHz, and a transmit and receive data rate of 128 kbps.

With a sample rate of 8 kHz, the frame time is 125 μs. Therefore, one 8-bit PCM word from each channel is transmitted and/or received during each 125-μs frame. For 16 bits to occur in 125 μs, a 128-kHz transmit and receive data clock is required.

$$\frac{1 \text{ channel}}{8 \text{ bits}} \times \frac{1 \text{ frame}}{2 \text{ channels}} \times \frac{125 \text{ μs}}{\text{frame}} = \frac{125 \text{ μs}}{16 \text{ bits}} = \frac{7.8125 \text{ μs}}{\text{bit}}$$

$$\text{bit rate} = \frac{1}{t_b} = \frac{1}{7.8125 \text{ μs}} = 128 \text{ kbps}$$

The transmit and receive enable signals (FSX and FSR) for each codec are active for one-half of the total frame time. Consequently, 8-kHz, 50% duty cycle transmit and receive data enable signals (FXS and FXR) are fed directly to one codec and fed to the other codec 180° out of phase (inverted), thereby enabling only one codec at a time.

To expand to a four-channel system, simply increase the transmit and receive data clock rates to 256 kHz and change the enable signals to an 8-kHz, 25% duty cycle pulse.

## Supervisory Signaling

With the 2914 combo chip, *supervisory signaling* can be used only in the fixed-data-rate mode. A transmit signaling frame is identified by making the FSX and FSR pulses twice their normal width. During a transmit signaling frame, the signal present on input SIGX is substituted into the least significant bit position ($b_1$) of the encoded PCM word. At the receive end, the signaling bit is extracted from the PCM word prior to decoding and placed on output SIGR until updated by reception of another signaling frame.

## Asynchronous Operation

Asynchronous operation is when the master transmit and receive clocks are derived from separate independent sources. The 2914 combo chip can be operated in either the synchronous or asynchronous mode. The 2914 has separate digital-to-analog converters and voltage references in the transmit and receive channels, which allows them to be operated completely independent of each other. With either synchronous or asynchronous operation, the master clock, data clock, and time-slot strobe must be synchronized at the beginning of each frame. In the variable data rate mode, CLKX and DCLKX must be synchronized once per frame but may be different frequencies.

## Transmit Filter Gain

The analog input to the transmit section of the 2914 is equipped with an uncommitted operational amplifier that can operate in the single-ended or differential mode. Figure 9–42 shows a circuit configuration commonly used to provide input gain. To operate with unity gain, simply strap VFXI− to GSX and apply the analog input to VFXI+.

## Receive Output Power Amplifier

The 2914 is equipped with an internal balanced output amplifier that may be used as two separate single ended outputs or as a single differential output. Figure 9–43 shows the gain setting configuration for the output amplifier operating in the differential mode. To operate with a single-ended output and unity gain, simply pin strap PWRO− to GSR and take the output from PWRO+.

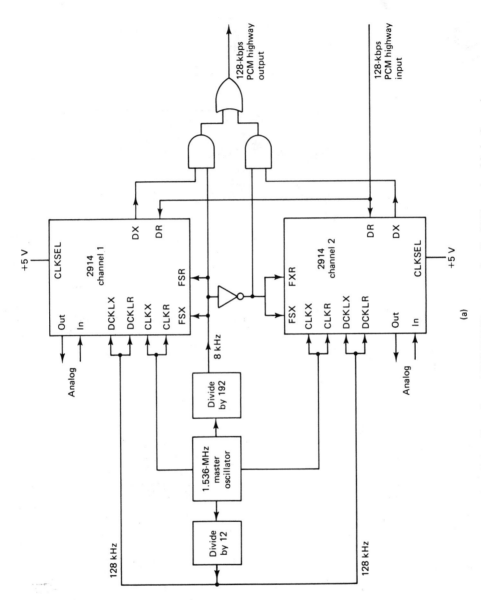

(a)

**FIGURE 9-41** Two-channel PCM-TDM system using the 2914 combo chip in the variable-data-rate mode with a master clock frequency of 1.536 MHz: (a) block diagram; (b) timing diagram.

202

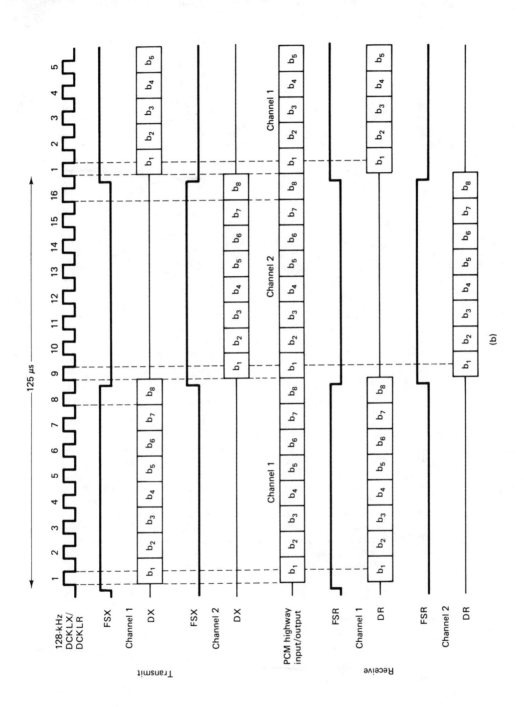

(b)

203

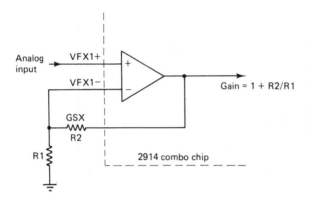

**FIGURE 9–42** Transmit filter gain circuit.

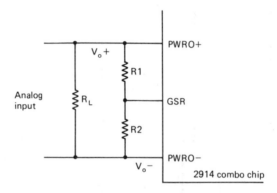

**FIGURE 9–43** Receive output power amplifier. PWRO+ and PWRO− are low-impedance complementary outputs. The voltages at the nodes are $V_o$+ at PWRO+ and $V_o$− at PRWO−. R1 and R2 comprise a gain-setting resistor network with the center tap connected to the GSR input. A value greater than 10 k$\Omega$ for R1 and a value less than 100 k$\Omega$ for R2 is recommended because (1) the parallel combination of R1 + R2 and RL set the total load impedance to the analog sink, and (2) the total capacitance at the GSR input and the parallel combination of R1 and R2 define a time constant that has to be minimized to avoid inaccuracies. VA represents the maximum available digital miliwatt output response (VA = 3.006 V rms).

$$V_o = -A(\text{VA})$$

where

$$A = \frac{1 + R1/R2}{4 + R1/R2}$$

For design purposes, a useful form is R1/R2 as a function of A.

$$R1/R2 = \frac{4A - 1}{1 - A}$$

## PULSE TRANSMISSION

All digital carrier systems involve the transmission of pulses through a medium with a finite bandwidth. A minimum-bandwidth system would require an infinite number of filter sections, which is impossible. Therefore, practical digital systems generally utilize filters with bandwidths that are approximately 30% or more in excess of the ideal Nyquist bandwidth. Figure 9–44(a) shows the typical output spectrum from a *bandlimited* communications channel when a narrow pulse is applied to its input. The figure shows that bandlimiting a pulse causes the energy from the pulse to be spread over a significantly wider bandwidth in the form of *secondary lobes*. The secondary lobes are called *ringing tails*. The output frequency spectrum

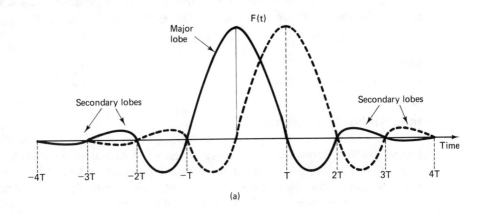

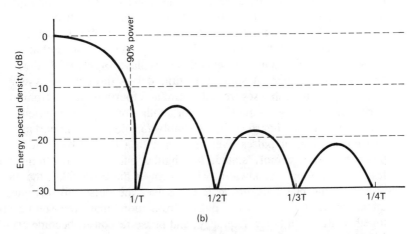

**FIGURE 9–44** Pulse response: (a) typical pulse response of a bandlimited filter; (b) spectrum of square pulse with duration *T*.

corresponding to a rectangular pulse is referred to as a (sin $x$)/$x$ response and is given as

$$F(\omega) = (T)\frac{\sin(\omega T/2)}{\omega T/2}$$

Figure 9–44(b) shows the approximate percentage of the total spectrum power at various bandwidths. It can be seen that approximately 90% of the signal energy is contained within the first *spectral null* (i.e., $F = 1/T$). Therefore, the signal can be confined to a bandwidth $B = 1/T$ and still pass most of the energy from the original waveform. In theory, only the amplitude at the middle of each pulse interval need to be preserved. Therefore, if the bandwidth is confined to $B = 1/2T$, the maximum signaling rate achievable through a low-pass filter with a specified band-width without causing excessive distortion is given as the Nyquist rate and is equal to twice the bandwidth. Mathematically, the Nyquist rate is

$$R = 2B$$

where   $R$ = signaling rate = $1/T$
        $B$ = specified bandwidth

## Intersymbol Interference

Figure 9–45(a) shows the input signal to an ideal minimum bandwidth lowpass filter. The input signal is a random, binary non-return-to-zero (NRZ) sequence. Figure 9–45(b) shows the output of a perfect filter (i.e., a filter that does not introduce any phase or amplitude distortion). Note that the output signal reaches its full value for each transmitted pulse at precisely the center of each sampling interval. However, if the lowpass filter is imperfect (which in reality it will be), the output response will more closely resemble that shown in Figure 9–45(c). At the sampling instants (i.e., the center of the pulses), the signal does not always attain the maximum value. The ringing tails of several pulses have *overlapped*, thus interfering with the *major pulse lobe*. Assuming no time delays through the system, energy in the form of spurious responses from the third and fourth impulses from one pulse appears during the sampling instant ($T = 0$) of another pulse. This interference is commonly called *intersymbol interference* or simply *ISI*. ISI is an important consideration in the transmission of pulses over circuits with a limited bandwidth and a nonlinear phase response. Simply stated, rectangular pulses will not remain rectangular in less than an infinite bandwidth. The narrower the bandwidth, the more rounded the pulses. If the phase distortion is excessive, the pulse will *tilt* and, consequently, affect the next pulse. When pulses from more than one source are multiplexed together, the amplitude, frequency, and phase responses become even more critical. ISI causes *crosstalk* between channels that occupy adjacent time slots in a time-division-multiplexed carrier system. Special filters called *equalizers* are inserted in the transmission path to "*equalize*" the distortion for all frequencies, creating a

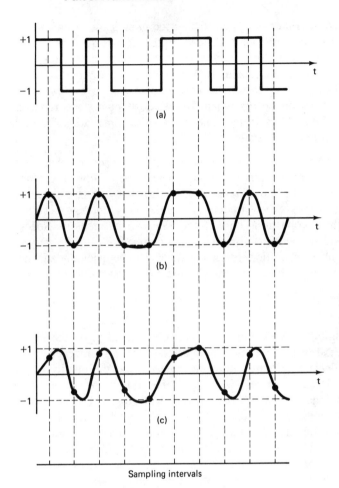

FIGURE 9–45  Pulse response: (a) NRZ input signal; (b) output from a perfect filter; (c) output from an imperfect filter.

uniform transmission medium and reducing transmission impairments. The four primary causes of ISI are:

*1. Timing inaccuracies.* In digital transmission systems, transmitter timing inaccuracies cause intersymbol interference if the rate of transmission does not conform to the *ringing frequency* designed into the communications channel. Generally, timing inaccuracies of this type are insignificant. Since receiver clocking information is derived from the received signals, which are contaminated with noise, inaccurate sample timing is more likely to occur in receivers than in transmitters.

*2. Insufficient bandwidth.* Timing errors are less likely to occur if the transmission rate is well below the channel bandwidth (i.e., the Nyquist bandwidth is significantly below the channel bandwidth). As the bandwidth of a communications channel is reduced, the ringing frequency is reduced and intersymbol interference is more likely to occur.

*3. Amplitude distortion.* Filters are placed in a communications channel to bandlimit signals and reduce or eliminate predicted noise and interference. Filters are also used to produce a specific pulse response. However, the frequency response of a channel cannot always be predicted absolutely. When the frequency characteristics of a communications channel depart from the normal or expected values, *pulse distortion* results. Pulse distortion occurs when the peaks of pulses are reduced, causing improper ringing frequencies in the time domain. Compensation for such impairments is called amplitude equalization.

*4. Phase distortion.* A pulse is simply the superposition of a series of harmonically related sine waves with specific amplitude and phase relationships. Therefore, if the relative phase relations of the individual sine waves are altered, phase distortion occurs. Phase distortion occurs when frequency components undergo different amounts of time delay while propagating through the transmission medium. Special delay equalizers are placed in the transmission path to compensate for the varying delays, thus reducing the phase distortion. Phase equalizers can be manually adjusted or designed to automatically adjust themselves to varying transmission characteristics.

## Eye Patterns

The performance of a digital transmission system depends, in part, on the ability of a repeater to regenerate the original pulses. Similarly, the quality of the regeneration process depends on the decision circuit within the repeater and the quality of the signal at the input to the decision circuit. Therefore, the performance of a digital transmission system can be measured by displaying the received signal on an oscilloscope and triggering the time base at the data rate. Thus all waveform combinations are superimposed over adjacent signaling intervals. Such a display is called an *eye pattern* or *eye diagram*. An eye pattern is a convenient technique for determining the effects of the degradations introduced into the pulses as they travel to the regenerator. The test setup to display an eye pattern is shown in Figure 9–46. The received pulse stream is fed to the vertical input of the oscilloscope, and the symbol clock is fed to the external trigger input, while the horizontal time base is set approximately equal to the symbol rate.

Figure 9–47 shows an eye pattern generated by a symmetrical waveform for *ternary* signals in which the individual pulses at the input to the regenerator have a

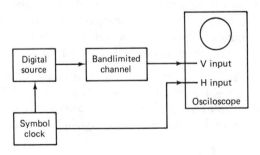

**FIGURE 9–46** **Eye diagram measurement setup.**

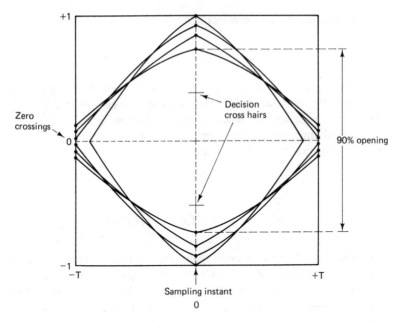

**FIGURE 9–47    Eye diagram.**

cosine-squared shape. In an *m*-level system, there will be $m - 1$ separate eyes. The horizontal lines labeled $+1$, $0$, and $-1$ correspond to the ideal received amplitudes. The vertical lines, separated by the signaling interval, $T$, correspond to the ideal *decision times*. The decision levels for the regenerator are represented by *crosshairs*. The vertical hair represents the decision time, while the horizontal hair represents the decision level. The eye pattern shows the quality of shaping and timing and discloses any noise and errors that might be present in the line equalization. The eye opening (the space in the middle of the eye pattern) defines a boundary within which no waveform *trajectories* can exist under any code-pattern condition. The eye opening is a function of the number of code levels and the intersymbol interference caused by the ringing tails of any preceding or succeeding pulses. To regenerate the pulse sequence without error, the eye must be open (i.e., a decision area must exist), and the decision crosshairs must be within the open area. The effect of pulse degradation is a reduction in the size of the ideal eye. In Figure 9–47 it can be seen that at the center of the eye (i.e., the sampling instant) the opening is about 90%, indicating only minor ISI degradation due to filtering imperfections. The small degradation is due to the nonideal Nyquist amplitude and phase characteristics of the transmission system. Mathematically, the ISI degradation is

$$20 \log \frac{H}{h}$$

where    $H$ = ideal vertical opening
$h$ = degraded vertical opening

For the eye diagram shown in Figure 9–47,

$$20 \log \frac{90}{100} = 0.9 \text{ dB ISI degradation}$$

In Figure 9–47 it can also be seen that the overlapping signal pattern does not cross the horizontal zero line at exact integer multiples of the symbol clock. This is an impairment known as data transition jitter. This jitter has an effect on the symbol timing (clock) recovery circuit and, if excessive, may significantly degrade the performance of cascaded regenerative sections.

# QUESTIONS

1. What is the difference between Digital Data Service and Dataphone Digital Service?
2. A frequency range of 3 to 10 kHz is to be sampled for transmission. What is the minimum sampling rate required for the original signal to be reproduced at the receive end?
3. The voltage in a S/H circuit is *continuously* adjusting to the analog input voltage.   (T, F)
4. The analog sample is converted to a digital number during the aperture time.   (T, F)
5. What is the cause of quantization noise?
6. What is the major advantage of digital communications?
7. If the resolution of an ADC is increased, (*more, the same amount, fewer*) bits would be used for each sample.
8. What is meant by a linear ADC?
9. In μ-law compression, the maximum input voltage appears (*amplified, unaffected, attenuated*) at the output.
10. In μ-law compression, a voltage less than the maximum appears (*amplified, unaffected, attenuated*) at the output.
11. In μ-law compression, the gain increases as the amplitude of the input signal _____ .
12. In μ-law compression of an analog signal, what is compressed?
13. All D channel banks use analog compression.   (T, F)
14. Since the formula for μ-law compression involves analog values, why is digital compression ascribed the name μ = 255?
15. Explain fully how a DS-1 signal (1.544 Mbps) is achieved.
16. All D channel banks output signals from 24 multiplexed voice band channels.   (T, F)
17. A muldem is an acronym for _____ .
18. As the line bit rate increases, the number of consecutive zeros allowed decreases.   (T, F)
19. What does AMI stand for?
20. Slope overload occurs when the staircase voltage of the modulator is changing faster than the analog signal.   (T, F)
21. Granular noise is the same as quantization noise.   (T, F)

22. Increasing the step size would increase:
    (a) The probability of slope overload.   (T, F)
    (b) Granular noise.   (T, F)
23. Digital pulses outputted by the delta modulator vary in amplitude proportional to the difference in the previous signal amplitude samples.   (T, F)
24. What determines the sampling rate in a delta modulator?
25. What does "CVSD" stand for?
26. CVSD is a form of adaptive delta modulation.   (T, F)
27. In the CVSD discussed, if five consecutive 1's are transmitted, the modulator has been trying to reach the analog voltage by increasing the staircase voltage in five equal steps.   (T, F)
28. Describe idle channel noise in a PCM system.
29. How is idle channel noise reduced with PCM?
30. What is midrise quantization?
31. What is midtread quantization?
32. What is a vocoder?
33. What is the purpose of a vocoder?
34. List three types of vocoders and briefly describe their differences.
35. Briefly describe the CCITT time-division-multiplexing system.
36. What is a combo chip?
37. Describe the fixed-data-rate mode of operation.
38. What is another name for the fixed-data-rate mode?
39. Describe the variable-data-rate mode of operation.
40. What is another name for the variable-data-rate mode?
41. Describe how differential PCM is accomplished.
42. What is the advantage of differential PCM?
43. List and describe five methods of achieving frame synchronization with PCM transmission.
44. Describe the differences between bit and word interleaving.
45. Describe the results of bandlimiting a pulse.
46. What is a ringing tail?
47. Describe intersymbol interference.
48. List the four primary causes of intersymbol interference.
49. Describe an eye pattern and how it is used with pulse transmission.
50. Briefly describe the test setup for displaying an eye pattern with a standard oscilloscope.

# PROBLEMS

1. Analog compression is used with $\mu = 255$.
   (a) If $V_{max} = 4.032$ V, what is the corresponding compressed voltage value?
   (b) If $V_{min} = 2$ mV, what is the corresponding compressed voltage value?

(c) If 6 bits are used to encode $V_{max}$, what is the voltage value of the LSB?

(d) If a computer is available,

(1) Manipulate the $\mu$-law formula so that given $V_{out}$, $V_{in}$ may be computed.

(2) Write a program to print out $V_{in}$, $V_{out}$ and gain for $V_{out}$ values ranging from 0.064 to 4.032 V in increments of 0.064 V.

The following questions may be answered from the computer printout or they may be calculated individually.

(e) What was the gain if the compression output was 0.064 V?

(f) What was the gain if the compression output was 4.032 V?

(g) Is the amplification linear?

(h) What is the uncompressed value for $V_{max} - V_{min}$?

(i) What is the compressed value for $V_{max} - V_{min}$?

(j) For $V_{out}$ from 0.192 to 0.32 V, the difference is 0.128 V. What is the voltage difference for the two corresponding $V_{in}$ values?

(k) What is the difference in the two $V_{in}$ values corresponding to $V_{out}$ from 3.968 to 4.032 V?

(l) Are smaller or larger amplitudes compressed more?

2. If PCM used 6 bits for the code (1 bit for the sign, 5 bits for magnitude) and the LSB = 0.2 V:

(a) How many different voltage levels could be represented by this scheme?

(b) What is the maximum positive output voltage that can be obtained with this scheme?

(c) What is the voltage represented by 1 0 0 1 1 1?

(d) What voltage is represented by 0 1 0 1 0 1?

(e) What is the code for +0.88 V?

(f) What is the code for −2.15 V?

3. If the maximum input voltage to a PCM system was 12.7 V and 7 bits (1 bit for the sign, 6 bits for magnitude) were used for encoding:

(a) What is the voltage of the LSB?

(b) What is the maximum possible error voltage?

(c) What is the total number of output voltages that can be obtained?

(d) What is the voltage represented by 1 0 0 1 1 0 1?

(e) What is the voltage represented by 0 1 1 0 1 1 1?

(f) What is the code for +4.2 V?

(g) What is the code for −3.6 V?

4. PCM/TDM system multiplexes 24 voice band channels. Each sample is encoded into 7 bits and 1 bit is added to each frame as a framing bit. The sampling rate is 9000 samples/second.

(a) What is the time of one frame?

(b) What is the time of one bit?

(c) What is the bit transmission rate?

5. PCM/TDM system multiplexes 32 voice band (0 to 4 kHz) channels. Each sample is encoded into 8 bits. There is one frame bit.

(a) What is the time of one frame?

(b) What is the time of one bit?

(c) What is the bit transmission rate?

6. A 12-bit code has been digitally compressed into 8 bits. The LSB = 0.03 V. For an analog signal of 1.74 V:

(a) What is the 12-bit code?

(b) What is the compressed 8-bit code?

(c) What is the value of the received decoded voltage?

(d) What is the percentage error in the transmission?

7. If the received compressed 8-bit code is 1 1 0 1 1 1 0 1, to what 12-bit code is it converted?

8. LSB = 0.02 V for a 12-digit linear code. What is the input voltage range that would be converted to:

(a) 1 0 0 0 0 0 0 0 0 0 0 1?

(b) 0 0 0 0 0 0 0 0 0 0 0 1?

(c) 1 0 0 0 0 0 0 0 0 0 0 0?

(d) 0 0 0 0 0 0 0 0 0 0 0 0?

(e) 1 0 0 0 0 1 0 1 0 0 0 1?

(f) 1 0 0 0 0 1 0 1 0 0 1 0?

9. For each of the following 12-bit codes, find the input voltage range and the corresponding digitally compressed 8-bit code. LSB = 0.02 V.

(a) 1 0 0 0 0 0 0 0 1 0 0 0

(b) 1 0 0 0 0 0 0 0 1 0 0 1

(c) 1 0 0 0 0 0 0 0 1 0 1 0

(d) 1 0 0 0 0 0 0 0 1 0 1 1

(e) 1 0 0 0 0 0 0 0 1 1 0 0

(f) 1 0 0 0 0 0 0 0 1 1 0 1

(g) 1 0 0 0 0 0 0 0 1 1 1 0

(h) 1 0 0 0 0 0 0 0 1 1 1 1

(i) What is the maximum quantization error for each of these?

(j) When the receiver converted the received 8-bit codes, expanded them back to 12 bits, to what voltage values would these 12 bits be converted?

(k) In the transmission of the codes above, no error would be introduced as a result of companding.   (T, F)

10. Assuming that the previously transmitted 1 had a positive polarity, draw the RZ, AMI waveform for the following bit sequence:

1 1 0 0 1 1 0 0

11. The following bit sequence is to be transmitted on a T2 carrier. Make the necessary substitutions to prevent excessive strings of 0's.

+ 00000000 1 0000000 1 00000 1 0000000 1

12. The following bit sequence is to be transmitted on a T3 carrier. Make the necessary substitutions to prevent excessive strings of 0's.

−0000 1 0000 1 00 1 1 00000 1 0000 1 00 1 0000 1 0000 1 000

# TEN

## *ASYNCHRONOUS PROTOCOL*

Asynchronous transmission is characterized by the use of start and stop bits to frame each transmitted character (Chapter 5). For the typical system described, ASCII will be used as the character code. Asynchronous modems such as the 202S and the 202T are used to achieve half- and full-duplex operation (Chapter 7). As stated in Chapter 2, line protocol is a set of rules governing the transmission of information. The line protocol described will be that used in a multipoint system.

The primary function of the *line control unit* (LCU) is to control the flow of all transmission. No remote station may transmit unless it is first allowed to do so by the LCU. A *poll* is an interrogation of the remote station by the LCU to see if it has any information to send. A *selection* is an interrogation of the remote station by the LCU to see if it is ready to receive a message from the primary. Each remote station has its own unique *Transmitter Start Code* (TSC). When a station receives its own TSC, it is being polled. A TSC may consist of two ASCII characters, such as DC3 A. The first control character simply indicates that what follows is a unique remote TSC. Other systems may require that each two-character ASCII TSC be followed by DEL as a TSC identifier. The former symbology will be used in the examples in this text. Although DC3 and DEL stand for Device Control 3 and DELETE in the ASCII tables, remember that these are only 7-bit codes and any device or system may use these codes to represent or do whatever the system designer wishes. Each remote station has one or more *Call Directing Codes* (CDCs). These are not necessarily unique. When a remote station receives one of its CDCs, it is being asked by the LCU if it is ready to receive information—it is being selected. If several remotes had the same CDC, the LCU could select and send the same message to all of them simultaneously. If more than one station but not all possess a common CDC, that CDC is called a *group address*. If all stations have a common CDC, that CDC is called a *broadcast address*. Obviously, TSCs, by their nature, must be unique since the LCU would produce chaos if it could poll several remotes simultaneously and they all had something to send.

## REMOTE TERMINAL EQUIPMENT

The following equipment may be found at a remote terminal: a display unit (CRT), keyboard, and/or a read-only printer (ROP). A *station controller* (Staco) serves as an interface between this equipment and the communications link.

**Display (D).** A CRT is normally capable of displaying 24 lines with 80 characters/line. This implies that there must be $24 \times 80 = 1920$ bytes of memory associated with the CRT to store the ASCII code for each character shown. Additional display memory may provide storage for three such pages (72 lines). The first line of each page is preceded by one, two, or three dots to identify the page viewed. The information displayed on the CRT can be highlighted and/or protected. If a character space is *protected*, the *keyboard operator* cannot, in normal operation, overwrite that character.

**Keyboard (K).** Figure 10–1 shows a typical asynchronous keyboard. The keyboard may be a stand-alone device or it may be connected with the display (KD) and/or printer (KDP). When in KD configuration, whatever is typed is displayed on the CRT. The keys at the left are for cursor control on the CRT. The HOME position on the CRT is the first row, first column. Note that the uppercase characters for the letters A through Z are data link control characters. These characters may be obtained by depressing CONTROL and then hitting the appropriate key. The terminal may be in one of three modes, SEND, RECEIVE, or LOCAL (upper left on keyboard). If the terminal is on-line with no information to send, it would normally be in the REC mode. The Staco monitors the terminal mode and responds accordingly when polled or selected by the LCU. If the station is polled while the terminal is in the REC mode, the Staco would respond with: "I have no message to send. I am ready to receive." In response to a selection, the Staco would respond with: "I am ready to receive." The terminal is placed in the LOCAL mode if the operator is composing a message on the CRT. At this time, the typed information is simply entered into display memory and displayed on the CRT, but it is not transmitted. If the Staco is polled or selected while the terminal is in LOCAL, the Staco would respond with: "I have no message to send. I am not ready to receive." The LCU must receive positive acknowledgment before transmitting any information to the remote terminal. In the case just cited, if the LCU transmitted information while the terminal was in LOCAL, the transmitted information would be displayed on

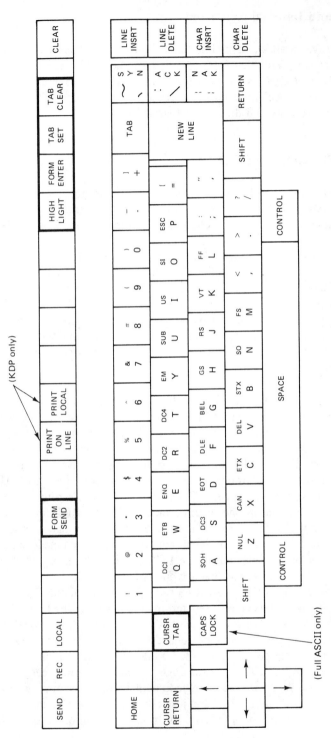

**FIGURE 10–1  Typical asynchronous keyboard.**

the CRT, destroying the message that the terminal operator was trying to compose. After the operator completed typing a message, he or she would enter the SEND mode. The message itself would not be transmitted until that station was polled by the LCU. If the station was selected while the terminal was in the SEND mode, the Staco would respond with: "I am not ready to receive. I have a message to send."

Depressing the PRINT LOCAL causes all information in display memory from the current cursor location to the control character that would halt transmission to be printed on the printer. If the PRINT ON LINE key is depressed while the terminal is in the REC mode, all received information is simultaneously printed and displayed on the CRT. If the PRINT ON LINE key is depressed before the operator enters the SEND mode, information in display memory will be printed as it is being transmitted.

The FORM ENTER key is used in the preparation of a form that will be used repeatedly. After the form is completed, depressing the FORM ENTER key a second time will get the operator out of the FORM ENTER mode. The form characters are protected and nonhighlighted. If the operator now fills out the form, the data used for form completion appear highlighted on the screen and are not protected. There are many KDs and each KD has various options. A KD may be optioned to transmit only the unprotected data in display memory. With this option, only the information entered to complete the form would be transmitted. This assumes that the primary already has a copy of the form and needs only the entered data. By depressing the FORM SEND key and then the SEND key, both the form and the entered data (protected and unprotected information) would be transmitted. The terminal operator may also want the primary to store the form for later retrieval. In this case, the operator would depress the FORM SEND key and then the SEND key with just the form displayed on his CRT.

Figure 10–2 shows a block diagram of the terminal equipment. Figure 10–3 shows the signal connections of the Staco to the terminal and to the modem. Note the similarity between the Staco-to-modem and the Staco-to-terminal connections.

**EXAMPLE**

If the terminal is in the SEND mode, the RTS line from terminal to Staco is active. When the Staco receives a poll, it activates the RTS line to the modem. After a predetermined delay, Staco receives a CTS from the modem. At this time, Staco activates the CTS to the terminal and the terminal starts transmitting information.

Asynchronous systems are used for small operations. Each terminal generally has only a keyboard, display and a printer. Terminals cannot communicate with each other directly, but the systems normally have a *store and forward* capability. A remote terminal, when polled, can send a message to the primary. As part of the message, the terminal specifies that the message is not intended for the primary but for some other terminal. The primary would store the message and, after selecting the designated terminal, would forward the message on to that terminal.

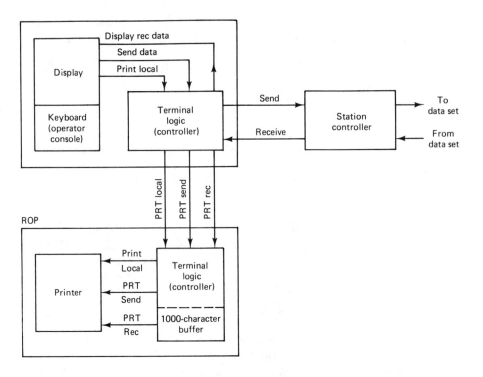

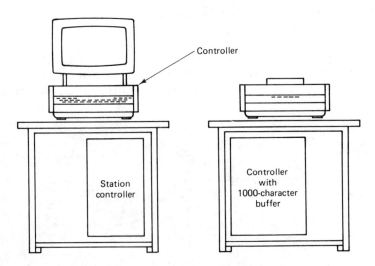

**FIGURE 10–2  Block diagram of terminal equipment.**

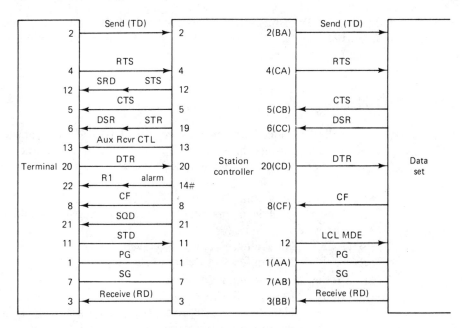

**FIGURE 10–3   Staco interface.**

## DATA LINK CONTROL CHARACTERS

These control codes are entered from the keyboard during data preparation and by the LCU at the primary.

**SOH (Start of Heading).**   This code is optional. When used, it must precede the heading portion of the message. The heading may contain destination information, message numbers, date, time, priority, and security classification.

**STX (Start of Text).**   This code is entered immediately after the heading information and defines the end of the heading and the start of the text. In normal operation, the Stacos monitor the data link looking for their TSC or CDC. If text is being transmitted between the LCU and a previously polled/selected station, the text contents may inadvertently contain a bit sequence that represents the TSC or CDC of a third station. If this third station now started to transmit a message or an acknowledgment, two stations would be transmitting simultaneously and communications would be disrupted. There is a need for all Stacos not previously selected to ignore all text contents. When a Staco not previously selected or polled detects a STX or a SOH on the data link, it enters a "blind" mode—it ignores all transmitted characters until it detects an EOT. At this time, the Staco will "unblind" itself and again monitor the data link for its CDC or TSC.

**EOT (End of Transmission).**　　This code must follow the last text character. Receipt of an EOT deselects all Stacos. To ensure that all remote stations are "unblinded," the LCU prefixes all polls and selections with EOT.

## SYSTEM RESPONSES

Responses to polls and selections generally consist of two ASCII characters. The system used normally permits the user to program whatever characters he wishes for this function. For our purpose, we will define these characters as follows:

1. Staco response to a poll:
    (a) If the terminal is in the SEND mode, the message would be sent. Otherwise:
    (b) \ ACK　　No Traffic, Ready to Receive.
    (c) \\　　　　No Traffic, Not Ready to Receive.
2. Staco response to a selection:
    (a) \ ACK　　Ready to Receive.
    (b) \\　　　　Not Ready to Receive. Terminal could be in LOCAL or out of service; the printer could be out of paper.
    (c) **　　　　Not Ready to Receive, Traffic to Send.
3. Optional Staco responses to a received message from the LCU (the station is capable of making these responses without being selected):
    (a) \ ACK　　Entire message received with no parity error.
    (b) \\　　　　Entire message not received.
    (c) **　　　　Station received the message with at least one parity error.

## DIALOGUE

In addition to line protocol, means need to be established by which transmitted text can be positioned properly on the CRT. Table 10–1 lists two-character escape sequences that can be entered from the keyboard, transmitted as part of the text sequence, and are designed to cause specific actions at the terminal.

| EXAMPLE | |
|---|---|
| Line | Col. 1 |
| 1 | FORM 75B |
| 6 | NAME ************ AGE ** |
| 11 | SEX * HT * FT ** IN |
| 14 | SHOE SIZE *** |

This entire form is to be protected. All asterisks will be unprotected. They identify the locations where the operator may enter information. All unused parts of the form will be protected. Spaces shown on the form (i.e., after M of FORM) will also be protected.

```
E D                            LCU polls a station whose TSC is D.
O C D
T ꓱ

S E                         E  Station was in the SEND mode.
T F O R M S 7 5 B O            Requests Form 75B.
X         P         T
```

CPU would find this program among that station's applications program. After it was found:

```
E                           Station with the TSC of D has a CDC of BD.
O B D
T                           That station is selected.

      A                     Station is ready to receive.
      \C
      K
```

```
S E   D D E
T S R E E S W F O R M S 7 5 B N N N N N N A M E S
X C   L L C           P       L L L L L       P

E                           E   S       S E
S X · · · · · · · · · · · · S W A G E   S X · ·
C                           C   P       P C

E   N N N N N       S E     E     S     S E
S W L L L L L S E X S S X · S W H T S S X ·
C             P C   C       P   P C

E   S     S E       E   S   N N N       S S
S W F T   S S X · · S W I N L L L S H O E S I Z E S
C P   P C C   C   P         P       P

E       E   N N N N N N N N N N E   E
S X · · · S W L L L L L L L L L L S @ O
C         C                     C   T
```

Explanation of the transmitted form:

| | |
|---|---|
| ESC R | Clears the screen and HOMEs the cursor. |
| Two DEL (or more) | Just a delay to allow the screen time to clear. |
| ESC W | The field that follows will be protected. |

**TABLE 10–1    Escape Sequence Terminal Controls**

| ESC sequence | Precedes transmission of: | Reception of, unless elected not to |
| --- | --- | --- |
| ESC 0 | Each character where there is a tab setting | Sets tab stop at cursor location |
| ESC 1 | | Sets tab stop at cursor location, on that line and all lines below to end of memory |
| ESC 2 | | Clears all tab stops to right and below cursor location; stops to left and below remain, as do stops above |
| ESC 3 | Each sequence of highlighted characters | Highlights characters that follow |
| ESC 4 | Normal characters that follow highlighted characters | Does not highlight characters that follow |
| ESC W | Each sequence of protected characters | Protects characters that follow and puts display into protected data mode wherein receipt of on-line controls that effect unprotected data will also effect protected |
| ESC X | Normal characters that follow protected characters | Does not protect characters that follow and removes display from protected data mode wherein receipt of on-line controls that affect unprotected data will not affect protected data |
| ESC @ | | Moves cursor to first tab stop on right, or to start of next line if no tab stop on right, or the first unprotected space following protected data whether on that line or next |
| ESC 7 | | Moves cursor up one line |
| ESC B | | Moves cursor down one line |
| ESC C | | Moves cursor right one character |
| ESC G | | Moves cursor to start of line |
| ESC H | | Advances display to first segment and moves cursor to start of 1st line |
| ESC J | | Clears data (unprotected only if not in protected data mode) to right of cursor, and all across all lines below cursor to end of memory |
| ESC L | | Moves unprotected data down one line to create a line of spaces on the line the cursor is on and returns the cursor to the start of the line |
| ESC M | | Clears unprotected data from the line the cursor is on, moves all unprotected data displayed below up one line, and returns the cursor to the start of the line |
| ESC ∧ | | Moves unprotected data one position to the right to create a character of space at the cursor position. |

**Table 10–1    (Continued)**

| ESC sequence   Precedes transmission of: | Reception of, unless elected not to |
| --- | --- |
| ESC P | Clears unprotected character at the cursor position and moves all unprotected data displayed on the right one position to the left |
| ESC R | Advances display to first segment, moves cursor to start of first line and clears all data from display memory, regardless of whether it is protected, unprotected, highlighted, unhighlighted or tab stop |
| ESC S | Moves all displayed data up one line |
| ESC T | Moves all displayed data down one line |
| ESC U | Causes next whole segment to be displayed |

| | |
| --- | --- |
| ESC X | The field that follows is unprotected. The operator can type information into these locations. |
| NL | New line. The last 10 NLs protect through line 24. For problem simplification, only one page of display memory is assumed. |
| ESC @ | Places the cursor at the first unprotected space on the CRT. This is done as a courtesy to the operator. |

## QUESTIONS

1. An on-line terminal can be in one of three modes. What are the three modes?

2. When the operator at a remote site presses the SEND key, the message on the display is immediately transmitted.  (T, F)

3. If the operator wants to type protected information on the display, which key must first be depressed?

4. What is meant by protected information?

5. If the remote terminal is solely printing out what is shown on the CRT, it is in the (*SEND*, *RECEIVE*, *LOCAL*) mode.

6. If a terminal is selected by the LCU, it is being asked to transmit if it has a message ready.  (T, F)

7. If terminal is selected by the LCU while it is in the SEND mode, Staco (*sends "OK to transmit" to the primary, sends the terminal's message to the primary, sends "Not ready to receive, have a message to send" to the primary, does nothing to keep the primary in suspense*).

8. What is the normal line monitoring mode for the remote terminal?

9. Which two ASCII characters are capable of "blinding" all unselected stations?

10. Which ASCII character "unblinds" the station controllers?

11. A remote station can have more than one TSC.  (T, F)

12. A remote station can have more than one CDC.  (T, F)

13. Data link control characters may be entered directly from an asynchronous keyboard.  (T, F)

14. Within a message, how is a field designated as protected?

15. If the remote terminal operator is composing a protected form, after the form is completed, how does the operator get out of the protected mode?

# PROBLEM

1. The operator at a remote station has requested Form 8561B from the primary. The format for this form is shown below. All characters included in the form are protected and all asterisks are unprotected. Write the character sequence that would display this form on the operator's CRT. Numbers under characters indicate the column the character is in.

```
LINE
 1      F O R M   8 5 6 1 B
        1     4   6       10

 2

 3      N A M E   *  ...  * S / S   * * * − * * − * * * *
        1     4   6       3032      36                  46

 4

 5      S T R E E T   *  ...  *
        1             8      37

 6

 7      C I T Y   *  ...  *   S T A T E   * *
        1         6       30 32      36 38 39

 8

 9      Z I P   * * * * *
        1   3   5       9
```

# ELEVEN

## *BINARY SYNCHRONOUS COMMUNICATIONS*

*Binary Synchronous Communications* (BSC or Bisync) is a synchronous, character-oriented protocol. ASCII and EBCDIC are generally used as the transmission codes with this protocol. ASCII will be used in all examples of transmission dialogue. Bisync is a *character-oriented* protocol because ASCII characters are used for data link control functions. This is in contrast with Synchronous Data Link Control (SDLC), a *bit-oriented* protocol, which uses bits or groups of bits to perform the data link control functions. Bisync is a *synchronous* protocol because SYN characters are sent with each transmission from which the receive DTE can achieve character synchronization.

   A picture of a typical synchronous keyboard is shown in Figure 11–1. Note the absence of data link control character keys that are present on an asynchronous keyboard. This keyboard has two PA keys and 12 PF keys. The function of these keys will be explained as the need arises.

## DATA LINK CONTROL CHARACTERS

(If a particular character is not as ASCII character, the ASCII character used will be shown in parentheses.)

   **SYN—Synchronous Idle.**   This character is transmitted to establish and maintain character synchronization. Unless operating in the transparent text mode (see latter part of this chapter), this character is automatically stripped by the station controller. In a multipoint system, information is transmitted in blocks of 256 charac-

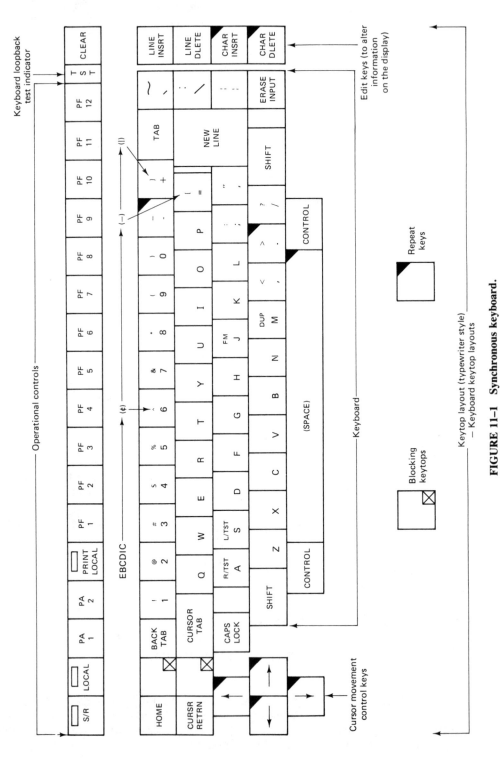

**FIGURE 11–1  Synchronous keyboard.**

226

ters. In point-to-point systems, the length of one block can be considerably longer. If the length of a single transmission is 1 s or longer, SYN characters are inserted directly into the message to maintain character synchronization. This would be necessary in some point-to-point transmissions.

**EOT—End of Transmission.** This character indicates the end of the message and resets all stations on line to the control mode. In this mode, the secondaries are prepared to receive a poll or selection from the primary.

In a multipoint system, the primary starts all polls and selections with

```
P  S  S  E  P
A  Y  Y  O  A
D  N  N  T  D
```

The leading PAD is 55H or AAH (H = hexadecimal) and is used to establish bit synchronization. This PAD is not always used and will not be shown in any subsequent examples. Station controllers manufactured by IBM always supply this leading PAD. This is not true of station controllers built by other manufacturers. The SYN SYN (could be 2 or 5) pattern establishes character synchronization, and hence the name Bisync by which this protocol is often called. The EOT does not indicate end of transmission, but ensures that all secondaries are in the control mode. The trailing PAD (1 to 7) is FFH. It follows the last character of every transmission to allow that character to be demodulated by the receive modem and be sent to the DTE before the modem receiver turns off (see action of RLSD and RD of the RS 232C interface). All transmissions start with the SYN SYN pattern.

**ETB—End of Transmission Block.** In a multipoint environment, a message may be more than one block long. ETB terminates the current block but infers that there are additional blocks to the message which will follow. The final block is terminated with ETX and not ETB or EOT.

**ETX—End of Text.** This character is used to indicate the end of the current message.

ETB and ETX are always immediately followed by a BCC. A BCC must be *immediately acknowledged* by the receiver (except after ITB). Therefore, ETB BCC or ETX BCC always produce *line turnaround* (the station that was receiving will now transmit).

**ITB—(US—Unit Separator)—End of Intermediate Block.** The function of this control character is similar to ETB. Although ITB must also be followed by a BCC, it differs from ETB in that its BCC will not be acknowledged immediately. The transmission of ITB will not produce a line turnaround. Eventually, a following block will be terminated by ETX. The receiving station must now acknowledge the BCCs. A positive acknowledgment indicates that the BCC following the ETB or ETX and all of the BCCs that followed previous ITBs are correct. If an error was detected in any of these BCCs, retransmission of all blocks will be required since the receiver has no way of informing the transmitter which specific BCCs were bad. The use of ITB speeds up message transmittal by eliminating line turn-arounds required for acknowledgments. If, however, only one block is bad, all blocks must be retransmitted. ITB is used more in point-to-point systems where a single message may be quite lengthy. In a multipoint system, the length of a transmission is generally limited by the capacity of the terminal buffers.

**ACK0 (DLE 0), ACK 1 (DLE 1)—Positive Acknowledgments.**     These are positive acknowledgments of BCCs and are sent alternately. ACK 1 is always the positive acknowledgment to the first and odd blocks of text (all text blocks start with STX); ACK 0 is used to acknowledge even blocks of text and selections. A positive acknowledgment merely indicates no error in the BCC; it does not mean acceptance or compliance with the contents of the message.

**WACK (DLE;)—Wait, Positive Acknowledgments.**     This character is used to provide positive acknowledgment to a BCC; it also requests the primary to delay further transmission to this station or device. A WACK is never sent by the primary to the secondary.

**NAK—Negative Acknowledgment.**     This response indicates an error was detected in the BCC and that retransmission of that block (see ITB) is requested.

**ENQ—Enquiry (format character).**     It is transmitted by the primary to indicate line turnaround. The secondary must transmit in response.

**SOH—Start of Heading.**     SOH is generally followed by auxiliary information such as priority, time of message, and so on. SOH% R is used to begin a *Status and Sense message*; SOH% / is used for *Remote Test Request messages*.

**STX—Start of Text.**     STX terminates a heading (SOH) if present and always precedes a block of text characters.

**RVI—(DLE <)—Reverse Interrupt.**     This will be explained later.

## MESSAGE FORMAT

The following information applies to multipoint, dedicated systems (Figures 11–2 and 11–3) which use Bisync protocol. Although there are systems that have as many as 50 devices controlled by a single station controller, the system under discus-

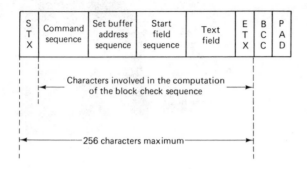

FIGURE 11-2   Multipoint dedicated system.

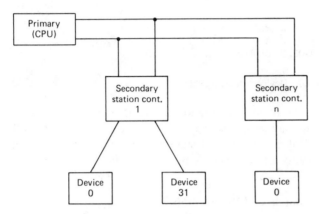

FIGURE 11-3   Multipoint dedicated system.

sion will be considered to have a maximum of 32 devices that can be handled by a single station controller.

## General Poll

### FORMAT

```
S S E P S S S        E P
Y Y O A Y Y P P ' '  ' ' N A
N N T D N N A A      Q D
```

SPA: Station Polling Address (Table 11-1).

A general poll will always have two ' 'characters following the SPA. A *general poll* is an invitation from the CPU to a particular station for any device at that station to send a message to the CPU. As the various devices at that station enter the SEND mode, the station controller keeps track of this sequence and will allow transmission in this order. In a general poll, the station controller determines which device will transmit if more than one are in the SEND mode.

**EXAMPLE**

```
S  S  E  P  S  S                    E  P
Y  Y  O  A  Y  Y  D  D  ''    ''    N  A
N  N  T  D  N  N                    Q  D
```

General poll of station 4.

## Specific Poll

**FORMAT**

```
S  S  E  P  S  S  S  S  D  D  E  P
Y  Y  O  A  Y  Y  P  P  A  A  N  A
N  N  T  D  N  N  A  A        Q  D
```

DA: Device Address (Table 11–1).

A *specific poll* is an invitation by the CPU to a particular device at a particular station to transmit. The CPU may send a specific poll in place of a general poll if there is only a small cluster of devices at that station. A specific poll of a device is required when the CPU wants a Status and Sense message from that device.

**TABLE 11–1   Station and Device Addresses**

| Station or device number | SPA | SSA | DA | Station or device number | SPA | SSA | DA |
|---|---|---|---|---|---|---|---|
| 0 | sp | - | sp | 16 | & | 0 | & |
| 1 | A | / | A | 17 | J | 1 | J |
| 2 | B | S | B | 18 | K | 2 | K |
| 3 | C | T | C | 19 | L | 3 | L |
| 4 | D | U | D | 20 | M | 4 | M |
| 5 | E | V | E | 21 | N | 5 | N |
| 6 | F | W | F | 22 | O | 6 | O |
| 7 | G | X | G | 23 | P | 7 | P |
| 8 | H | Y | H | 24 | Q | 8 | Q |
| 9 | I | Z | I | 25 | R | 9 | R |
| 10 | [ | \| | [ | 26 | ] | : | ] |
| 11 | . | ' | . | 27 | $ | # | $ |
| 12 | < | % | < | 28 | * | @ | * |
| 13 | ( | — | ( | 29 | ) | ' | ) |
| 14 | + | > | + | 30 | ; | = | ; |
| 15 | ! | ? | ! | 31 | ∧ | '' | ∧ |

**EXAMPLE**

```
S  S  E  P  S  S              E  P
Y  Y  O  A  Y  Y  D  D  H  H  N  A
N  N  T  D  N  N              Q  D
```

Specific poll of device 8 at station 4.

## Selection

### FORMAT

```
S  S  E  P  S  S  S  S         E  P
Y  Y  O  A  Y  Y  S  S  D  D   N  A
N  N  T  D  N  N  A  A  A  A   Q  D
```

SSA: Station Selection Address (Table 11–1).

A *selection* means the CPU has a message for the device addressed and wants to determine if that device is in position to receive a message (i.e., not in local).

**EXAMPLE**

```
S  S  E  P  S  S              E  P
Y  Y  O  A  Y  Y  T  T  F  F  N  A
N  N  T  D  N  N              Q  D
```

Selection of device 6 at station 3.

No BCC is transmitted with a poll or selection. This necessitates the repetition of SPA, DA, and SSA. If an error occurred with the first SSA, it is unlikely that exactly the same error will occur with the second SSA. Unless both SSAs are identical, no station will respond. A retransmission of the poll or selection, hopefully, will not suffer the same problem.

If there is no response to a general poll (for any reason), the CPU waits 4 s and then retransmits the general poll. If no response, after another 4-s delay, a third general poll is sent. If there is still no response, the CPU will continue with its general poll of other stations. When it is time for the "problem" station to be polled, the CPU will again try three times at 4-s intervals. After the CPU has polled the station on 10 occasions, three times on each occasion at 4-s intervals, and it still has not raised a response, the CPU will drop that station from the poll sequence. Manual intervention will be required to replace that station on the poll sequence. The polling options are established at the CPU through software.

**Block Check Characters (BCC).**   The characters following the first SOH or STX in a block are involved in the computation of the BCC. The last character of a block involved in this computation is either ETB, ETX, or ITB.

**EXAMPLE**

```
S S S     S           E   B P
Y Y O     T           T   C A
N N H|    X           X | C D

S S S                 E   B P
Y Y T                 T   C A
N N X|                B | C D
```

Block characters involved in the computation of the BCC are shown grouped.

## Message Format

1. Commands
   (a) Write:

```
E   W
S 1 C   (see Table 11-2 for WCC)
C   C
```

   Will overwrite previous information in the data buffers with received information.
   (b) Erase-Write:

```
E   W
S 5 C
C   C
```

   Clears all data buffers before entering received data. Also homes the cursor.
   (c) Erase All Unprotected:

```
E
S ?
C
```

   Clears all unprotected fields in the buffer.
   (d) Copy:

```
E   C D
S ? C   (see Table 11-7 for CCC)
C   C A
```

**TABLE 11–2   Write Control Characters**

| Start printer | Sound KD alarm | Put KD in LOCAL | Reset ACs to unmod | WCC | | | |
|:---:|:---:|:---:|:---:|:---:|:---:|:---:|:---:|
| | | | | NL char/line | 40 char/line | 64 char/line | 80 char/line |
| | | | | sp | & | - | 0 |
| | | | × | A | J | / | 1 |
| | | × | | B | K | S | 2 |
| | | × | × | C | L | T | 3 |
| | × | | | D | M | U | 4 |
| | × | | × | E | N | V | 5 |
| | × | × | | F | O | W | 6 |
| | × | × | × | G | P | X | 7 |
| × | | | | H | Q | Y | 8 |
| × | | | × | I | R | Z | 9 |
| × | | × | | [ | ] | ¦ | : |
| × | | × | × | . | $ | , | # |
| × | × | | | < | * | % | @ |
| × | × | | × | ( | ) | — | ' |
| × | × | × | | + | ; | > | = |
| × | × | × | × | ! | ∧ | ? | " |

Transfers specified contents of the buffer of a specified device into the buffer of the selected device on the same station cluster.

(e) Read Modified:

```
E
S b
C
```

Transmits data fields flagged as modified from the buffer.

(f) Read All:

```
E
S 2
C
```

Transmits the contents of the buffer, including nulls.

**WCC—Write Control Character.**    A single control character chosen to supply the required conditions.

**EXAMPLE**

$$WCC = 6$$

*Conditions*: Sounds KD alarm and puts KD in LOCAL and prints with 80 characters/line.

    **CCC—Copy Control Character (Table 11–7).**    A single character identifying the type of information to be copied. The DA following this character identifies the device that will be copied from.

**EXAMPLE**

$$CCC = 9$$

Want to start printer and copy only the unprotected data and ACs. Print at 80-character/line format.

   2. *Device order*: Used by the line control unit (LCU) to format the message.
     (a) Set Buffer Address:

```
D B B
C A A    (see Table 11-3 for BA1 and BA2)
1 1 2
```

Identifies the buffer address at which the operation will begin or continue. The buffer address is identified by two characters: BA1, BA2 (buffer addresses 1 and 2). These addresses identify the starting row and column.

**EXAMPLE**

```
D        BA1 = F BA2 = K
C F K
1        Identifies row 6, column 3
```

     (b) Start Field:

```
G A
S C
```

    Indicates the start of a new field. The AC (Attribute Character, see Table 11–4) immediately following the GS (ASCII Group Separator) identifies the characteristics of this field. The field will extend up to the location where the next AC is stored. On the screen, the field will wrap-around if an AC is stored in row 23, column 7 and no other AC exists below it (to and including row 24, column 80), the field continues to HOME and the top of the screen until the next AC is found.

**TABLE 11–3   Address Matrix**

|   | B A 1 | B A 2 | C A 1 | C A 2 | A 1 | A 2 |
|---|---|---|---|---|---|---|

*Column*

| Row | 1 | 2 | 3 | 4 | 5 | 6 | 7 | 8 | 9 | 10 | 11 | 12 | 13 | 14 | 15 | 16 |
|---|---|---|---|---|---|---|---|---|---|---|---|---|---|---|---|---|
| 1 | SS / PP | $S_A$ / $P^A$ | $S_B$ / $P^B$ | $S_C$ / $P^C$ | $S_D$ / $P^D$ | $S_E$ / $P^E$ | $S_F$ / $P^F$ | $S_G$ / $P^G$ | $S_H$ / $P^H$ | $S_I$ / $P^I$ | $S_J$ / $P^J$ | S / $P^.$ | $S_\vee$ / $P^V$ | $S_($ / $P^[$ | $S_+$ / $P^{+}$ | $S_|$ / $P^|$ |
| 2 | A& | AJ | AK | AL | AM | AN | AO | AP | AQ | AR | A] | A$ | A* | A) | A: | A∧ |
| 3 | B– | B/ | BS | BT | BU | BV | BW | BX | BY | BZ | B: | B. | B% | B⌐ | B∧ | B? |
| 4 | C0 | C1 | C2 | C3 | C4 | C5 | C6 | C7 | C8 | C9 | C: | C# | C@ | C' | C= | C'' |
| 5 | $E_P^S$ | EA | EB | EC | ED | EE | EF | EG | EH | EI | E: | E. | E< | E( | E+ | E! |
| 6 | F& | F/ | FK | FL | FM | FN | FO | FP | FQ | FR | F: | F$ | F* | F) | F: | F∧ |
| 7 | G– | G/ | GS | GT | GU | GV | GW | GX | GY | GZ | G: | G. | G% | G⌐ | G∧ | G? |
| 8 | H0 | HI | H2 | H3 | H4 | H5 | H6 | H7 | H8 | H9 | H: | H# | H@ | H' | H= | H'' |
| 9 | $I_P^S$ | IA | IB | IC | ID | IE | IF | IG | IH | II | I: | I. | I< | I( | I+ | I! |
| 10 | .& | .J | .K | .L | .M | .N | .O | .P | .Q | .R | .J | .$ | *. | .) | .: | < |
| 11 | ∨↓ | ∇/ | ∨S | ∨T | ∨U | ∨V | ∨W | ∨X | ∨Y | ∨Z | ∨J | ∨, | ∨% | ∨⌐ | ∧ | ∨? |
| 12 | (0 | (1 | (2 | (3 | (4 | (5 | (6 | (7 | (8 | (9 | (: | (# | (@ | (' | (= | (: |
| 13 | $S_P$ | !A | !B | !C | !D | !E | !F | !G | !H | !I | !: | !. | !< | !( | !+ | !: |
| 14 | && | &J | &K | &L | &M | &N | &O | &P | &Q | &R | &] | &$ | &* | &) | &; | &∧ |
| 15 | J– | J/ | JS | JT | JU | JV | JW | JX | JY | JZ | J: | J, | J% | J⌐ | J∧ | J? |
| 16 | K0 | K1 | K2 | K3 | K4 | K5 | K6 | K7 | K8 | K9 | K: | K# | K@ | K' | K= | K'' |
| 17 | $M_P^S$ | MA | MB | MC | MD | ME | MF | MG | MH | MI | MI | M. | M< | M( | M+ | M! |
| 18 | N& | NJ | NK | NL | NM | NN | NO | NP | NQ | NR | N] | N$ | N* | N) | N; | N∧ |
| 19 | 0– | O/ | OS | OT | OU | OV | OW | OX | OY | OZ | O: | O. | O% | O⌐ | O∧ | O? |
| 20 | P0 | P1 | P2 | P3 | P4 | P5 | P6 | P7 | P8 | P9 | P: | P# | P@ | P' | P= | P'' |
| 21 | $R_P^S$ | RA | RB | RC | RD | RE | RF | RG | RH | RI | R] | R. | R< | R( | R+ | R! |
| 22 | J& | J/ | JK | JL | JM | JN | JO | JP | JQ | JR | ]: | J$ | J* | J) | J: | J∧ |
| 23 | $– | $/ | $S | $T | $U | $V | $W | $X | $Y | $Z | $] | $. | $% | $⌐ | $∧ | $? |
| 24 | *0 | *1 | *2 | *3 | *4 | *5 | *6 | *7 | *8 | *9 | *: | *# | *@ | *; | *= | *'' |

235

**TABLE 11–3** (continued)

|  | | 17 | 18 | 19 | 20 | 21 | 22 | 23 | 24 | 25 | 26 | 27 | 28 | 29 | 30 | 31 | 32 |
|---|---|---|---|---|---|---|---|---|---|---|---|---|---|---|---|---|
| B B | A A 1 2 | | | | | | | | | | | | | | | | |
| C C | A A 1 2 | | | | | | | | | | | | | | | | |
| A A | A A 1 2 | | | | | | | | | | | | | | | | |
| | **Row** | | | | | | | | | **Column** | | | | | | | |
| | 1 | S&/P | SJ/PJ | SK/PK | SL/PL | SM/PM | SN/PN | SO/PO | SP/PP | SQ/PQ | SR/PR | S]/P] | S$/P$ | S*/P* | S)/P) | S;/P; | S\/P\ |
| | 2 | A– | A/ | AS | AT | AU | AV | AW | AX | AY | AZ | A! | A, | A% | A⌐ | A> | A? |
| | 3 | B0 | B1 | B2 | B3 | B4 | B5 | B6 | B7 | B8 | B9 | B: | B# | B@ | B' | B= | B'' |
| | 4 | D S/P | DA | DB | DC | DD | DE | DF | DG | DH | DI | D] | D. | D< | D< | D+ | D! |
| | 5 | E& | EJ | EK | EL | EM | EN | EO | EP | EQ | ER | E] | E$ | E* | E) | E; | E\ |
| | 6 | F— | F/ | FS | FT | FU | FV | FW | FX | FY | FZ | F. | F, | F% | F⌐ | F> | F? |
| | 7 | G0 | G1 | G2 | G3 | G4 | G5 | G6 | G7 | G8 | G9 | G: | G# | G@ | G' | G= | G'' |
| | 8 | S/P | IA | IB | IC | ID | IE | IF | IG | IH | II | I] | I. | I< | I( | I+ | I! |
| | 9 | |& | |J | |K | |L | |M | |N | |O | |P | |Q | |R | |] | |$ | |* | |) | |; | |\ |
| | 10 | .| | ./ | .S | .T | .U | .V | .W | .X | .Y | .Z | .! | .: | .% | .⌐ | .^ | .? |
| | 11 | ∇0 | <1 | <2 | <3 | <4 | <5 | <6 | <7 | <8 | <9 | <: | <# | <@ | <' | <= | <'' |
| | 12 | + S/P | +A | +B | +C | +D | +E | +F | +G | +H | +I | +] | +. | +< | +( | ++ | +! |
| | 13 | !& | !J | !K | !L | !M | !N | !O | !P | !Q | !R | !] | !$ | !* | !) | !; | !\ |
| | 14 | &— | &/ | &S | &T | &U | &V | &W | &X | &Y | &Z | &! | &, | &% | &⌐ | &> | &? |
| | 15 | J0 | J1 | J2 | J3 | J4 | J5 | J6 | J7 | J8 | J9 | J: | J# | J@ | J' | J= | J'' |
| | 16 | S L/P | LA | LB | LC | LD | LE | LF | LG | LH | LI | L] | L. | L< | L( | L+ | L! |
| | 17 | M& | MJ | MK | ML | MM | MN | MO | MP | MQ | MR | M] | M$ | M* | M) | M; | M\ |
| | 18 | N– | N/ | NS | NT | NU | NV | NW | NX | NY | NZ | N! | N, | N% | N⌐ | N> | N? |
| | 19 | O0 | O1 | O2 | O3 | O4 | O5 | O6 | O7 | O8 | O9 | O: | O# | O@ | O' | O= | O'' |
| | 20 | Q S/P | QA | QB | QC | QD | QE | QF | QG | QH | QI | Q] | Q. | Q< | Q( | Q+ | Q! |
| | 21 | R& | RJ | RK | RL | RM | RN | RO | RP | RQ | RR | R] | R$ | R* | R) | R; | R\ |
| | 22 | |– | |/ | |S | |T | |U | |V | |W | |X | |Y | |Z | |! | |, | |% | |⌐ | |> | |? |
| | 23 | $0 | $1 | $2 | $3 | $4 | $5 | $6 | $7 | $8 | $9 | $: | $# | $@ | $' | $= | $'' |
| | 24 | ) S/P | )A | )B | )C | )D | )E | )F | )G | )H | )I | )] | ) | )< | )x | )+ | )! |

236

TABLE 11–3  (continued)

|  |  |  |
|---|---|---|
| B A 1 | B A 2 |  |
| C A 1 | C A 2 |  |
| A A 1 | A A 2 |  |

Column

| Row | 33 | 34 | 35 | 36 | 37 | 38 | 39 | 40 | 41 | 42 | 43 | 44 | 45 | 46 | 47 | 48 |
|---|---|---|---|---|---|---|---|---|---|---|---|---|---|---|---|---|
| 1 | S⌐ / P⌐ | S⁄ / P | S_S / P | S_T / P | S_U / P | S_V / P | S_W / P | S_X / P | S_Y / P | S_Z / P | S; / P | S / P' | S% / P | S / P⌐ | S<> / P | S? / P' |
| 2 | A0 | A1 | A2 | A3 | A4 | A5 | A6 | A7 | A8 | A9 | A: | A# | A@ | A' | A= | A'' |
| 3 | C^S_P | CA | CB | CC | CD | CE | CF | CG | CH | CI | CJ | C. | C< | C( | C+ | C) |
| 4 | D& | DJ | DK | DL | DM | DN | DO | DP | DQ | DR | DJ | D$ | D* | D) | D; | D\ |
| 5 | E⌐ | E/ | ES | ET | EU | EV | EW | EX | EY | EZ | E! | E, | E% | E! | E△ | E? |
| 6 | F0 | FI | F2 | F3 | F4 | F5 | F6 | F7 | F8 | F9 | F. | F# | F@ | F' | F= | F' |
| 7 | H^S_P | HA | HB | HC | HD | HE | HF | HG | HH | HI | HI | H. | H< | H( | H+ | HI |
| 8 | I& | IJ | IK | IL | IM | IN | IO | IP | IQ | IR | II | I$ | I* | I) | I; | I< |
| 9 | I⌐ | I/ | IS | IT | IU | IV | IW | IX | IY | IZ | I= | I. | I% | I⊥ | I△ | I? |
| 10 | .0 | .1 | .2 | .3 | .4 | .5 | .6 | .7 | .8 | .9 | .: | .# | .@ | .' | .= | .'' |
| 11 | S^S_P | (A | (B | (C | (D | (E | (F | (G | (H | (I | (I | (. | (< | (( | (+ | (< |
| 12 | +& | +J | +K | +L | +M | +N | +O | +P | +Q | +R | +! | +$ | +* | +! | +; | +< |
| 13 | !⌐ | !⁄ | !S | !T | !U | !V | !W | !X | !Y | !Z | !; | !, | !% | !⊥ | !△ | !? |
| 14 | &0 | &1 | &2 | &3 | &4 | &5 | &6 | &7 | &8 | &9 | &: | &# | &@ | & | &= | &'' |
| 15 | K^S_P | KA | KB | KC | KD | KE | KF | KG | KH | KI | KI | K. | K< | K( | K+ | K! |
| 16 | L& | LJ | LK | LL | LM | LN | LO | LP | LQ | LR | LI | L$ | L* | L) | L; | L\ |
| 17 | M⌐ | M/ | MS | MT | MU | MV | MW | MX | MY | MZ | MI | M, | M% | M⊥ | M△ | M? |
| 18 | N0 | N1 | N2 | N3 | N4 | N5 | N6 | N7 | N8 | N9 | N: | N# | N@ | N' | N= | N'' |
| 19 | P^S_P | PA | PB | PC | PD | PE | PF | PG | PH | PI | PI | P. | P< | P( | P+ | PI |
| 20 | Q& | QJ | QK | QL | QM | QN | QO | QP | QQ | QR | QI | Q$ | Q* | Q) | Q; | Q\ |
| 21 | R⌐ | R/ | RS | RT | RU | RV | RW | RX | RY | RZ | RI | R, | R% | R⊥ | R△ | R? |
| 22 | J0 | J1 | J2 | J3 | J4 | J5 | J6 | J7 | J8 | J9 | J: | J# | J@ | J' | J= | J'' |
| 23 | *^S_P | *A | *B | *C | *D | *E | *F | *G | *H | *I | *I | *. | *< | *( | *+ | *! |
| 24 | )& | )J | )K | )L | )M | )N | )O | )P | )Q | )R | )I | )$ | )< | )) | ); | )< |

**TABLE 11–3** (continued)

| | | | | | | | | Column | | | | | | | | |
|---|---|---|---|---|---|---|---|---|---|---|---|---|---|---|---|---|
| Row | 49 | 50 | 51 | 52 | 53 | 54 | 55 | 56 | 57 | 58 | 59 | 60 | 61 | 62 | 63 | 64 |
| 1 | $S_0$ / $^SP$ | $S_1$ / P | $S_2$ / P | $S_3$ / P | $S_4$ / P | $S_5$ / P | $S_6$ / P | $S_7$ / P | $S_8$ / P | $S_9$ / P | S: / P | S# / P | S@ / P | S / P' | S= / P | S, / P |
| 2 | B& | BA | BB | BC | BD | BE | BF | BG | BH | BI | B\| | B. | B< | B( | B+ | B! |
| 3 | C& | CJ | CK | CL | CM | CN | CO | CP | CQ | CR | C\| | C$ | C* | C) | C; | C\ |
| 4 | D— | D/ | DS | DT | DU | DV | DW | DX | DY | DZ | D\| | D, | D% | D↓ | D△ | D? |
| 5 | E0 | E1 | E2 | E3 | E4 | E5 | E6 | E7 | E8 | E9 | E: | E# | E@ | E' | E= | E' |
| 6 | $G\&^{S}_{P}$ | GA | GB | GC | GD | GE | GF | GG | GH | GI | G\| | G. | G< | G( | G+ | G! |
| 7 | H& | HJ | HK | HL | HM | HN | HO | HP | HQ | HR | H\| | H$ | H* | H) | H; | H\ |
| 8 | ⊥— | I/ | IS | IT | IU | IV | IW | IX | IY | IZ | = : | I, | I% | ⊥' | ⊥△ | I? |
| 9 | I0 | I1 | I2 | I3 | I4 | I5 | I6 | I7 | I8 | I9 | \|: | \|# | I@ | \|' | \|= | \|' |
| 10 | $<^{S}_{P}$ / B | <A | <B | <C | <D | <E | <F | <G | <H | <I | <\| | <. | << | <' | <+ | <! |
| 11 | (& | (J | (K | (L | (M | (N | (O | (P | (Q | (R | (\| | ($ | (* | 0↓ | (; | (\ |
| 12 | +— | +/ | +S | +T | +U | +V | +W | +X | +Y | +Z | +\| | +, | +% | +↓ | +△ | +? |
| 13 | !0 | !1 | !2 | !3 | !4 | !5 | !6 | !7 | !8 | !9 | \|: | !# | !@ | !' | != | !' |
| 14 | $J^{S}_{P}$ | JA | JB | JC | JD | JE | JF | JG | JH | JI | J\| | J. | J< | J( | J+ | J! |
| 15 | K& | KJ | KK | KL | KM | KN | KO | KP | KQ | KR | K\| | K$ | K* | K) | K; | K\ |
| 16 | L— | L/ | LS | LT | LU | LV | LW | LX | LY | LZ | L\| | L, | L% | L↓ | L△ | L? |
| 17 | M0 | M1 | M2 | M3 | M4 | M5 | M6 | M7 | M8 | M9 | M: | M# | M@ | M' | M= | M' |
| 18 | $O^{S}_{P}$ | OA | OB | OC | OD | OE | OF | OG | OH | OI | O\| | O. | O< | O( | O+ | O! |
| 19 | P& | PJ | PK | PL | PM | PN | PO | PP | PQ | PR | P\| | P$ | P* | P) | P; | P\ |
| 20 | Q— | Q/ | QS | QT | QU | QV | QW | QX | QY | QZ | Q\| | Q. | Q% | Q↓ | Q△ | Q? |
| 21 | R0 | R1 | R2 | R3 | R4 | R5 | R6 | R7 | R8 | R9 | R: | R# | R@ | R' | R= | R' |
| 22 | $\$^{S}_{P}$ | $A | $B | $C | $D | $E | $F | $G | $H | $I | $\| | $. | $< | $( | $+ | $! |
| 23 | *& | *J | *K | *L | *M | *N | *O | *P | *Q | *R | *\| | *$ | ** | *) | *; | *\ |
| 24 | )— | )/ | )S | )T | )U | )V | )W | )X | )Y | )Z | )\| | ). | )% | )↓ | )△ | )? |

238

**TABLE 11-3** *(continued)*

| | B B A A A A 1 2 | C C A A 1 2 | | | | | | | | | | | | | | | |
|---|---|---|---|---|---|---|---|---|---|---|---|---|---|---|---|---|---|

| Row | 65 | 66 | 67 | 68 | 69 | 70 | 71 | 72 | 73 | 74 | 75 | 76 | 77 | 78 | 79 | 80 |
|---|---|---|---|---|---|---|---|---|---|---|---|---|---|---|---|---|---|
| 1 | $A^{S}_{P}$ | AA | AB | AC | AD | AE | AF | AG | AH | AI | A| | A. | A< | A( | A+ | A! |
| 2 | B& | BJ | BK | BL | BM | BN | BO | BP | BQ | BR | B] | B$ | B* | B) | B; | B\ |
| 3 | C– | C/ | CS | CT | CU | CV | CW | CX | CY | CZ | C| | C, | C% | C_ | C> | C? |
| 4 | D0 | D1 | D2 | D3 | D4 | D5 | D6 | D7 | D8 | D9 | D: | D# | D@ | D' | D= | D" |
| 5 | $F^{S}_{P}$ | FA | FB | FC | FD | FE | FF | FG | FH | FI | F| | F. | F< | F( | F+ | F! |
| 6 | G& | GJ | GK | GL | GM | GN | GO | GP | GQ | GR | G] | G$ | G* | G) | G; | G\ |
| 7 | H– | H/ | HS | HT | HU | HV | HW | HX | HY | HZ | H: | H, | H% | H_ | H> | H? |
| 8 | I0 | I1 | I2 | I3 | I4 | I5 | I6 | I7 | I8 | I9 | I: | I# | I@ | I' | I= | I" |
| 9 | $^{S}_{P}$ | ·A | ·B | ·C | ·D | ·E | ·F | ·G | ·H | ·I | ·| | : | ·< | ·( | ·+ | ·! |
| 10 | <& | ∇J | <K | <L | <M | <N | <O | <P | <Q | <R | ∇| | <$ | <* | ∇) | ∇; | <\ |
| 11 | (– | (/ | (S | (T | (U | (V | (W | (X | (Y | (Z | (: | (, | (% | (_ | (> | (? |
| 12 | +0 | +1 | +2 | +3 | +4 | +5 | +6 | +7 | +8 | +9 | +: | +# | +@ | +' | += | +" |
| 13 | $\&^{S}_{P}$ | &A | &B | &C | &D | &E | &F | &G | &H | &I | &[ | &. | &< | &( | &+ | &! |
| 14 | J& | JJ | JK | JL | JM | JN | JO | JP | JQ | JR | JJ | JS | J* | J) | J; | J\ |
| 15 | K– | K/ | KS | KT | KU | KV | KW | KX | KY | KZ | K: | K, | K% | K_ | K> | K? |
| 16 | L0 | L1 | L2 | L3 | L4 | L5 | L6 | L7 | L8 | L9 | L: | L# | L@ | L' | L= | L" |
| 17 | $N^{S}_{P}$ | NA | NB | NC | ND | NE | NF | NG | NH | NI | N[ | N. | N< | N( | N+ | N! |
| 18 | O& | OJ | OK | OL | OM | ON | OO | OP | OQ | OR | O] | O$ | O* | O) | O; | O\ |
| 19 | P– | P/ | PS | PT | PU | PV | PW | PX | PY | PZ | P| | P. | P% | P_ | P> | P? |
| 20 | Q0 | QI | Q2 | Q3 | Q4 | Q5 | Q6 | Q7 | Q8 | Q9 | Q: | Q# | Q@ | Q' | Q= | Q" |
| 21 | $I^{S}_{P}$ | IA | IB | IC | ID | IE | IF | IG | IH | II | I| | I. | I< | I( | I+ | I! |
| 22 | $& | $J | $K | $L | $M | $N | $O | $P | $Q | $R | $] | $$ | $* | $) | $; | $\ |
| 23 | *– | */ | *S | *T | *U | *V | *W | *X | *Y | *Z | *: | *. | *% | *_ | *> | *? |
| 24 | )0 | )I | )2 | )3 | )4 | )5 | )6 | )7 | )8 | )9 | ): | )# | )@ | )' | )= | )" |

*Column*

239

**TABLE 11–4   Attribute Characters**

| Protected | Numeric | Hidden | Blinked | Intensified | Modified | AC | Protected | Numeric | Hidden | Blinked | Intensified | Modified | AC |
|---|---|---|---|---|---|---|---|---|---|---|---|---|---|
| | | | | | | sp | X | | | | | | - |
| | | | | | X | A | X | | | | | X | / |
| | | | | X | | H | X | | | | X | | Y |
| | | | | X | X | I | X | | | | X | X | Z |
| | | | X | | | [ | X | | | X | | | ¦ |
| | | | X | | X | . | X | | | X | | X | , |
| | | X | | | | < | X | | X | | | | % |
| | | X | | | X | ( | X | | X | | | X | — |
| X | | | | | | & | X | X | | | | | 0 |
| X | | | | | X | J | X | X | | | | X | 1 |
| X | | | | X | | Q | X | X | | | X | | 8 |
| X | | | | X | X | R | X | X | | | X | X | 9 |
| X | | | X | | | ] | X | X | | X | | | : |
| X | | | X | | X | S | X | X | | X | | X | # |
| X | X | | | | | * | X | X | X | | | | @ |
| X | X | | | | X | ) | X | X | X | | | X | ' |

The GS simply indicates that the next received character is an AC. When received, the GS is discarded. The AC is stored in a *protected* character position in the buffer but is displayed on the screen as a *space*.

    *AC*. The contents of the device buffers are broken up into *fields*. Each field starts with a field AC which denotes the properties of that field.

    The types of fields described by attribute characters are as follows:

(1) *Protected*: accessible only by the remote LCU. The terminal operator cannot enter or alter data in this field.

(2) *Unprotected*: accessible to the operator and LCU.

(3) *Numeric*: allows only 0–9, DUP, minus, and period to be entered in this field by the operator.

(4) *Hidden*: data in this field are stored in the display buffers but are not visible on the screen or printed on the printer.

(5) *Modified*: fields in which data have been altered by the operator or which has been declared as such by the AC.

    *Note*: The write or erase-write command is followed by a WCC. A WCC may declare all ACs as unmodified. Normally, when data are transmitted, only modified data are sent (data which the operator has entered). This eliminates unnecessary transmission of a form and simply sends out the information that was entered in completing the form. When required, however, the data field of the form itself may be declared as modified. Then, when transmission takes place, both the form and the data required to complete the form are transmitted as modified data.

(6) *Intensified* (*highlighted*): characters in this field are displayed at a higher than normal intensity on the display (X 1.5).

(7) *Blinked*: characters in this field are displayed alternately between full and half intensity.

(8) *Protected and Numeric*: these two attributes combine to provide *AUTOSKIP*. Any time the cursor falls on an autoskip AC, the cursor automatically skips to the first location of the next unprotected field. Where conditions exist, it will wrap-around.

    (c) Insert Cursor:

```
D
C      Places cursor at the current buffer address.
3
```

---

**EXAMPLE**

```
D     D
C F K C    Places cursor at row 6, column 3.
1     3
```

---

    (d) Program Tab:

```
H
T
```

Advances the buffer pointer to the first location of the next unprotected field. If currently on an attribute of an unprotected field, the cursor advances one position. If the pointer is in an unprotected field and allows either text or another HT, the rest of the field is replaced by nulls.

    (e) Repeat to Address:

```
D B B
C A A C
4 1 2
```

Stores a specified character into every buffer position from the present position up to but excluding the specified buffer position. Will wrap-around. If the buffer address is the same as the current address, the character will be placed in all locations.

> **EXAMPLE**
>
> $$\begin{array}{ccc} & D & \\ C & \substack{S \\ P} & N\ T \\ & 4 & \end{array}$$

Assuming that the current address was HOME, T would be placed in every location up to but not including row 1, column 22.

    (f) Erase Unprotected to Address:

```
D B B     Self-explanatory.
C A A
2 1 2     Will wrap-around.
```

If the buffer address equals the current address, nulls are set in all unprotected locations.

## AIDS—Attention Identifiers

Refer to the layout of the synchronous keyboard in Figure 11–1. When a PA or a PF key is depressed, the ASCII code corresponding to the AID (Table 11–5) of the depressed key is transmitted (when polled).

**TABLE 11–5   Attention Identifiers (AIDS)**

| Key depressed | AID | Message response to poll |
|---|---|---|
| S/R | , | Read Modified |
| PA1 | % | Short Read |
| PA2 | < | Short Read |
| PA3 | , | Short Read |
| R/TST | 0 | Read Modified |
| PF1 | 1 | Read Modified |
| PF2 | 2 | Read Modified |
| PF3 | 3 | Read Modified |
| PF4 | 4 | Read Modified |
| PF5 | 5 | Read Modified |
| PF6 | 6 | Read Modified |
| PF7 | 7 | Read Modified |
| PF8 | 8 | Read Modified |
| PF9 | 9 | Read Modified |
| PF10 | · | Read Modified |
| PF11 | # | Read Modified |
| PF12 | @ | Read Modified |
| CLEAR | — | Short Read |
| None, KD | - | Response to Read |
| None, P | Y | All Command Only |
| Magnetic Stripe Reader | W | Read Modified |

The PA (*Program Access*) keys request a preformatted form from the CPU. No text can be transmitted by the secondary when these keys are depressed.

The PF (*Program Form*) keys can cause a variety of options to occur: the CPU can transmit a preformatted form; the CPU can initiate a copy command; anything. Text can be transmitted with these AIDS.

Each AID is just a single character. When the CPU receives that character, it goes to the application programs for that station to determine the action necessitated by that character. Therefore, the function of each PF key depends on what has been programmed *for that station* and *for that key*.

## Format for Secondary Messages

1. Read Modified Message (in response to poll or read modified command if PA or Clear key is depressed):

| S S | | A E B |
|-----|-------|-------|
| T P | D A | I T C |
| X A | | D X C |

**EXAMPLE**

Operator of device 1 at station 2 depressed the PA1 key. After being polled, the station would transmit:

```
S S S        E B P
Y Y T B A % T C A
N N X        X C D
```

2. Read All Message (in reply to a Read All command):

| S S | | A C C | data in | E B |
|-----|-----|-------|---------|-----|
| | D | | G A | |
| T P | | I A A | fields + | T C |
| | A | | S C | |
| X A | | D 1 2 | nulls | X C |
| | | | for each field | |

**EXAMPLE**

```
    S S E P S S        E P
    Y Y O A Y Y / / C C N A  ───────────────►
    N N T D N N        Q D
```
Selection of station 1, device 3: a KD.

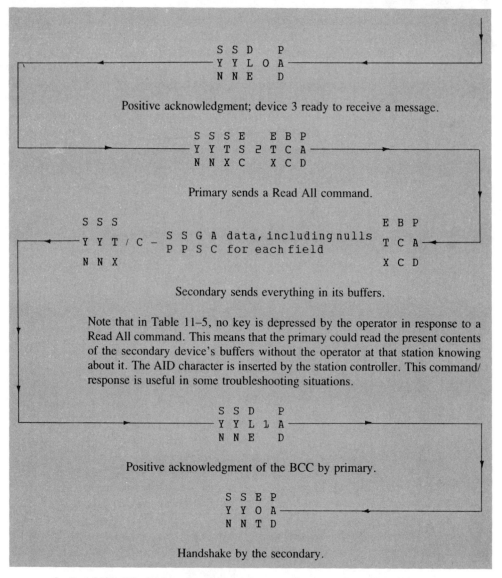

```
            S  S  D     P
          Y  Y  L  O  A
            N  N  E     D
```

Positive acknowledgment; device 3 ready to receive a message.

```
            S  S  S  E     E  B  P
          Y  Y  T  S  2  T  C  A
            N  N  X  C     X  C  D
```

Primary sends a Read All command.

```
S  S  S                                    E  B  P
  Y  Y  T / C  ─  S  S  G  A  data, including nulls  T  C  A
N  N  X          P  P  S  C  for each field          X  C  D
```

Secondary sends everything in its buffers.

Note that in Table 11–5, no key is depressed by the operator in response to a Read All command. This means that the primary could read the present contents of the secondary device's buffers without the operator at that station knowing about it. The AID character is inserted by the station controller. This command/response is useful in some troubleshooting situations.

```
            S  S  D     P
          Y  Y  L  1  A
            N  N  E     D
```

Positive acknowledgment of the BCC by primary.

```
            S  S  E  P
          Y  Y  O  A
            N  N  T  D
```

Handshake by the secondary.

3. Read Modified Message (in reply to a poll or to a Read Modified command if either the S/R or PF key is depressed)

```
S  S     A  C  C | D  A  A  data in        | E  B
         D                                   
T  P  A  I  A  A | C  A  A  fields          | T  C
X  A     D  1  2 | 1  1  2  less nulls      | X  C
                 | for each modified field  |
```

**EXAMPLE**

```
S S E P S S       E P
Y Y O A Y Y B B D D N A
N N T D N N         Q D
```

Specific poll of device 4 at station 2.

If the S/R key had been depressed by the operator:

```
S S S              D        data in field less   E B P
Y Y T B D ' B T    C A J    nulls for each       T C A
N N X              1        modified field       X C D
```

Cursor was left at row 3, column 4; first modified character is in row 2, column 2.

```
S S D   P
Y Y L 1 A
N N E   D
```

Primary sends positive acknowledgment of BCC.

```
S S E P
Y Y O A
N N T D
```

Handshake by the secondary.

### Points to keep in mind

(a) All transmissions containing a BCC must be acknowledged. BCCs after an ETB or ETX must be acknowledged immediately.

(b) Negative acknowledgments require retransmission of the last block received.

(c) If the secondary is sending the message, the secondary will handshake the primary with

```
S S E P
Y Y O A
N N T D
```

after it receives a final positive acknowledgment of its BCC (after ETX). If the primary is sending the message, it does not handshake the secondary after it receives a final positive acknowledgment.

(d) The secondary cannot transmit or receive unless it is first polled or selected, respectively, by the primary.

(e) Although there is an ASCII code for ACK, DLE is used as the transmission code for this purpose.

(f) In the transmission of certain forms, the buffer address (BA1 and BA2), the cursor address (CA1 and CA2), or the attribute address (AA1 and AA2) need to be specified. The same coding is used for all of these addresses. The buffer address is the specified location in memory; the cursor address is the current address of the cursor. However, where an attribute address is specified, it does not mean the address where the attribute character is stored but the *first location following this address*.

---

**EXAMPLE**

Given:

$$\begin{smallmatrix} G \\ S \end{smallmatrix} \; Y$$

Assume that the attribute Y is stored in row 3, column 5. If the attribute address was called for by any format, the address for row 3, column 6 should be given.

---

(g) *Difference between NULL and SPACE*. If a device's memory is erased, NULLS (ASCII = 00H) are stored in all locations and displayed on the screen as blank spaces. A SPACE (ASCII = 20H) is considered a separation character. It would also be displayed on the screen as a blank space. During the transmission of modified data, if a SPACE was entered in an unprotected field (operator hit the space bar), it would be transmitted. A NULL in the same location would not be transmitted.

(h) A field is declared unprotected and unmodified. If it is unprotected, the operator can write into that field. Once the operator enters characters into that field, since he or she changed the characters from what they were, those characters are considered modified data.

---

**EXAMPLE**

A form is already prepared and stored in computer memory at the primary. The secondary can obtain this form by depressing the PF6 key (this key was programmed for this function). Device 6 at station 2 is the secondary in question. The form is as follows:

CLASS *(5 unprotected spaces) COURSE *(10 unprotected spaces).
(* denotes a protected space.)
The C of CLASS will be on row 5, column 20.
The C of COURSE will be on row 5, column 40.

When the form is sent:

*Command*: sound tone, reset attributes, 64 characters/line.
*Form*: protected and intensified.

---

*Operator-entered information*: blinked.

*Dialogue*. PF6 key is depressed by the operator.

```
S S E P S S           E P
Y Y O A Y Y B B " "   N A ──────────────────────────┐
N N T D N N           O D                            │
```

General poll of station 2. Station controller detects that device 6 has its PF6 key depressed.

```
 S S S                E B P
┌──────◄───── Y Y T B F 6  S S  T C A ──────◄────────┘
│            N N X         P P  X C D
│                          P P
│                          X C
```

B = SPA for station 2; F = DA for device 6; 6 = AID for PF6; two SPs = cursor address (cursor is in the HOME position on the screen); no text is sent since screen is clear.

```
│            S S D   P
└──────►──── Y Y L 1 A ──────────────►────────┐
             N N E   D                         │
```

Primary acknowledges BCC. DLE 1 is used for the first block of text.

```
│            S S E P                           │
┌──────◄──── Y Y O A ──────────◄───────────────┘
│            N N T D
```

Handshake by the secondary station.

The primary has recovered the desired form from memory and is now ready to transmit it to the secondary.

```
│            S S E P S S           E P
└──────►──── Y Y O A Y Y S S F F F N A ──────►────────┐
             N N T D N N           Q D                │
```

Primary selects device 6 at station 2. S = SSA for station 2; F = DA for device 6.

```
│            S S D   P                                 │
┌──────◄──── Y Y L O A ──────────◄─────────────────────┘
│            N N E   D
```

Positive acknowledgment indicating that device 6 is ready to receive. DLE 0 is used since the station has not yet received any text information. The following sequence is the form transmitted by the primary station.

```
        S   S   S   E           D
 ──►    Y   Y   T   S   5   V   C   E   K   G   Y   C   L   A   S   S   G   [   S   S   S   S   S   G   8
        N   N   X   C           1                   S                           S   P   P   P   P   P   S


        D                               D           D           D   E   B   P
        C   E   X   C   O   U   R   S   E   G   [   C   E   8   G   8   C   E   R   C   T   C   A  ──►
        1                               S   1           S   1                       3   X   C   D
```

Explanation of message:

**ESC5**    Erase-write command. Clears screen, homes cursor.

**V**    WCC. Sounds KD alarm, resets ACs to unmodified, 64 characters/line.

**DC1**    Device order. The next two characters will represent a buffer address.

**EK**    Row 5, column 19.

**GS**    Indicated field will start at row 5, column 19. The type of field will be specified by the attribute character that follows.

**Y**    Attribute character. Field is protected and intensified. Remember, the attribute character is displayed as a protected space. Row 5, column 19 is now a protected space.

**CLASS**    This is the text. C is stored in row 5, column 20 and the last S will be in row 5, column 24.

**GS**    Indicated field will start at the next location—row 5, column 25.

**[**    Attribute character. Characters entered in this field will be blinked. The attribute character will also provide the protected space after CLASS.

**5 SPs**    Spaces allowed for the operator to enter his information.

**GS**    Indicated field will start at row 5, column 31.

**8**    Attribute character. Field is protected, numeric, and intensified. Protected and numeric = autoskip. When the operator enters the fifth character after CLASS, the cursor will move to the next location, which is an autoskip field. Once the cursor finds itself in this field, it moves on to the first unprotected location of the next unprotected field. The operator would then be automatically ready to enter the course.

**DC1**    Set buffer address to . . .

**EX**    Row 5, column 40.

**COURSE**    Text that would start at row 5, column 40. The E ends up on row 5, column 45. COURSE is being written in a field identified as protected and numeric. The field is protected only as far as the device operator is concerned. The operator cannot overwrite the word COURSE. In

|     |     |
| --- | --- |
|     | fact, the operator cannot enter anything in this field. Similarly, numeric applies only to what the operator may enter. COURSE is not numeric, but the operator does not enter COURSE; it is part of the form. If the field is protected, it is entirely incidental whether the field is numeric or not as far as the operator is concerned; the operator cannot enter anything into this field anyway. |
| GS | Indicated field will start at the next location: row 5, column 46. |
| [ | Attribute character. Field is blinked. Row 5, column 46 is a protected space for the attribute character. This also provides the required protected space after COURSE. |
| DC1 | Set buffer address to . . . |
| E8 | Row 5, column 57. |
| GS | New field starts at row 5, column 57. This means that the operator can enter information (which will be blinked) from row 5, column 47 to row 5, column 56, which are the 10 required spaces. This is shown as an alternative method for providing these spaces in contrast with the way this was achieved after CLASS. |
| 8 | Attribute character. Protected, numeric, and intensified. Since there is no further form information, this protects the rest of the form. Remember, protected will wrap-around, so this AC protects the form from row 5, column 57 to row 24, column 80 and then back around from row 1, column 1 to row 5, column 18. Also, since this is an autoskip field, if the operator entered the tenth character in the preceding blinked field, the cursor would move into this autoskip field and end up at row 5, column 26. If programmed correctly, often the CPU can determine which form was completed simply from the final position of the cursor. |
| DC1 | Set buffer address to . . . |
| ER | Row 5, column 26. |
| DC3 | Insert cursor. This is done as a courtesy to the operator so that he or she may start completing the form without any unnecessary waste of time. |

```
                          S  S  D      P
◄─────────────────────────Y  Y  L  1  A──────────────◄──
                          N  N  E      D
```

Station acknowledges BCC. Primary goes on to poll station 3.

(i) *Difference between unprotected and modified.* An unprotected field is a field in which the operator can enter data. In the previous message form, 10 spaces were left unprotected after COURSE for operator data. If the operator entered MATH, these four spaces immediately become modified: the contents of those locations were changed by the operator.

After entering MATH, if the operator hit the TAB key, the cursor would automatically move to the first unprotected space of the next field. The remaining spaces in the previous field would contain nulls which were not modified. If only modified data were transmitted, only MATH in the field after COURSE would be transmitted.

After entering MATH, if the operator repeatedly hit the space bar to get to the first unprotected space of the next field, the character for SPACE would have been entered in the six locations following MATH and the contents of these six locations would also have become modified data. If only modified data were transmitted, MATH and six 20H characters (20H = ASCII space) would be transmitted.

What if the attribute character after COURSE defined the next field to be modified and unprotected? If only MATH was entered by the operator and then he hit the TAB key, MATH, and six NULLS would be transmitted as modified data. Although the operator changed only four characters in this field, the entire field was declared as modified by the attribute character and the entire field would be transmitted whether the operator entered anything into it or not.

Similarly, if the COURSE field was declared protected and modified by the attribute character, the operator could not overwrite COURSE. However, COURSE would be transmitted whenever modified data were transmitted.

4. Status and Sense Message:

| S | | S | S | | S | S | E | B | P |
|---|---|---|---|---|---|---|---|---|---|
| O | % | R | T | P | D A | S | S | T | C | A |
| H | | X | A | | 1 | 2 | X | C | D |

Transmission of RVI—Reverse Interrupt—by the receiving station constitutes a positive acknowledgment of a received BCC. In addition, the RVI requests the sending station to halt transmission at its earliest convenience because the receiving station has something to send of higher priority. This application of an RVI is used mainly in a two-point system where a block of data may be much larger than 256 characters. In a multipoint data link system, the length of a message from the primary is limited by the capacity of the terminal buffers at the secondary station. In a multipoint system, the RVI still serves as a positive acknowledgment, but it also indicates to the primary that the secondary is experiencing some form of difficulty. To determine the nature of the problem, the primary must send a specific poll. In response to this poll, the secondary returns a Status and Sense message. SOH% R informs the primary that what follows is the Status and Sense message.

SS1 and SS2 is a two-character code which describe the type of problem encountered (Table 11–6). An RVI is never sent by the primary to the secondary.

**TABLE 11–6   Status and Sense Codes**

| SS1 | SS2 | |
|-----|-----|---|
| S<br>P | A | ESC missing.<br>ETX missing or out of place.<br>Contains invalid buffer address or device address.<br>Example: Specific poll of a device address which that station does not have. |
| S<br>P | - | Contains invalid BCC sequence. |
| H | A | Device being copied. |
| H | S<br>P | Device busy for some other reason. |
| B | S*<br>P | Device no longer busy. |
| D | A | Device to be copied not available because the buffer is locked. With some systems, if a protected alphanumeric attribute is in the home position, that buffer cannot be copied. An attempted copy command with the buffer locked will always give a buffer locked status and sense message. |
| S<br>P | J | Device to be copied not available for some other reason. |
| B | S*<br>P | Device now available. |
| B | &[b] | Device not available because it is out of paper or the cover is up. |
| S<br>P | & | Device not available for other reasons. |

[a] Sent only in reply to a specific poll except as noted by an asterisk.

[b] Sent in reply to a general poll as well as a specific poll.

**EXAMPLE**

Assume that the primary has selected a particular device at a station and received a positive acknowledgment.

$$
\begin{array}{cccc c ccc}
S & S & S & D & & E & B & P \\
Y & Y & T & C & E\ E\ {}^G_S \ \ldots\ldots\ & T & C & A \\
N & N & X & 1 & & X & C & D \\
\end{array}
$$

Message sent by the primary.

$$
\begin{array}{cccc}
S & S & D & P \\
Y & Y & L & < & A \\
N & N & E & & D \\
\end{array}
$$

Secondary responds with an RVI.

```
S  S  E  P  S  S  S        E  P
Y  Y  O  A  Y  Y  P  P  D  D  N  A
N  N  T  D  N  N  A  A        Q  D
```

Primary polls the device to determine the nature of the problem.

```
S  S  S        S  S     D  S     E  B  P
Y  Y  O  %  R  T  P     A  A  T  C  A
N  N  H        X  A  A  P        X  C  D
```

Since the message from the primary did not start with a command field, the status and sense code (SP A) was sent back.

The primary would now have to correct its error, reselect, and retransmit the original message.

5. Remote Test Request Message (in reply to a poll or a read modified command if R/TST depressed).

| S |   | S | D | A | A | data in fields | E | B | P |
|---|---|---|---|---|---|----------------|---|---|---|
| O | % | / T | C | A | A | less nulls | T | C | A |
| H |   | X | 1 | 1 | 2 |  | X | C | D |
|   |   |   | For each modified field |   |   |   |   |   |   |

## The Copy Command

The copy command is sent by the primary to a station controller indicating which device will be copied from. The information to be copied will be sent to a previously selected device. The sole restriction is that the device copied from and the device copied onto must be at the same station. CCC stands for Copy Control Character (Table 11–7). This character identifies what will be copied.

The sequence of events involved with copying is initiated by the device that will be copied from. If copying from one device onto another is required quite often, a PA key at the device that will be copied from can be programmed to initiate a copy command onto the desired device. Once programmed, depressing that particular PA key will cause a copy only onto the device for which it was programmed. The operator at the device would compose (or have) the message to be copied on the screen and then press the appropriate PA key, because what is on the screen is intended for another device and not the primary station. If a PF key was used, the screen text would be transmitted to the primary. When polled, the station controller would send out a Read Modified message. The primary would then select the device that will be copied onto. If that device is ready to receive (powered up and not in LOCAL), the station controller returns a positive acknowledg-

**TABLE 11–7  Copy Control Characters**

| Start printer | Sound KD tone | ACs | Protected data | Unprotected data | NL char/line | 40 char/line | 64 char/line | 80 char/line |
|---|---|---|---|---|---|---|---|---|
| | | | | | Copy | | Print | |
| | X | | | | sp | & | - | 0 |
| | | X | | X | A | J | / | 1 |
| | | X | X | | B | K | S | 2 |
| | | X | X | X | C | L | T | 3 |
| | X | X | | | D | M | U | 4 |
| | X | X | | X | E | N | V | 5 |
| | X | X | X | | F | O | W | 6 |
| | X | X | X | X | G | P | X | 7 |
| X | | X | | | H | Q | Y | 8 |
| X | | X | | X | I | R | Z | 9 |
| X | | X | X | | [ | ] | ¦ | : |
| X | | X | X | X | . | $ | , | # |
| X | X | X | | | < | * | % | @ |
| X | X | X | | X | ( | ) | – | ' |
| X | X | X | X | | + | ; | > | = |
| X | X | X | X | X | ! | / | ? | " |

ment to the primary. At this time, the primary sends a copy command to the station controller. Up to this time, the operator at the device that will be copied from does not know if the copy request has been sent to the primary; the operator at the device that will be copied onto has no inkling of any copying operation being planned.

On receiving a copy command, the station controller returns a WACK to the primary indicating positive acknowledgment and also requests no transmission from the primary since it will be involved in a copy operation. The station controller then initiates the actual copying with the information going from the device being copied from, to the station controller, to the device being copied onto. While the copying is being carried out, the primary is free to poll or select other stations in the system.

**EXAMPLE**

Device 3 at station 5 wants to copy everything on its display onto device 8. The PA1 key has been programmed to inform the primary to send out the appropriate copy command.

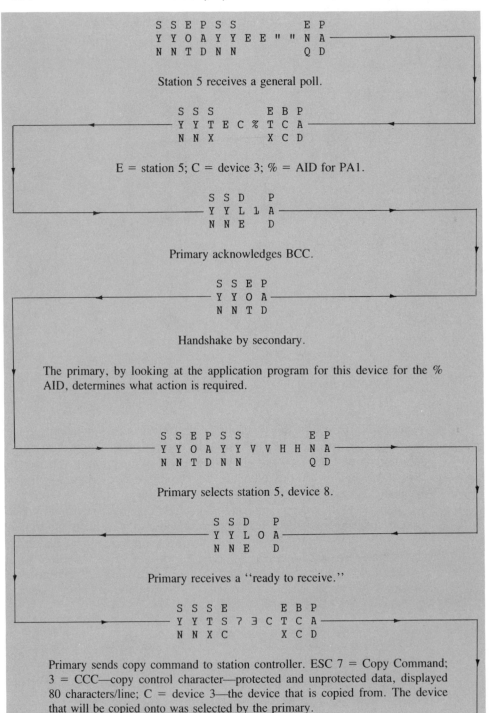

```
S  S E P S S              E P
Y  Y O A Y Y E E " " N A ──────────────────►
N  N T D N N              Q D
```

Station 5 receives a general poll.

```
      S S S       E B P
    ─ Y Y T E C % T C A ─
      N N X       X C D
```

E = station 5; C = device 3; % = AID for PA1.

```
      S S D   P
    ─ Y Y L 1 A ─
      N N E   D
```

Primary acknowledges BCC.

```
      S S E P
    ─ Y Y O A ─
      N N T D
```

Handshake by secondary.

The primary, by looking at the application program for this device for the % AID, determines what action is required.

```
      S S E P S S           E P
    ─ Y Y O A Y Y V V H H N A ─────────►
      N N T D N N           Q D
```

Primary selects station 5, device 8.

```
      S S D   P
    ─ Y Y L O A ─
      N N E   D
```

Primary receives a "ready to receive."

```
      S S S E       E B P
    ─ Y Y T S 7 ϶ C T C A ─
      N N X C       X C D
```

Primary sends copy command to station controller. ESC 7 = Copy Command; 3 = CCC—copy control character—protected and unprotected data, displayed 80 characters/line; C = device 3—the device that is copied from. The device that will be copied onto was selected by the primary.

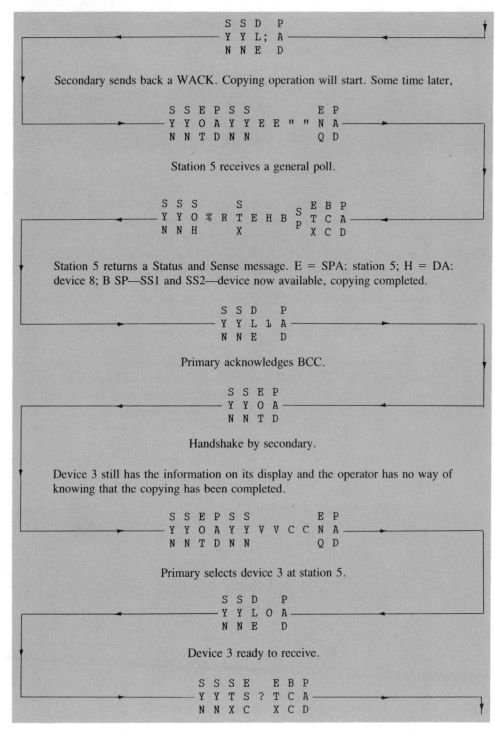

```
                    S  S  D     P
                    Y  Y  L; A
                    N  N  E     D
```

Secondary sends back a WACK. Copying operation will start. Some time later,

```
       S  S  E  P  S  S              E  P
       Y  Y  O  A  Y  Y  E  E  "  "  N  A
       N  N  T  D  N  N              Q  D
```

Station 5 receives a general poll.

```
   S  S  S        S           E  B  P
   Y  Y  O  %  R  T  E  H  B  S  T  C  A
   N  N  H        X        P  X  C  D
```

Station 5 returns a Status and Sense message. E = SPA: station 5; H = DA: device 8; B SP—SS1 and SS2—device now available, copying completed.

```
                    S  S  D     P
                    Y  Y  L  1  A
                    N  N  E     D
```

Primary acknowledges BCC.

```
                    S  S  E  P
                    Y  Y  O  A
                    N  N  T  D
```

Handshake by secondary.

Device 3 still has the information on its display and the operator has no way of knowing that the copying has been completed.

```
       S  S  E  P  S  S              E  P
       Y  Y  O  A  Y  Y  V  V  C  C  N  A
       N  N  T  D  N  N              Q  D
```

Primary selects device 3 at station 5.

```
                    S  S  D     P
                    Y  Y  L  O  A
                    N  N  E     D
```

Device 3 ready to receive.

```
       S  S  S  E     E  B  P
       Y  Y  T  S  ?  T  C  A
       N  N  X  C     X  C  D
```

ESC? = Erase all unprotected command. Primary erases all unprotected data from the screen of device 3 as a positive indication that the copying has been completed. The erase-write command with various options of WCC could have also been used.

```
S  S  D    P
Y  Y  L  1  A ──────────────────────────
N  N  E    D
```

Secondary returns positive acknowledgment of BCC.

## Transparent Text Mode

If Bisync is to be used for the transmission of binary data, floating-point numbers, unique specialized codes, or machine language computer programs, it is possible for these bit patterns to be ASCII codes for certain data link control characters such as SYN, STX, ETB, ENQ, DLE, ETX, or EOT. These bit patterns would have to be *transparent* to the receiver—the receiver would have to know that these bit patterns are part of the text and not intended to be data link control characters. How, then, is the receiver to know when a particular bit pattern is part of the text and when that same bit pattern actually is a control character? The transmitter would have to transmit in transparent text mode. The transmitter enters transparent text mode by transmitting DLE STX instead of just STX. From here on, if a bit pattern is to be a control character, it *must* be prefixed by DLE. To terminate a block, the transmitter must send DLE ETB. When operating in transparent text mode, the receiver, whenever it receives a DLE, strips it and interprets the next character as a data link control character. If ETX was received without being preceded by DLE, the bit pattern for ETX would simply be treated as part of the data stream and not as a control character. To transmit DLE as part of the data stream, DLE DLE would have to be transmitted. The first DLE would be stripped.

DLE STX    Initiates transparent mode for the text following.

DLE ETB    Terminates a block of transparent text. The next block could be non-transparent and the message then be terminated by ETX alone.

**EXAMPLE**

Assume the primary has already selected the secondary, received a "ready to receive," and is now transmitting the message. Acknowledgment of BCCs is not shown.

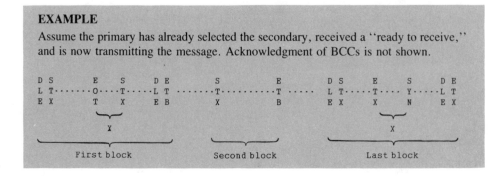

Three blocks of text are shown, which make up one message. The first and last blocks use transparent text mode. The control characters, labeled ''X,'' contained within transparent text and not preceded by DLE will be treated by the receiver as part of the bit stream of the message and not as control characters.

## EXAMPLES OF DIALOGUE IN A MULTIPOINT ENVIRONMENT

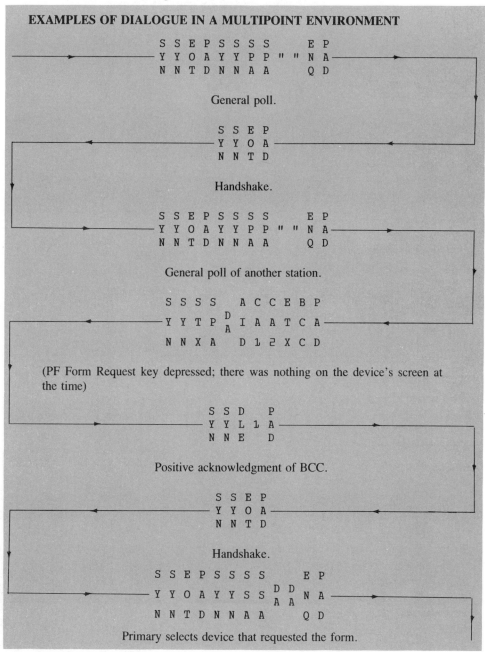

```
          S S E P S S S       E P
          Y Y O A Y Y P P " " N A
          N N T D N N A A       Q D
```

General poll.

```
          S S E P
          Y Y O A
          N N T D
```

Handshake.

```
          S S E P S S S       E P
          Y Y O A Y Y P P " " N A
          N N T D N N A A       Q D
```

General poll of another station.

```
          S S S S   A C C E B P
                   D
          Y Y T P  A I A A T C A
          N N X A   D 1 2 X C D
```

(PF Form Request key depressed; there was nothing on the device's screen at the time)

```
          S S D   P
          Y Y L 1 A
          N N E   D
```

Positive acknowledgment of BCC.

```
          S S E P
          Y Y O A
          N N T D
```

Handshake.

```
          S S E P S S S       E P
                           D D
          Y Y O A Y Y S S  A A N A
          N N T D N N A A       Q D
```

Primary selects device that requested the form.

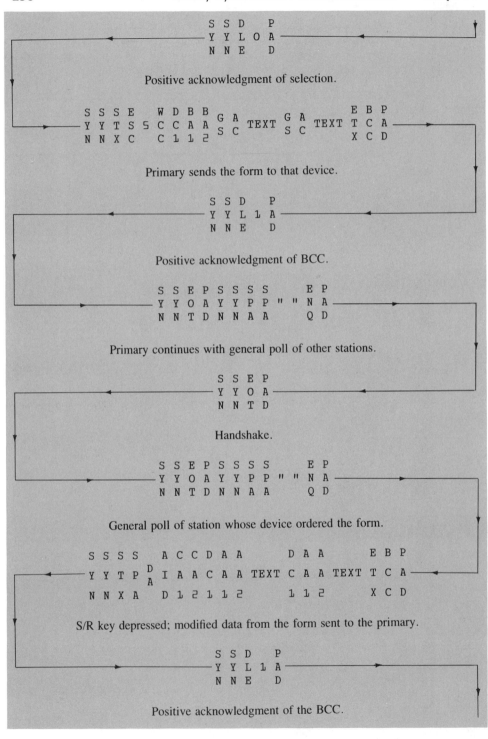

```
                          S S D   P
                          Y Y L O A
                          N N E   D
```

Positive acknowledgment of selection.

```
   S S S E   W D B B G A          G A          E B P
   Y Y T S 5 C C A A S C  TEXT    S C  TEXT    T C A
   N N X C   C 1 1 2 C            C            X C D
```

Primary sends the form to that device.

```
                          S S D   P
                          Y Y L 1 A
                          N N E   D
```

Positive acknowledgment of BCC.

```
   S S E P S S S     E P
   Y Y O A Y Y P P " " N A
   N N T D N N A A     Q D
```

Primary continues with general poll of other stations.

```
                        S S E P
                        Y Y O A
                        N N T D
```

Handshake.

```
   S S E P S S S     E P
   Y Y O A Y Y P P " " N A
   N N T D N N A A     Q D
```

General poll of station whose device ordered the form.

```
   S S S S   A C C D A A       D A A       E B P
   Y Y T P D I A A C A A  TEXT C A A  TEXT T C A
   N N X A A D 1 2 1 1 2       1 1 2       X C D
```

S/R key depressed; modified data from the form sent to the primary.

```
                          S S D   P
                          Y Y L 1 A
                          N N E   D
```

Positive acknowledgment of the BCC.

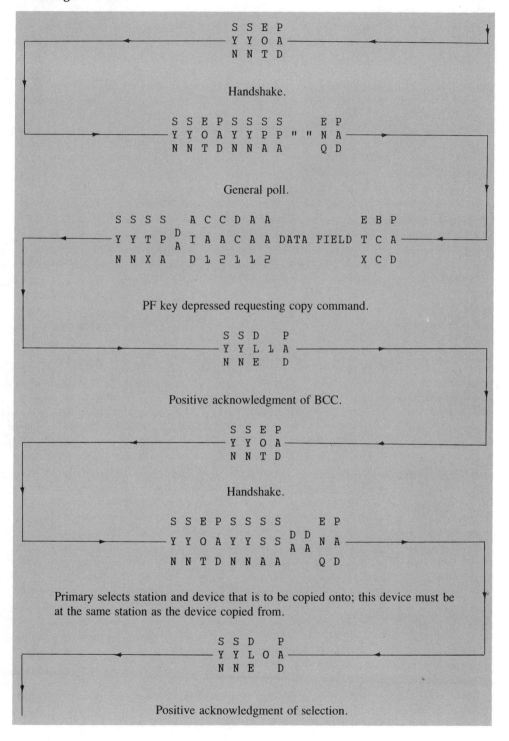

```
          S  S  E  P
          Y  Y  O  A
          N  N  T  D
```

Handshake.

```
S  S  E  P  S  S  S        E  P
Y  Y  O  A  Y  Y  P  P  "  "  N  A
N  N  T  D  N  N  A  A        Q  D
```

General poll.

```
S  S  S  S     A  C  C  D  A  A                    E  B  P
Y  Y  T  P   D I  A  A  C  A  A  DATA FIELD  T  C  A
           A
N  N  X  A     D  1  2  1  1  2                    X  C  D
```

PF key depressed requesting copy command.

```
          S  S  D     P
          Y  Y  L  1  A
          N  N  E     D
```

Positive acknowledgment of BCC.

```
          S  S  E  P
          Y  Y  O  A
          N  N  T  D
```

Handshake.

```
S  S  E  P  S  S  S        E  P
Y  Y  O  A  Y  Y  S  S  D  D  N  A
                       A  A
N  N  T  D  N  N  A  A        Q  D
```

Primary selects station and device that is to be copied onto; this device must be at the same station as the device copied from.

```
          S  S  D  P
          Y  Y  L  O  A
          N  N  E     D
```

Positive acknowledgment of selection.

```
S S S E   C D E B P
Y Y T S ? C A T C A
N N X C   C   X C D
```

Copy command is sent to the station identifying the device that is to be copied from.

```
S S D   P
Y Y L ; A
N N E   D
```

Wait acknowledgment. Positive acknowledgment to BCC; printer cannot start printing until the motor comes up to speed.

```
S S E P S S S     E P
Y Y O A Y Y P P " " N A
N N T D N N A A   Q D
```

Primary resumes general poll.

```
S S S     S S D S S E B P
Y Y O % R T P A S S T C A
N N H     X A   1 2 X C D
```

Later, in response to a general poll, a Status and Sense message is sent by the station to indicate that the device is no longer busy—copying has been completed.

```
S S D   P
Y Y L 1 A
N N E   D
```

Positive acknowledgment of the BCC.

```
S S E P
Y Y O A
N N T D
```

Handshake.

```
S S E P S S S D D E P
Y Y O A Y Y S S A A N A
N N T D N N A A     Q D
```

Primary selects the device that was copied from.

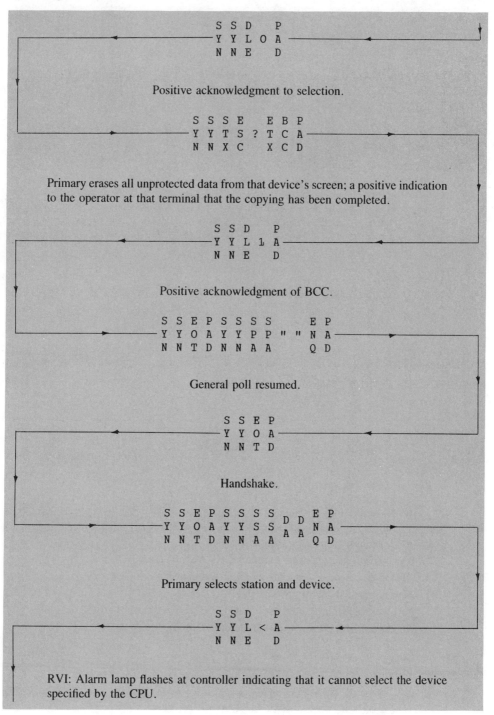

```
                    S  S  D     P
                    Y  Y  L  O  A
                    N  N  E     D
```

Positive acknowledgment to selection.

```
          S  S  S  E     E  B  P
          Y  Y  T  S  ?  T  C  A
          N  N  X  C     X  C  D
```

Primary erases all unprotected data from that device's screen; a positive indication to the operator at that terminal that the copying has been completed.

```
                    S  S  D     P
                    Y  Y  L  1  A
                    N  N  E     D
```

Positive acknowledgment of BCC.

```
    S  S  E  P  S  S  S        E  P
    Y  Y  O  A  Y  Y  P  P  "  "  N  A
    N  N  T  D  N  N  A  A        Q  D
```

General poll resumed.

```
              S  S  E  P
              Y  Y  O  A
              N  N  T  D
```

Handshake.

```
    S  S  E  P  S  S  S        E  P
    Y  Y  O  A  Y  Y  S  S  D  D  N  A
    N  N  T  D  N  N  A  A  A  A  Q  D
```

Primary selects station and device.

```
                    S  S  D     P
                    Y  Y  L  <  A
                    N  N  E     D
```

RVI: Alarm lamp flashes at controller indicating that it cannot select the device specified by the CPU.

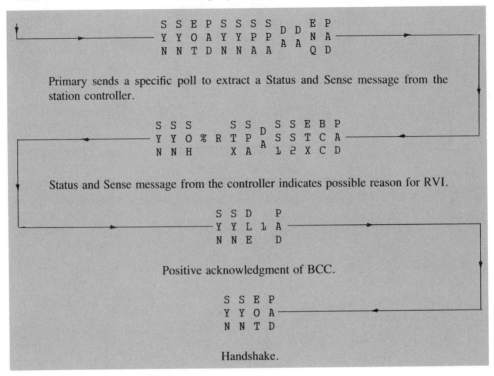

Primary sends a specific poll to extract a Status and Sense message from the station controller.

Status and Sense message from the controller indicates possible reason for RVI.

Positive acknowledgment of BCC.

Handshake.

# QUESTIONS

1. Which data link control character(s):
   (a) Immediately precedes each block of text?
   (b) Is used to indicate the end of the message?
   (c) May be used to terminate a block of text which is not the last block of the message?
   (d) Terminates heading information?

2. Which ASCII characters are transmitted for ACK 0 and ACK 1, respectively?

3. What is the response to the first block of text correctly received?

4. What response should be given if the BCC received is bad?

5. WACK is neither a positive nor a negative acknowledgment of a BCC.   (T, F)

6. When required RVI can be sent in place of an ACK 0 or an ACK 1.   (T, F)

7. In the course of a dialogue, the RVI is sent by the transmitting station to the receiving station.   (T, F)

8. When does the primary send an RVI to the secondary?

9. When a station receives an ITB followed by a BCC, it must immediately respond with a ACK 0, ACK 1, WACK, NAK, or an RVI.   (T, F)

10. What response is given to a selection to indicate "ready to receive"?
11. Given the following sequence, draw brackets around all characters that would be involved in the computation of the BCC.

```
S S S                     S                     E B P
Y Y O.....................T.....................T C A
N N H                     X                     X C D
```

12. Given the following sequence:

```
S S S                     I E
Y Y T.....................T T .....................
N N X                     B X
```

   (a) What does ITB indicate?
   (b) What is the purpose of ETX?
   (c) What response is immediately given as a response to the correct reception of this sequence?

13. Given the following sequence:

```
S E     D                               E B P
          G   S         G S S G
T S 5 9 C K L   Z   S E X         O T C A
          S   P         S P P S
X C     1                               X C D
```

   (a) Where is the cursor after ESC 5 is executed:
   (b) Identify all of the following that have been accomplished by the write control character: (1) start printer, (2) sound KD alarm; (3) KD in LOCAL; (4) ACs reset to unmodified.
   (c) How many characters can you have on one line?
   (d) In which row and column will the S of SEX be stored?
   (e) Can the terminal operator enter a character in the space immediately before SEX?
   (f) How many spaces are allowed the operator to enter the type of sex?
   (g) On this form, the operator could enter information into (*none*, *1*, *2*, *3*, *more than 3*) locations.
   (h) Assuming that the operator only enters M (for male) properly, since the only visible data transmitted to the CPU are modified data, what portions of the completed form will be sent back?
   (i) In which row and column is the attribute character for the SEX field stored?

14. NULLS and SPACES appear the same way on the display. (T, F)
15. Which key will the operator have to depress in response to a Read All command?
16. A field in which the operator cannot enter information is called __________ .
17. What is the name for the information that the operator enters?
18. What three characters identify a Status and Sense message?
19. Transparent text is initiated by __________ .
20. If a field is identified as protected, neither the terminal operator nor the transmitter of the form can enter information into that field. (T, F)

**21.** A handshake is given by (*the secondary only, the primary only, either the primary or the secondary*).

**22.** All Status and Sense messages can be sent only in response to a specific poll.   (T, F)

**23.** In a copy command, the device address supplied is the address of the device that will be copied from.   (T, F)

**24.** What keys are present on an asynchronous keyboard that are not present on a synchronous keyboard?

**25.** What problems would be encountered if selections were not specific?

**26.** What is the difference between a poll and a selection?

# PROBLEMS

**1.** Give the ASCII code in hex for ITB.

**2.** Write out the entire ASCII *character* sequence for a general poll of station 11.

**3.** Write out the entire ASCII character response by station 11 if no device at that station had any messages to send and no PF or PA keys had been depressed.

**4.** Given:

```
S  S  E  P  S  S              E  P
Y  Y  O  A  Y  Y  $  $  J  J  N  A
N  N  T  D  N  N              Q  D
```

**(a)** The sequence above is a (*general poll, specific poll, selection*).
**(b)** Which station is involved in this message?
**(c)** Which device (if any) at this station is involved in this message?

**5.** Write out the entire ASCII character sequence from the primary to the secondary to send the following form:

```
Row   7    PART (10)
Row  10    QUANTITY (4)
PART and QUANTITY will start in column 10.
```

(1) Clear the screen, sound the KD bell, reset ACs to unmodified, and use 80 characters/ line.
(2) The form is protected.
(3) Operator-entered data will be intensified.
(4) Use autoskip and numeric where applicable.
(5) Insert cursor at the first location where the operator can enter data.

**6.** Write out the entire ASCII character sequence from the primary to the secondary to send the following form:

```
Row    NAME S (16 spaces) S  S/S  S 11 spaces)    (these 11 spaces include two -'s)
             P             P       P
```

(1) The N of NAME will start in column 7.
(2) Alphanumeric information can be entered into the spaces after NAME.
(3) Only numbers will be entered after S/S.
(4) Command: 64-character line, put KD in LOCAL, reset ACs to unmodified, ring alarm bell.
(5) Use auto skip where applicable.
(6) Operator-entered data: intensified and modified.
(7) Insert cursor for the operator at the first location where operator can enter information.
(8) Everything else on the form should be protected.

# TWELVE

## *SYNCHRONOUS DATA LINK CONTROL AND HIGH LEVEL DATA LINK CONTROL*

### SYNCHRONOUS DATA LINK CONTROL

SDLC is a *bit-oriented protocol*. This means that transmission acknowledgments, polls, and other data link controls are accomplished by a single bit or a group of bits. In contrast, Bisynch is a character-oriented protocol which accomplishes these functions with a data word. With SDLC, text is transmitted in EBCDIC. Bit 7 is always understood to be the least significant bit and bit 0 is the most significant bit. This convention will be adhered to in this text. As shown in Figure 12–1, bit 7 is always transmitted first.

$$b_0 \dots\dots\dots\dots b_7 \longrightarrow$$

| MSB | LSB |
|---|---|
| High-order bit | Low-order bit |

**FIGURE 12–1   EBCDIC transmission sequence.**

### SDLC FRAME

Information is transmitted on a SDLC data link in units called *frames*. A frame always starts and ends with a flag. The bit sequence for a flag is 01111110 = 7EH. The frame format and the number of bits in each field are shown in Figure 12–2.

There are several variations of how a terminating flag of one frame and the starting flag of the next frame may be transmitted. These variations are shown in Figure 12–3.

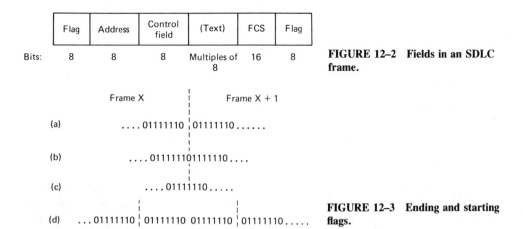

Bits:  8     8     8     Multiples of   16    8

**FIGURE 12–2  Fields in an SDLC frame.**

**FIGURE 12–3  Ending and starting flags.**

(a) Two totally separate flags.
(b) The last 0 of the terminating flag is also the first 0 of the starting flag.
(c) The terminating flag of one frame is also the starting flag of the next frame.
(d) Additional flags inserted between the terminating flag of one frame and the starting flag of the next frame.

The transmitted address is always the address of the secondary. If the primary is transmitting, the address would mean: "This message is for. . . ." If the secondary is transmitting, the address would mean: "This message is from. . . ." The primary does not have an address. Secondaries may have individual, group, and/or broadcast addresses. A *group address* is an address common to two or more secondaries in the data link but not to all. A *broadcast address* is common to all secondaries. FFH is used as the broadcast address. If the primary sent information addressed to FFH, all secondary stations in the data link would receive the information. The address 00H is not assigned to any station. This address is called a *void address* and may be used by the primary to troubleshoot the system.

The *frame check sequence* (FCS) is computed using CRC-16 (Chapter 4). The CRC is computed on the bit stream following the starting flag to and including the text if present (Figure 12–4). CRC-16 generates 16 bits for error detection.

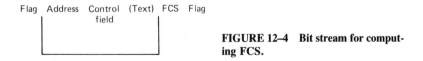

**FIGURE 12–4  Bit stream for computing FCS.**

The control field has three formats: information, supervisory, and unnumbered (unsequenced). The type of format used is determined by the two least significant bits of this field:

```
b₆   b₇
X    0    Information (text always present)
0    1    Supervisory (text never present)
1    1    Unnumbered (text may be present)
```

## Information Format

Figure 12–5 shows the bit pattern of an information control field. Bit 3 is the *Poll/Final* (P/F) *bit*. If the transmission is from the primary to the secondary, this bit is a poll bit. If P = 0, the frame is not a poll and the secondary is not invited to answer. If P = 1, the frame is a polling frame and the secondary is required to answer after it receives this frame. If the transmission is from the secondary to the primary, this bit is a final bit. If F = 0, the frame sent by the secondary is not the last frame and the primary will not respond. If F = 1, the frame sent is the last frame that the secondary wishes to transmit in this sequence and the primary will respond appropriately.

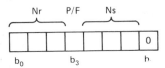

**FIGURE 12–5   Information control field.**

The Nr (number received) transmitted by the primary is related to the Ns (number sent) transmitted by the secondary; the Nr transmitted by the secondary is related to the Ns transmitted by the primary. If information format is used, text is sent in numbered frames. The sender identifies which frame is being sent by sending Ns, a 3-bit binary number. If Ns = 010, frame 2 is being sent. Whoever receives the information transmits the Nr. Nr identifies the number of the next frame expected. If the secondary transmitted an Ns = 2 and received an Nr = 3, the primary is acknowledging successful reception through frames Nr − 1 and is expecting to receive frame 3 next.

Since each frame has an FCS, does each frame have to be acknowledged immediately? No. A maximum of seven frames can be transmitted without requesting confirmation. Confirmation is requested by making the P/F bit a 1. Acknowledgment (Nr) can be made either with an information frame if the receiving end has information to transmit or with a supervisory frame if there is no text to send.

## Supervisory Format

The supervisory control field (Figure 12–6) is identified by 01 in its two least significant bits. The secondary clears a RNR by sending an information frame with F = 1 or

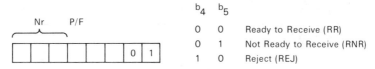

FIGURE 12–6  Supervisory control field.

a RR or REJ frame with F = 0 or 1. The primary clears a RNR by sending an information frame with P = 1 or a RR or REJ frame with P = 0 or 1.

---

**EXAMPLE**

Assume that the primary is transmitting to the secondary and all Nr and Ns counters are initially zeroed.

| Primary | | | Secondary | | | |
|---------|---|---|-----------|---|---|---|
| Nr | P | Ns | Nr | F | Ns | |
| 000 | 0 | 000 0 | | | | Primary transmits three information frames. |
| 000 | 0 | 001 0 | | | | The third frame is a poll. |
| 000 | 1 | 010 0 | | | | |
| | | | 011 | 0 | 000 0 | Secondary acknowledges with Nr = 3. |
| | | | 011 | 1 | 001 0 | It sends two information frames back. |
| 010 | 0 | 011 0 | | | | Primary acknowledges errorless reception of two frames with |
| 010 | 0 | 100 0 | | | | Nr = 2. Send six information frames. |
| 010 | 0 | 101 0 | | | | |
| 010 | 0 | 110 0 | | | | |
| 010 | 0 | 111 0 | | | | |
| 010 | 1 | 000 0 | | | | |
| | | | 110 | 0 | 010 0 | Since Nr = 6, the secondary acknowledges all frames through |
| | | | 110 | 0 | 011 0 | Nr − 1 = 5 and requests retransmission of frame 6. Since |
| | | | 110 | 1 | 100 0 | the primary does not know whether only frame 6 had an |
| | | | | | | FCS error or whether any subsequent frames also had an |
| | | | | | | error, the primary must retransmit everything from the sixth |
| | | | | | | frame on. |
| 101 | 0 | 110 0 | | | | Primary acknowledges secondary's frame 4 with Nr = 5. |
| 101 | 0 | 111 0 | | | | Continues to transmit its text starting with frame 6. |
| 101 | 0 | 000 0 | | | | Note the counter rollover. |
| 101 | 1 | 001 0 | | | | |
| | | | 010 | 1 | 00 01 | Secondary confirms all frames with Nr = 2. |
| | | | | | | Since the secondary has no text to send, supervisory format |
| | | | | | | is used. |
| 101 | 0 | 010 0 | | | | Primary sends five additional frames. |
| 101 | 0 | 011 0 | | | | |
| 101 | 0 | 100 0 | | | | |
| 101 | 0 | 101 0 | | | | |
| 101 | 1 | 110 0 | | | | |
| | | | 101 | 1 | 00 01 | Secondary acknowledges through frame 4 and requests |
| | | | | (RR) | | retransmission of frame 5 and what follows. Since the Nr |
| | | | | or | | indicates this, note the redundancy produced by bits 4 and |
| | | | 101 | 1 | 10 01 | 5 of the reject format. |
| | | | | (REJ) | | |

**TABLE 12–1**

| Transmitted Ns | |
|---|---|
| 000 | If this many frames were transmitted before requesting acknowledgment, |
| 001 | and an Nr = 0 was returned, there would be confusion as to whether |
| 010 | the first frame had an FCS error and everything had to be retransmitted |
| 011 | or if everything was received without error and the next expected |
| 100 | frame number was 000. |
| 101 | |
| 110 | |
| 111 | |

If more than seven frames are sent before requesting acknowledgment, ambiguity can be introduced (Table 12–1).

## Unnumbered Format

If the two least significant bits of the control field are both 1's, an unnumbered format is being used (Table 12–2). As the name implies, frames sent with the control field in this format are not numbered. The control field contains neither Nr nor Ns. If Ns is not present, these frames cannot be acknowledged with an Nr.

**TABLE 12–2**

| Binary Configuration | | | Acronym[a] | Sent by the Primary | Sent by the Secondary | Text Permitted |
|---|---|---|---|---|---|---|
| $b_0$ | | $b_7$ | | | | |
| 000 | P/F | 0011 | UI | X | X | X |
| 000 | P | 0111 | SIM | X | | |
| 000 | F | 0111 | RIM | | X | |
| 100 | P | 0011 | SNRM | X | | |
| 000 | F | 1111 | DM | | X | |
| 010 | F | 0011 | RD | | X | |
| 010 | P | 0011 | DISC | X | | |
| 011 | P/F | 0011 | UA | X | X | |
| 100 | F | 0111 | FRMR | | X | X |

[a] Obsolete acronyms and their meaning are shown in parentheses:

| | |
|---|---|
| UI | Unnumbered information (NSI—Nonsequenced Information) |
| SIM | Set Initialization Mode |
| RIM | Request Initialization (RQI—Request Initialization) |
| SNRM | Set Normal Response Mode |
| DM | Disconnect Mode (ROL—Request On-Line) |
| RD | Request Disconnect (RQD—Request Disconnect) |
| DISC | Disconnect |
| UA | Unnumbered Acknowledgment (NSA—Nonsequenced Acknowledgment) |
| FRMR | Frame Reject (CMDR—Command Reject) |

Similarly, since these control fields do not contain an Nr, this type of format cannot be used to acknowledge numbered frames (frames containing an Ns). Only those control characters used in a multipoint data link will be discussed.

RIM is transmitted by the secondary to notify the primary of the need for a SIM command. A SIM command sent to the secondary would initiate system-specified procedures for the purpose of initializing link-level functions. The primary and secondary Nr and Ns are reset by this command. The secondary is expected to answer with UA. After initialization, the secondary is either in the normal response mode or the disconnect mode. If it is in the *disconnect mode*, the secondary can transmit and receive only unnumbered commands. In this mode, the secondary is effectively off-line. In the *normal response mode*, the secondary is capable of transmitting and receiving information, supervisory, and unnumbered frames. The secondary cannot go into the normal response mode or the disconnect mode on its own but must be placed in that mode by the primary. Assume initially that the secondary is already in the disconnect mode. Assume also that the primary sends an information frame to the secondary. The secondary does receive it but cannot react to it. The secondary would reply with the unnumbered frame of DM. In effect, it is telling the primary: "I am in the disconnect mode and I cannot receive information frames." The primary would then transmit a SNRM frame which would place the secondary in the normal response mode. When the primary receives a UA from the secondary, it could then retransmit the information frame(s). If the secondary wished to return to the disconnect mode after receiving the primary's message, it would return a supervisory frame to provide the primary with a numbered acknowledgment with the F bit a 0. The following frame would be an unnumbered frame—RD. The primary would respond with DISC, which would return the secondary to the disconnect mode. The secondary would be expected to return a UA. SNRM also resets Nr and Ns of the primary and the secondary.

To transmit FRMR, the secondary must be in the normal response mode. The FRMR indicates to the primary that the previously received frame by the secondary is invalid. Reasons why a frame may be invalid:

1. The control field cannot be implemented by the secondary.

---

**EXAMPLE**

Secondary receives a control field of

$$011 \ P \ 11 \ 11 \quad (\text{LSB on right})$$

The two LSBs identify this as an unnumbered frame, but there is no command that corresponds to the remaining bits.

---

2. The information field is too long to fit into the receiving station's buffers.
3. The received control field does not allow an information field in that frame.

**EXAMPLE**

The secondary receives a supervisory control field. Since no text is permitted with this field, the secondary expects the next two bytes to be the FCS followed by a flag. If a flag is not detected until after 20 bytes have been received following the control field, it presumes that a text field has been included.

4. The Nr received from the primary is invalid.

**EXAMPLE**

Secondary transmits three numbered frames: Ns = 000, 001, and 010. In response, it receives an Nr = 011, implying that all frames were received correctly. The secondary sends four more frames: Ns = 011, 100, 101, and 110. In reply, it now receives an Nr of 001. Only an Nr of 011, 100, 101, 110, or 111 would make any sense. The secondary sends back a FRMR.

The text field allowed with FRMR is fixed and is used to identify the reason why the received frame was invalid. The format of a Frame Reject frame is shown in Figure 12–7.

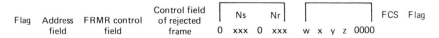

**FIGURE 12–7    Frame Reject frame format.**

The two bracketed bytes are shown with the order of the arrow to the left (leftmost bit is transmitted first). Ns and Nr are the current values of these two counters at the secondary.

w = 1     if the secondary received a command which is invalid or if it cannot be implemented

x = 1     if a prohibited information field is received

y = 1     if there was a buffer overrun (information field too long)

z = 1     if the received Nr disagrees with the transmitted Ns

The four o's following z are simply stuffed to complete this byte.

Once the secondary transmits a FRMR, it cannot release itself from this condition, nor will it act on the frame that caused this condition. It will repeat the FRMR whenever it responds except to an acceptable mode-setting command such as DISC, SIM, or SNRM which resets the FRMR condition. Table 12–3 shows a sample dialogue in a multipoint, full-full-duplex system (Figure 12–8). Station A is initially in NRM and station B is initially in DM.

**TABLE 12–3**

| | |
|---|---|
| Flag A  RR FCS Flag<br>7E   C1   11       7E | Primary polls Station A |
| Flag A    Info (Text) FCS Flag<br>7E   C1   00               7E | Station A sends first information frame;<br>Nr = Ns = 0, F = 0 |
| Flag A    Info (Text) FCS Flag<br>7E   C1   02               7E | Station A sends second information frame;<br>Nr = F = 0, Ns = 1 |
| Flag A    Info (Text) FCS Flag<br>7E   C1   14               7E | Station A sends third information frame,<br>Nr = 0, F = 1, Ns = 2 |
| Flag A    Info (Text) FCS Flag<br>7E   C1   60               7E | Primary sends text reply; Nr = 3,<br>P = 0, Ns = 0 |
| Flag A    Info (Text) FCS Flag<br>7E   C1   72               7E | Primary sends second information frame;<br>Nr = 3, P = 1, Ns = 1 |
| Flag A    REJ FCS Flag<br>7E   C1   39     7E | Station A had an FCS error with second<br>frame; Nr = 1, F = 1 |
| Flag A    Info (Text) FCS Flag<br>7E   C1   72               7E | Primary retransmits second frame; Nr =<br>3, P = 1, Ns = 1 |
| Flag A    RR FCS Flag<br>7E   C1   51     7E | Station A acknowledges the second frame;<br>Nr = 2, F = 1 |
| Flag B    Info (Text) FCS Flag<br>7E   C2   10               7E | Primary sends information frame to<br>Station B; Nr = Ns = 0, P = 1 |
| Flag B    DM FCS Flag<br>7E   C2   1F     7E | Station B indicates that it is in DM |
| Flag B    SNRM FCS Flag<br>7E   C2   93       7E | Primary places station B in NRM |
| Flag B    UA FCS Flag<br>7E   C2   73     7E | Station B returns an unnumbered<br>acknowledgment |
| Flag B    Info (Text) FCS Flag<br>7E   C2   00               7E | Primary sends station B the first<br>information frame; Ns = Nr = 0,<br>P = 0 |
| Flag B    Info (Text) FCS Flag<br>7E   C2   02               7E | Primary sends station B the second<br>information frame; Ns = 1, P = 0,<br>Nr = 0 |
| Flag B    Info (Text) FCS Flag<br>7E   C2   04               7E | Primary sends station B the third<br>information frame; Ns = 2, P = 0,<br>Nr = 0 |
| Flag B    Info (Text) FCS Flag<br>7E   C2   16               7E | Primary sends station B the fourth<br>information frame; Ns = 3, P = 1,<br>Nr = 0 |
| Flag B    RR FCS Flag<br>7E   C2   91     7E | Station B acknowledges all frames;<br>Nr = 4, F = 1 |
| Flag B    DISC (Text) FCS Flag<br>7E   C2   53               7E | Primary sends a disconnect command to<br>station B; erroneously sends text along<br>in the same frame |
| Flag B    FRMR DISC — — FCS Flag<br>7E   C2   97     53   80 02        7E | Station B replies with a frame reject; DISC<br>is the control field of the rejected<br>command<br>80 Nr = 4, Ns = 0: current counter of<br>station B<br>02: x = 1: text not allowed |
| Flag B    DISC FCS Flag<br>7E   C2   53       7E | Primary sends disconnect command to<br>station B |
| Flag B    UA FCS Flag<br>7E   C2   73       7E | Station B replies with an unnumbered<br>acknowledgment |

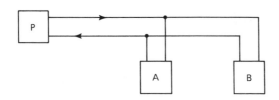

FIGURE 12–8    Multipoint F/FDX system.

## Zero Stuffing (Zero-Bit Insertion)

The flag bit pattern may occur in a frame where this pattern is not intended to be interpreted as a flag.

**EXAMPLES**

1. The transmission of an "=" as part of the text. The EBCDIC for "=" is 01111110.
2. A control field using information format where $N_r = 3$, $P/F = 1$, and $N_s = 7$. This control field would also yield a bit pattern of 01111110.

Anytime this pattern is received, it is interpreted as a flag. Therefore, the receiver would interpret the frame to be ending somewhere in the middle of the actual message. To eliminate this ambiguity, a zero is inserted automatically after five consecutive ones found between the starting and ending flags of a frame. This process is clearer if the sequence of events is understood:

1. Message is composed:

Control        (Text)
Field

2. The station controller inserts the station address:

Address        Control        (Text)
Field

3. The FCS is computed:

Address        Control        (Text)        FCS
Field

4. Zeros are now stuffed after any five contiguous ones found anywhere from the address to the FCS inclusive.
5. Starting and ending flags are now added before transmission.

The stuffed 0's are not included in the computation of the FCS since they are added after the FCS has been computed. In fact, the FCS may have 0's stuffed into it. This process prevents the occurrence of a flag pattern anywhere between the starting and ending flags.

At the receive terminal, the process is reversed. Once the starting flag is identified, any zero following five successive ones is stripped. Six successive ones followed by a zero will be interpreted as the terminating flag.

---

**EXAMPLE**

Station B transmits "C?" as the information field.
"o" is used to indicate a stuffed zero.
Order of the arrow is to the left.

```
01111110 01000011 011111o10 11111o110 11000011 111o10110
 Flag    Add. = B Cont. Fld  " = 7F    C = C3    ? = 6F

         11111o110 FCS 01111110 1111111111111...
            " = 7F         Flag    Idle Line 1's
```

---

**Message abort.**    Anytime after the starting flag is received, the reception of *seven to 14 consecutive 1's* indicates an abort condition. Zeros are not stuffed into an abort signal. An abort simply tells the receiving station to disregard everything sent thus far in the current frame. Clock synchronization is still maintained by the receiver during an abort condition. An abort condition is used to terminate a frame and immediately begin another frame of higher priority.

**Idle line state.**    If the data link is operational but there is no SDLC control or information transmission currently in progress, the idle line state exists. A station perceives the existence of an idle link when it receives *15 or more consecutive 1's.*

## Comparison of Bisynch and SDLC

Bisynch is a character-oriented protocol, whereas SDLC is a bit-oriented protocol. ASCII or EBCDIC can be used with Bisynch, whereas EBCDIC is always used with SDLC. Both Bisynch and SDLC are synchronous protocols. For error detection, ASCII uses a block check sequence derived from Longitudinal Redundancy Checking (LRC) which consists of one byte; EBCDIC uses a block or frame check sequence derived from Cyclic Redundancy Checking (CRC) which consists of two bytes. When operating in a multipoint environment, approximately 256 bytes are transmitted per frame/block for both protocols. Since Bisynch is a character-oriented protocol, it has a greater difficulty handling the occurrence of data bytes that have the bit patterns of control characters within the text. Bisynch must resort to the transparent text mode. SDLC does not have similar problems. The occurrence of flag patterns within a frame is the only thing that would present a misunderstanding, and this is remedied by the process of zero bit stuffing.

## Invert-on-Zero Coding

For the receiver to maintain bit synchronization, there must be frequent variations in the levels of the received signal—no prolonged receptions of constant 1's or constant 0's. Zero bit stuffing eliminates the possibility of receiving more than six consecutive 1's as part of any unaborted frame. The possibility of receiving a long string of 0's as part of a frame still exists. Data terminal may possess the option of non-return-to-zero inverted (NRZI) encoding of the data bits of a frame before these bits are applied to the RS 232C interface. With this form of encoding, there is no change in the signal level to transmit a 1; to transmit a 0, the signal level is always inverted from what it was. An example of NRZI encoding is shown in Figure 12–9. The signal level is assumed high before the reception of the first one of the actual bit stream.

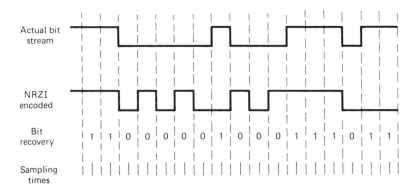

FIGURE 12–9    NRZI encoding.

## SDLC Loop Operation

An SDLC *loop* is operated in the half-duplex mode. The primary difference between the loop and bus configurations is that in a loop, all transmissions travel in the same direction on the communications channel. In a loop configuration, only one station transmits at a time. The primary transmits first, then each secondary station responds sequentially. In an SDLC loop, the transmit port of the primary station controller is connected to one or more secondary stations in a serial fashion; then the loop is terminated back at the receive port of the primary. Figure 12–10 shows an SDLC loop configuration.

In an SDLC loop, the primary transmits frames that are addressed to any or all of the secondary stations. Each frame transmitted by the primary contains an

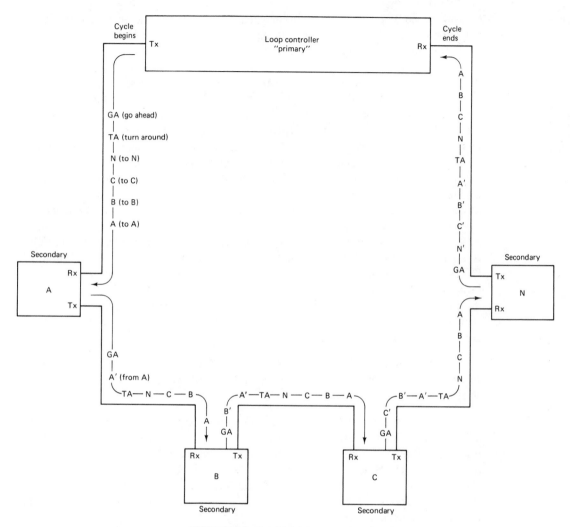

**FIGURE 12–10  SDLC loop configuration.**

address of the secondary station to which that frame is directed. Each secondary station, in turn, decodes the address field of every frame, then serves as a repeater for all stations that are down-loop from it. If a secondary detects a frame with its address, it accepts the frame, then passes it on to the next down-loop station. All frames transmitted by the primary are returned to the primary. When the primary has completed transmitting, it follows the last flag with eight consecutive 0's. A flag followed by eight consecutive 0's is called a *turnaround* sequence which signals the end of the primary's transmission. Immediately following the turnaround sequence, the primary transmits continuous 1's, which generates a *go-ahead* sequence (01111111). A secondary cannot transmit until it has received a frame addressed to it with the P bit set, a turnaround sequence, and then a go-ahead sequence. Once the primary has begun transmitting 1's, it goes into the receive mode.

The first down-loop secondary station that has received a frame addressed to it with the P bit set, changes the seventh 1 bit in the go-ahead sequence to a 0, thus creating a flag. That flag becomes the beginning flag of the secondary's response frame or frames. After the secondary has transmitted its last frame, it again becomes a repeater for the idle line 1's from the primary. These idle line 1's again become the go-ahead sequence for the next secondary station. The next down-loop station that has received a frame addressed to it with the P bit set detects the turnaround sequence, any frames transmitted from up-loop secondaries, and then the go-ahead sequence. Each secondary station inserts its response frames immediately after the last repeated frame. The cycle is completed when the primary receives its own turnaround sequence, a series of response frames, and then the go-ahead sequence.

**Configure command/response.**     The configure command/response (CFGR) is an unnumbered command/response that is used only in a loop configuration. CFGR contains a one-byte *function descriptor* (essentially a subcommand) in the information field. A CFGR command is acknowledged with a CFGR response. If the low-order bit of the function descriptor is set, a specified function is initiated. If it is reset, the specified function is cleared. There are six subcommands that can appear in the configure command's function field.

1. *Clear—00000000.* A clear subcommand causes all previously set functions to be cleared by the secondary. The secondary's response to a clear subcommand is another clear subcommand, 00000000.

2. *Beacon test (BCN)—0000000X.* The beacon test causes the secondary receiving it to turn on or turn off its carrier. If the X bit is set, the secondary suppresses transmission of the carrier. If the X bit is reset, the secondary resumes transmission of the carrier. The beacon test is used to isolate an open-loop problem. Also, whenever a secondary detects the loss of a receive carrier, it automatically begins to transmit its beacon response. The secondary will continue transmitting the beacon until the loop resumes normal status.

3. *Monitor mode—0000010X.* The monitor command causes the addressed secondary to place itself into a monitor (receive only) mode. Once in the monitor mode, a secondary cannot transmit until it receives a monitor mode clear (00000100) or a clear (00000000) subcommand.

4. *Wrap—0000100X.* The wrap command causes the secondary station to loop its transmissions directly to its receiver input. The wrap command places the secondary effectively off-line for the duration of the test. A secondary station does not send the results of a wrap test to the primary.

5. *Self-test—0000101X.* The self-test subcommand causes the addressed secondary to initiate a series of internal diagnostic tests. When the tests are completed,

the secondary will respond. If the P bit in the configure command is set, the secondary will respond following completion of the self-test at its earliest opportunity. If the P bit is reset, the secondary will respond following completion of the test to the next poll-type frame it receives. All other transmissions are ignored by the secondary while it is performing the self-tests. The secondary indicates the results of the self-test by setting or resetting the low-order bit (X) of its self-test response. A 1 indicates that the tests were unsuccessful, and a 0 indicates that they were successful.

6. *Modified link test—0000110X*. If the modified link test function is set (X bit set), the secondary station will respond to a TEST command with a TEST response that has an information field containing the first byte of the TEST command information field repeated *n* times. The number *n* is system implementation dependent. If the X bit is reset, the secondary station will respond to a TEST command, with or without an information field, with a TEST response with a zero-length information field. The modified link test is an optional subcommand and is only used to provide an alternative form of link test to that previously described for the TEST command.

# HIGH LEVEL DATA LINK CONTROL

In 1975, the International Standards Organization (ISO) defined a set of standards known as High Level Data Link Control. HDLC is a superset of IBM's SDLC. HDLC is divided into smaller subdivisions that, when combined, outline the frame structure, control standards, and class of operation of a bit-oriented Data Link Control (DLC). Since HDLC is a superset of SDLC, only the added capabilities as prescribed by the indicated ISO standards will be described.

## ISO 3309–1976(E)

This standard defines the frame structure, delimiting sequence, and transparency mechanism employed by HDLC. These are all identical to SDLC except that HDLC has extended address capability and the FCS is checked differently. In HDLC, the address field may be extended recursively. A logic 1 for $b_0$ identifies that byte as the last byte of the address field: a logic 0 indicates that the byte immediately following is part of an extended address. (*Note*: The ISO standard identifies the low-order bit as $b_0$. In SLDC, the low order bit is designated $b_7$). Basically, 7 bits are available for address encoding within each address byte. A given system cannot use both basic and exended addressing. Figure 12–11 shows an example of an HDLC frame using extended addressing with three address bytes.

HDLC uses a CRC, generated by a polynomial specified by CCITT V.41, as the FCS. The transmitting station computes a 16-bit CRC and transmits it just prior to the ending flag. The receive station computes the CRC in the same manner as SDLC except that the computation includes the received CRC. If the transmission is without error, the CRC computed at the receiver is always F0BBH.

| 01111110 | 0XXXXXXX 0XXXXXXX 1XXXXXXX | 01111110 |
|:---:|:---:|:---:|
| Beginning Flag | Three Byte Address Field | Ending Flag |

**FIGURE 12–11  HDLC frame using extended addressing.**

## ISO 4335–1979(E)

This standard defines the elements of procedures for HDLC. The control field, the information format, and the supervisory format have increased capability over SDLC. HDLC also has two additional operational modes.

**Operational modes.**  The three operational modes of SDLC—initialization mode, normal response mode, and disconnect mode—are described in SDLC under unnumbered formats. HDLC has the two following additional operational modes:

1. *Asynchronous Response Mode (ARM)*. In this mode, secondary stations are allowed to send unsolicited response frames. For a secondary station to transmit, it need not have received a frame with the P bit set. However, if a secondary receives a command with the P bit set, it must reply with a frame with the F bit set. Additional frames may be transmitted after the response to poll has been sent. Final indicates only a response to a poll and is not an indication of the last frame of a message.
2. *Asynchronous Disconnect Mode (ADM)*. ADM is identical to NDM except that a secondary may initiate a DM or RIM at any time.

*Control field*. HDLC has an extended mode capability. In the extended mode, the control field is increased to 16 bits. Seven bits are assigned to the Ns and 7 bits are assigned to the Nr. In the extended mode, there may be as many as 127 unacknowledged frames outstanding at any one time.

*Information format*. EBCDIC is the only code used with SDLC. The information field, prior to zero substitution, must have a multiple of 8-bit characters. HDLC is not byte oriented: a character or word within the information field of an HDLC frame may have any number of bits as long as all of the character and word lengths are the same. Consequently, the coding used with HDLC is less restricted than with SDLC.

*Supervisory format*. The supervisory format for HDLC, like SDLC, includes ready to receive (RR), ready not to receive (RNR), and reject (REJ). However, HDLC includes, in addition, a selective reject (SREJ). A selective reject calls for retransmission of only the frame identified by Nr. Only one SREJ may be established by a station at a time. Additional SREJs may be established after the first SREJ has been cleared by receipt of an I-frame with an Ns equal to the SREJ's Nr.

## ISO 6159–1980(E) (unbalanced) and ISO 6256–1981(E) (balanced)

ISO is currently merging these two standards into ISO 7809. These two standards identify the class of operations as either balanced or unbalanced. The defined class of operation identifies the procedures necessary to establish the link-level protocol.

*Unbalanced operation.* Circuit configuration consists of a primary station and one or more secondary stations. The data link operation may be half-duplex or full-duplex. Channel access is through polling using the NRM.

*Balanced operation.* Circuit configuration consists of only two stations. Each station can set up or disconnect the link. Channel access is through contention using ARM. Data link operation may be half-duplex or full-duplex.

## PROTOCOL CHIPS

### Intel's 8273 Programmable HDLC/SDLC Protocol Controller

*Characteristics*

1. May be used as either a primary or a secondary controller.
2. Compatible with the 8048, 8085, 8088, 8086, 80188, and 80186 microprocessors.
3. Capable of half duplex, full duplex, or loop operations.
4. Buffered operation: The address and the control fields are first loaded into the address and control buffers on this chip. During transmission, this chip provides the leading flag, transmits the address and control fields, and by way of DMA (direct memory access), reads memory and outputs the information field. It then adds the computed FCS and the trailing flag. It also accomplishes zero bit insertion. On reception, this chip will store the address and control fields in its internal buffer registers, check the FCS, strip the zero bits inserted for transparency, and store the information field in memory.
5. Nonbuffered operation. For transmission, same as buffered operation except that the address and control fields are stored in memory. This chip transmits the starting flag, fetches and transmits the address, control, and information fields from memory, and appends the FCS and the trailing flag. On reception, the address, control, and information fields are sent to memory.
6. Possesses a programmable NRZI encode/decode feature.
7. Can recognize and generate abort and idle characters.

## Intel's 8274 Multi-Protocol Serial Controller (MPSC)

*Characteristics*

1. Two independent full-duplex transmitters and receivers.
2. Fully compatible with the 8048, 8051, 8085, 8088, 8086, 80188, and 80186 CPUs.
3. Asynchronous operation:
    (a) 5 to 8 bit characters; odd, even or no parity; 1, 1.5, or 2 stop bits.
    (b) Error detection: framing, overrun, and parity.
4. Byte synchronous operation:
    (a) One or two sync characters.
    (b) Automatic CRC generation and checking.
    (c) IBM Bisync compatible.
5. Bit synchronous operation:
    (a) SDLC/HDLC flag generation/checking.
    (b) 8-bit address recognition.
    (c) Automatic zero bit insertion/deletion.
    (d) Automatic CRC generation/checking.

## Zilog's Z8440 Serial Input/Ouput Controller

This chip is designed for compatibility with the Z80 microprocessor and possesses characteristics very much similar to Intel's 8274 chip.

## Motorola's MC68564 Serial Input/Output and MC2652/MC68652 Multi-Protocol Communications Controller

These chips readily interface with the MC6800 and the 68000 family of microprocessors. Again, their general characteristics are similar to Intel's 8274 chip, although the MC68564 and the MC68562 are capable of handling a higher bit rate.

# QUESTIONS

1. SDLC is a _____ -oriented protocol.
2. Identify the code used with SDLC.
3. Which bit is the low-order bit (bit position transmitted first) in EBCDIC?
4. What are the hex characters for a flag?
5. Excluding flags, identify the four fields that comprise an SDLC frame.
6. In SDLC, the transmitted address is always the address of what station?

   **7.** An address common to all secondary stations is called a _____ address. What is the hex sequence for this address?

   **8.** An address common to more than one but not all of the secondary stations is called a _____ address.

   **9.** Identify the hex address that is never assigned to a secondary station.

  **10.** Identify the three control field formats that are used with SDLC.

  **11.** Which of the formats in Question 10 allows the transmission of text information?

  **12.** The P/F bit is included with all three of the control field formats.   (T, F)

  **13.** All frames containing information are numbered.   (T, F)

  **14.** Which control field format is used to acknowledge the reception of numbered frames, but does not allow the transmission of an information field?

  **15.** What is the purpose of the Nr bits?

  **16.** What is the purpose of the Ns bits?

  **17.** An information frame is identified by making bit position _____ of the _____ field a logic _____ .

  **18.** What is the maximum number of information frames that can be sent before an acknowledgment is required with SDLC?

  **19.** What are the low-order bits for unnumbered control characters?

  **20.** What follows the control character in a supervisory frame?

  **21.** What is the secondary's response to an invalid command from the primary?

  **22.** What is the secondary's response to a SNRM command from the primary?

  **23.** Supervisory control characters cannot be used to acknowledge previously received numbered frames.   (T, F)

  **24.** To distinguish a flag from an EBCDIC = sign, a process of _____ is used.

  **25.** The frame check sequence is made up of 8 bits.   (T, F)

  **26.** Which unnumbered commands reset the Nr and the Ns counters to all 0's?

  **27.** Which response indicates that the secondary station is in the disconnect mode?

  **28.** When in the disconnect mode, the secondary cannot transmit either of the three frame formats.   (T, F)

  **29.** What is the purpose of NRZI encoding?

  **30.** Flags are zero-inserted.   (T, F)

  **31.** What is the maximum number of outstanding (unacknowledged) frames allowed with HDLC in the basic control mode? In the extended control mode?

  **33.** What is the delimiting sequence used with HDLC?

  **33.** What supervisory condition is included with HDLC that is not included in SDLC?

  **34.** EBCDIC is the only code allowed in HDLC.   (T, F)

  **35.** How does HDLC achieve transparency?

  **36.** Which bit in the address field of an HDLC frame indicates that additional address bytes follow when in the extended addressing mode?

  **37.** In what mode is an SDLC loop operated?

  **38.** What is a turnaround sequence?

  **39.** What is a go-ahead sequence?

**40.** Briefly describe the operation of an SDLC loop.

**41.** Describe how the configure command/response is used.

# PROBLEMS

**1.** 1 1 1 1 1 1 1 0 1 1 1 1 1 1 0 1 1 1 1 1 1 0 1 1 1 0 0 0 0 0 0 0 0 1 1 1 1 1 0 1 0 1 1 0 0 1 1 1 0 1 0 1 1 0 0 1 1 0 1 1 1 1 1 0 0
0 1 0 0 1 1 1 1 1 0 0 1 0 1 1 0 0 1 0 1 0 1 1 0 0 0 1 1 0 0 1 1 1 0 1 1 1 1 1 1 0 1 1 1 1 1 1 1 1 1 1 1 1 1 1 1 1 1 1 1 1 1
**The message above is an SDLC data stream; the order of the arrow is to the left.**
   **(a)** What is the format of the frame above?
   **(b)** What is the hex address of the secondary involved in this transmission?
   **(c)** If an Ns is associated with this frame, what is it?
   **(d)** If an Nr is associated with this frame, what is it?
   **(e)** This frame is the last of a series of frames, or the only frame in this message.
      (T, F)
   **(f)** Excluding flags, how many characters are in this frame?

**2.** Stuff 0's where required. Order of the arrow is to the left.

                1 0 0 1 1 1 1 1 1 1 1 1 1 0 0 1 1 1 1 0 1 0

**3.** SDLC message in EBCDIC:

$$\begin{array}{ccc} & S & F \\ = & T\ p\ .\ .\ .\ .\ .\ .\ C & = \\ & X & S \end{array}$$

   **(a)** What is the address of the secondary?
   **(b)** The message is from a primary to a secondary station.   (T, F)
   **(c)** How many bytes follow p in the message above?

**4.** If the primary sent four successive frames starting with Ns = 2, the secondary would return a positive acknowledgment of all four frames by returning an Nr of _____ .

**5.** Complete the following chart for the control characters given.

| Control character | Format | P/F, $\overline{\text{P}}/\overline{\text{F}}$ | Ns | Nr | Command/response |
|---|---|---|---|---|---|
| $A6_{Hex.}$ | | | | | |
| $N_{EBCDIC}$ | | | | | |
| $53_{Hex}$ | | | | | |
| $68_{Hex}$ | | | | | |
| $22_{Hex}$ | | | | | |

**Enter P/F if this bit is a 1, but the transmission direction is not determined. Enter P or F if the transmission direction is known.**

**6.** Detemine the hex characters for the control field that are required to convey the following information:

   **(a)** Represent an information frame sending frame 3. It is the final frame and frame 5 is the next frame expected to be received.

   **(b)** Represent the primary sending a supervisory frame that is a poll. It is ready to receive and frame 6 is the next frame expected to be received.

   **(c)** Represent the secondary sending an unnumbered frame telling the primary that it is in the disconnect mode.

   **(d)** Represent the secondary sending information frame 7. It is a final frame and the next frame it expects to receive is frame 3.

**7.** Draw the level diagram that the bit stream shown in Figure P12–7 when encoded in NRZI, would produce.

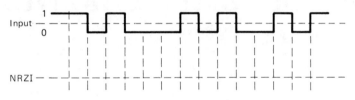

**FIGURE P12–7**

# THIRTEEN

## _PACKET SWITCHING AND LOCAL AREA NETWORKS_

## PUBLIC DATA NETWORK

A _public data network_ (PDN) is a switched data communications network similar to the public telephone network except that a PDN is designed for transferring data only. Public data networks combine the concepts of both _value-added networks_ (VANs) and _packet-switching networks_.

### Value-Added Network

A value-added network "_adds value_" to the services or facilities provided by a common carrier to provide new types of communication services. Examples of added values are error control, enhanced connection reliability, dynamic routing, failure protection, logical multiplexing, and data format conversions. A VAN comprises an organization that leases communications lines from common carriers such as AT&T and MCI and adds new types of communications services to those lines. Examples of value-added networks are GTE Telnet, DATAPAC, TRANSPAC, and Tymnet Inc.

## SWITCHING TECHNIQUES

Packet switching is the transmittal of small bundles of information through switches to the intended destination. This sounds logical. However, if you did not know what packet switching was before, probably very little was added to your knowledge

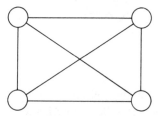

**FIGURE 13–1   Dedicated connections.**

of this subject. Let's take a step back and see how it came about and what the alternatives are. The rapid growth in the number of personal computers, small business computers, mainframe computers, and terminals has created a situation where more and more people have the need to exchange digital information with each other. One method of accomplishing this is by making direct connections between all of the terminals. If $n$ terminals are to be interconnected, the first terminal would require $n - 1$ connections; the second would require $n - 2$ additional connections; the third would require $n - 3$ additional connections; and so on. These connections are shown in Figure 13–1. Mathematically, the sum of a series of integers 1, 2, . . . , $k$ is computed from the formula $k(k + 1)/2$. Therefore, to interconnect all of our terminals, $(n - 1)n/2$ connections are required. If only a few devices are all connected to each other, the number of connections is not too high. However, as the $n$th terminal is added to the network, an additional $n - 1$ connections are required. Soon, this number of dedicated connections becomes totally impractical. The alternative is to use a switch to which all of the terminals are connected. If there are 100 terminals, then 100 connections would be required to connect each terminal to the switch. The switch would have the capability of linking any two terminals that are connected to it. A switched network is shown in Figure 13–2. For applications where a large number of terminals are involved, the cost of the switch is offset by the savings realized in the reduction of the number of connections. As a network becomes larger, the inclusion of more than one switch may be required. If so, interswitch trunks would be needed.

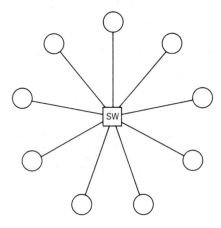

**FIGURE 13–2   Switched network.**

There are three basic switching techniques: circuit switching, message switching, and packet switching.

## Circuit Switching

The public telephone network in the United States uses circuit-switching principles. Once a call has been established, the network provides a dedicated line for the duration of the call. The time required to establish the connection is called the *setup time*, which may be quite long. Since one line is allocated for each call, blocking may occur. *Blocking* occurs when the user is denied a connection through the network because no facilities are available between the points of origin and destination. If circuit switching is used for the transmission of digital information, the terminals at both ends of the connection must be compatible—they must use a common protocol, bit rate, and character set. The equipment used to interface the data terminal equipment to the network is voice compatible and common calling procedures may be used. Once a call is established, information is transferred in real time. Inefficiencies in this type of switching arise after a line has been assigned. If the user does not use the full capacity of the line for the entire time, no other user may have access to it during this period.

## Message Switching

Message switching is a form of *store-and-forward* network. Each switch within the network has a message storing capacity. The user inputs information into the network and the network transfers the information to the next switch or to the final destination when it is convenient to do so. Consequently, the delay time from message transmission to reception varies and may be quite long. Blocking cannot occur. Once the information has entered the network, it is converted to a format role suitable for transmission within the network. At the terminating end, the information is converted to a format compatible with the receiving data terminal equipment. Therefore, the source and destination terminals need not be compatible. The basic unit of transmission is still the entire message. Message switching is more efficient than circuit switching since messages received at a time of peak load can be delayed and transmitted when the load has decreased. This efficiency is achieved at the cost of not transmitting messages in real time.

## Packet Switching

With packet switching, the user's data are divided into small segments called *packets*. Since a packet may be held in memory for a short period of time, packet switching is called a *hold-and-forward* network. The individual packets of a long message need not take the same path to the destination. Since the packets are small, the hold time is also small and the information is transmitted in near real time. The source and destination equipment need not be compatible since the network makes the necessary adjustments. Blocking cannot occur. The price paid for this high efficiency and near real-time transmission is the requirement of complex switching arrangements and complex protocols.

A circuit switch is called a *transparent switch*. It does nothing but link the transmitted information to the intended destination. It is totally transparent to the user. A message switch or a packet switch is called a *transactional switch* because it does more than connect the send station to the receiver. It can either store the data temporarily or change the data format. At the receive end, the data may again be changed so that they are compatible with the user's terminal equipment.

## THE INTERNATIONAL STANDARDS ORGANIZATION (ISO) PROTOCOL HIERARCHY

In order to facilitate the intercommunication of data-processing equipment, the ISO developed a standard protocol hierarchy (Figure 13–3). The intent of this layered structure of network protocols is to allow a network of open systems to be logically composed of smaller subdivisions called levels or layers. Each layer is further divided into smaller subsystems corresponding to the intersection of the system layers. The basic idea of layering is that each layer adds value to services provided by the set of lower layers in such a way that the highest layer is offered the full set of services needed to run distributed applications. Layering thus divides the total system problem into smaller pieces.

**Physical layer.**   The physical layer is the lowest and most fundamental layer of the ISO hierarchy. This layer specifies the physical, electrical, functional, and procedural standards for accessing the network. Several documents are being established to further define the physical layer. The functions specified in the physical layer are similar to those outlined in the EIA RS 232C interface (Chapter 6).

**Data link layer.**   The primary function of the data link layer is to ensure the reliable transfer of user data over the data link. This layer is responsible for the communication between nodes within the network. The data link layer is the final framing of the message envelope. It facilitates the orderly transmission of data between nodes and allows for error detection and retransmission.

| | |
|---|---|
| Level 7 | Application |
| Level 6 | Presentation |
| Level 5 | Session |
| Level 4 | Transport |
| Level 3 | Network |
| Level 2 | Data link |
| Level 1 | Physical |

**FIGURE 13–3   ISO protocol hierarchy.**

**Network layer.**   The network layer determines the most appropriate network configuration for the function provided by the network. Examples of network configurations are dial-up, leased, or packet. Flow control procedures are initiated at this level and switched services are initiated through a data call establishment procedure.

**Transport layer.**   The transport layer acts as an interface between the session and network layers. The transport layer is concerned with economic efficiency. One important function performed by this layer is controlling end-to-end data integrity, which includes routing, message segmenting, and error recovery.

**Session layer.**   The session layer is responsible for availability. A session is a temporary relationship. Activating a session makes available the appropriate resources such as buffer storage and processor capacity. Session control includes such things as network log-on and user authentication. The session layer coordinates interaction between end application processes.

**Presentation layer.**   The presentation layer defines protocols that communicate syntax (or representation). The presentation layer functions include formatting, encoding, encrypting/decrypting of messages, dialogue management, synchronization, interruption, and termination.

**Application layer.**   The application layer is analogous to the general manager of the network. It is responsible for the transfer of information between application processes and the end-to-end encryption. The application layer controls the sequence of events between the computer application and the user or another application. It also controls the sequence of activities within the application itself. The application layer is the layer that communicates directly with the user's application program or process.

## CCITT X.1: INTERNATIONAL USER CLASS OF SERVICE IN PUBLIC DATA NETWORKS

The CCITT X.1 standard divides 11 classes of service into three basic modes of operation: start/stop (S/S), synchronous (SY), and packet (PM). Data entering a public data network must be in one of these three forms.

**Start/stop mode (classes 1 and 2).**   In the start/stop mode, data are transferred from the user to the network in an asynchronous data format (each character is framed between a start and 1 or 2 stop bits). IBM's 83B protocol and the Bell System's 8A1/8B1 selective calling arrangement (Chapter 10) are examples of common asynchronous protocols. Call control signaling is done in International Alphabet

#5 (ASCII), and data transfer signaling is in accordance with CCITT X.4 with the following bit/second and bit/character structure:

Class 1: 300 bps, 11 bits/character

Class 2:   50 bps, 7.5 bits/character

              100 bps, 7.5 bits/character

              110 bps, 11 bits/character

              134.5 bps, 9 bits/character

              200 bps, 11 bits/character

**Synchronous mode (classes 3 through 7).**    In the synchronous mode, data are transferred from the user to the network in a synchronous data format (each transmission is preceded with a unique 8-bit synchronizing character). IBM's 3270 BSC (Chapter 11), Burrough's BASIC, and UNIVAC's UNISCOPE are examples of common synchronous protocols. Call control and data transmission rates are identical to private-line applications. The classes of service for synchronous modes are:

Class 3: 600 bps

Class 4:    2.4 kbps

Class 5:    4.8 kbps

Class 6:    9.6 kbps

Class 7:    48 kbps

**Packet mode (classes 8 through 11).**    In the packet mode, data are transferred from the user to the network in a frame format. Each frame is delimited by a beginning and an ending flag. The ISO HDLC frame format (Chapter 12) is a common data link control procedure. Within the frame, call control and data transfer are in accordance with CCITT X.25 at the following transmission rates:

Class  8:    2.4 kbps

Class  9:    4.8 kbps

Class 10:    9.6 kbps

Class 11:    48 kbps

Figure 13–4 is a typical data network interconnect arrangement showing all three modes of operation. The packet assembler/disassembler (PAD) provides the necessary assembly/disassembly functions to interface non-packet-mode users to the public data network. Asynchronous users must comply with additional standards, such as the CCITT X.28 and X.29 to emulate point-to-point connections for nonintelligent terminals operating within the data network.

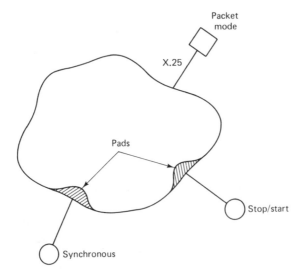

FIGURE 13–4   Public data network.

## CCITT X.25: THE USER-NETWORK INTERFACE

In 1976, the CCITT defined the X.25 standard as the international standard for packet network access. The X.25 is strictly a user-to-network interface standard. X.25 addresses only the lower three levels of the ISO protocol hierarchy (the physical, data link, and network levels). Whenever possible, the X.25 uses existing standards. For example, for the physical level, X.25 specifies existing DTE/DCE standards. The CCITT X.21, X.26, and X.27 standards are specified as the physical interfaces for X.25. These standards correspond to EIA standards RS 232C, RS 423A, and RS 422A, respectively. For the second level, the data link level, X.25, defines the ISO HDLC as the international standard and the American National Standards Institute (ANSI) 3.66, *Advanced Data Communications Control Procedures* (ADCCP), as the United States Standard. The data link level provides an envelope in which the data of the higher levels are embedded. Unfortunately, many of the features of these data link control procedures are unnecessary for packet networks. HDLC and ADCCP were designed for two-point private-line and multipoint polling applications. Consequently, the addressing and control procedures outlined by them are superfluous for packet networks. HDLC and ADCCP were selected because their block format and error detection mechanism are ideally suited to a packet environment. Level three, the network level, of the X.25 standard specifies three types of switching services offered in the packet mode on a switched data network. They are the *permanent virtual circuit*, the *virtual call*, and the *datagram*. All three modes transfer messages in packet format. The X.25 simply outlines the format the data must be in when they enter the data network.

**Permanent Virtual Circuit (PVC).** The PVC is logically equivalent to a leased line, but slower. A PVC guarantees the connection, on demand, between two predetermined users of the network. Since the users of the PVC are fixed, there is no need to include identification codes or addresses when initiating a call or transferring data.

**Virtual Call (VC).** The VC is logically equivalent to a dial-up circuit using the public telephone network. The VC provides a one-to-many access capability. Each user of the network has access to every other user through a common network of switches. Properties essential to a VC are sequenced data transfer, transparency, full-duplex capabilities, flow control, error control, and interface independence.

**Datagram (DG).** The datagram is not completely defined in the X.25 standard. Currently, work is being done to standardize the datagram concept. Until agreed upon, its usefulness is questionable. A DG is the functional opposite of a VC. Users send small messages into the network. Each message is an independent entity and is not acknowledged. The network does not guarantee that the messages will arrive at the destination in the same order in which they were sent. However, very often the message will fit into a single packet. This is referred to as a *single-packet-per-segment protocol*. The DG is the most unreliable of the three packet modes of service.

**X.25 packet formats.** The switched virtual call makes the most efficient use of the packet network. The virtual call has two packet frame formats, the call-request frame and the data-transfer frame.

*Call-request frame.* Figure 13–5 shows the field structure of the call-request frame. Each frame is delimited by a beginning and an ending flag. Following the

| Flag | Link address field | Link control field | Format identifier | Logical channel identifier | Packet type | Calling address length | Called address length | Called address |
|------|------|------|------|------|------|------|------|------|
| Bits: 8 | 8 or 8N | 8 or 16 | 4 | 12 | 8 | 4 | 4 | To 60 |

| Calling address | 0 | Facilities field length | Facilities field | Protocol ID | User data | Frame check sequence | Flag |
|------|------|------|------|------|------|------|------|
| Bits: To 60 | 2 | 6 | To 512 | 32 | To 96 | 16 | 8 |

**FIGURE 13–5 Call-request packet frame format.**

beginning flag is the link address field, then a link control field. A CRC-16 frame check sequence precedes the ending flag. The flag, address, control, and frame check sequence fields are defined by the ISO's link-level protocol, HDLC.

1. *Flag field*. The flag sequence is 01111110. The flag identifies the beginning and ending of the packet.

2. *Address field*. The address field contains the link-level address. Either the basic 8-bit or extended 8 + 8N-bit addressing modes may be used.

3. *Control field*. This field is used for local counting and error control and correction between the DCE and the DTE. Either the basic 8-bit or the extended 16-bit control format may be used.

4. *Format identifier*. The format identifier defines the nature of the packet—new call request, data packet, or a previously established call. The format identifier also identifies the numbering sequence limitations used (either 7 or 127).

5. *Logical channel identifier (LCI)*. The logical channel identifier designates a 12-bit logical channel number for the packet. Up to 4096 logical channels may be in use at any one time. The LCI, along with the called and calling addresses, is used to identify the source user and the destination user. When a user gains access to the network, it is assigned an LCI. When a user disconnects from the network, it relinquishes the assigned LCI. Once a call has been established, data transfer packets need only to include the LCI and not the calling or called addresses.

6. *Packet type*. This 8-bit sequence identifies the function and the content of the packet—new request, call clear, call reset, and so on. This field also identifies flow control and data restrictions.

7. *Calling address length*. This 4-bit field identifies (in binary) the number of digits included in the address of the calling party. Fifteen is the largest number of digits that may be used.

8. *Called address length*. Same as the calling address length except that it identifies the number of digits included in the address of the called party. Again, the maximum number of digits that may be used is 15.

9. *Called address*. This field contains the destination's address in binary-coded decimal (BCD) notation. The number of BCD codes in this field was previously specified in the called address length field. Each BCD number is made up of 4 bits. Fifteen 4-bit BCD codes represent a maximum of 60 bits for this field.

10. *Calling address*. Same as the called address field except that this field contains the address of the calling party.

11. *Facilities length field*. This field identifies, in binary, the number of 8-bit octets included in the following facilities field.

| Flag | Link address field | Link control field | Format identifier | Logical channel identifier | Send packet sequence number P(s) | 0 | Receive packet sequence number P(r) | 0 | User data | Frame check sequence | Flag |
|------|------|------|------|------|------|------|------|------|------|------|------|
| Bits: 8 | 8 | 8 | 4 | 12 | 3/7 | 5/1 | 3/7 | 5/1 | To 1024 | 16 | 8 |

**FIGURE 13–6  Data-transfer packet frame format.**

12. *Facilities field*. This field contains information concerning optional network facilities, such as reverse billing, a closed user group, one-way transmit connections, and one-way receive connections.

13. *Protocol identifier*. This 32-bit field can be used by the subscriber for unique, user-level protocol functions, such as log-on procedures and user identification practices.

14. *User data field*. Up to 96 bits of user information can be transmitted with a data request packet. These are nonsequenced data and usually contain such information as the user's password number.

*Data-transfer packet*. Figure 13–6 shows the field format for a data-transfer packet. It is similar to the call-request packet except that it contains considerably less overhead. The DLC frame format is HDLC. A data-transfer packet includes a send packet field and a receive packet field that were not included in the call-request packet.

1. *Send packet sequence number*. This field contains a sequential numbering scheme used to identify packets within a logical channel identifier. The P(s) count is used essentially the same way as the Ns is used in HDLC and SDLC. Each successive packet transmitted is assigned the next P(s) number in the sequence. The P(s) can be a 3-bit or a 7-bit binary number. Consequently, the maximum number of unacknowledged packets that may be outstanding is either 7 or 127. The number of P(s) bits used is identified in the format identifier field. The send packet field is 8 bits long; unused bits are reset to a logic 0 condition.

2. *Receive packet sequence number*. This field contains the number P(r) of the last packet received on this LCI without error. The P(r) is used to ACK/NAK previously received data packets.

## LOCAL AREA NETWORKS

The decrease in the price of computers, word processors, high-speed terminals, and so on, has allowed businesses to install this equipment in numerous offices. Different types of computers are purchased to satisfy different goals. Local area networks (LANs) are designed to interconnect this equipment, with the obvious advantages of resource sharing and simplified managerial controls. LANs are privately owned and operated and are generally used to connect stations in the same building or in buildings in the same locale. Most LANs are 1 to 2 miles in length. A gateway is provided by which this network connects to the outside world. This eliminates the problem of interfacing various equipment with different networks.

## Network Considerations

**Topology.**   The topology or architecture of a network identifies how the various stations will be interconnected. The star, bus, ring, and mesh topologies are the most common, each with its own advantages and disadvantages. These topologies are shown in Figure 13–7.

**Connecting medium.**   Although communications may be established on twisted pairs, coaxial cable, and fiber optics, all present local area network systems use coaxial cable as the transmission medium.

**Information transmission.**   The question now arises as to how information will be placed on this cable. There are two basic approaches: baseband and broadband transmission. *Baseband* transmission uses the cable as a *single channel* device. All stations connected to this channel must transmit the same type of information at

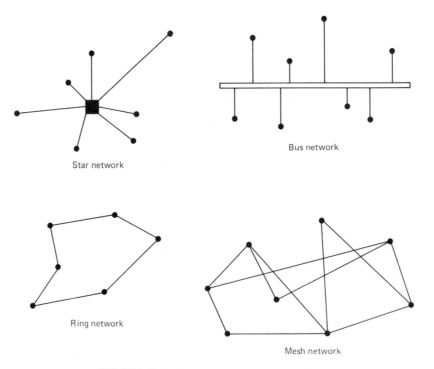

Star network

Bus network

Ring network

Mesh network

**FIGURE 13–7   Local area network topologies.**

the same speed. *Broadband* transmission is a *multichannel* operation. Each channel occupies a different frequency band and the information on each channel may be multiplexed. This permits voice, digital data, and video information to be transmitted simultaneously. Because of the need of modems, amplifiers, and more complex transmitters and receivers, a broadband system is more complex than a baseband system.

**Channel access.** Channel access defines how a station gets control of the channel to transmit. Dedicated channels have been discounted because of high inefficiency due to nonuse of the channel for a large portion of the time. Contrasted with dedicated channels is random access. Although many means are available for achieving random access, *carrier sense, multiple access with collision detection (CSMA/CD)* and *token passing* have been receiving the most attention. With CSMA/CD, a station "listens" to detect if any transmissions are currently being sent by other stations. If so, that station will wait until the transmission ends and then transmit. If another station started to transmit at the same time, a collision of data will occur. In such a case, both stations will immediately stop transmitting and back-off for a random period of time before attempting to retransmit. A detailed explanation of CSMA/CD as used by Ethernet is given in a subsequent discussion. CSMA/CD is used by most baseband networks. Token passing is suited for a ring topology network. An electrical token is placed on the channel. As it circulates around the ring from station to station, if a particular station wishes to transmit, it removes the token, places its message on the channel, then reinserts the token after its message. The station possessing the token is the only one permitted to transmit. As the message circulates, it is removed at its intended destination. With token passing, each station is assured of having access to the channel in turn and of being able to transmit all of its messages. This guarantee of successful message transmission is not available with CSMA/CD.

## Systems

This is not intended to be an inclusive list of all existing LANs. The networks shown demonstrate the various channel accesses used with either baseband or broadband networks.

*Ethernet*: developed by Xerox Corporation in conjunction with Digital Equipment Corporation and Intel Corporation. It is a baseband system using CSMA/CD.

*Wangnet*: developed by Wang Computers. It uses CSMA/CD with a broadband network.

*Localnet*: developed by Sytek. It also uses CSMA/CD with a broadband network.

*Domain*: developed by Apollo Computers. It uses a token-passing ring structure in a broadband network.

## IEEE 802 Local Area Network Committee

Although the same accessing technique may be used by different manufacturers, this does not mean that all used the same codes, signal levels, and so on. In 1980, the 802 Local Area Network Committee was established to standardize the means of connecting digital computer equipment and peripherals with the local environment. In 1983, the IEEE standards board approved standard 802.3—CSMA/CD accessing on a bus topology—and standard 802.4—token passing accessing on a bus topology. These standards, together with standard 802.2, which describes the logical link control, have been adopted by ISO. Standard 802.5 describes token passing accessing in a ring topology.

# ETHERNET

Ethernet is a baseband system which uses a bus topology and is designed for use in a local area network. Since there are many computer manufacturers, the problem is to design a system in which different types of computers can communicate with each other. In a demonstration in 1982, the computers of 10 major companies were linked on a 1500-ft Ethernet cable for electronic mail, word processing, and so on.

Before explaining the operation of Ethernet, a description of this system's hardware and software is provided.

## Software

Information is transmitted from one station (computer) to another in the form of packets. The format for a packet is shown in Figure 13–8. The preamble is used for bit synchronization. The source and destination addresses and the field type make up the header information. The CRC is computed on the header information and the data field. The packet information is converted to Manchester code and transmitted at a 10-Mbps rate. In the Manchester encoding, each bit cell is divided into two parts. The first half contains the complement of the bit value and the second contains the actual bit value. This is illustrated in Figure 13–9. This code ensures that a signal transition exists in every bit.

The destination of a packet may be a single station or a group of stations. A

| Preamble | Destination address | Source address | Type field | Data field | CRC |
|----------|---------------------|----------------|------------|------------|-----|
| Bytes:        8 | 6 | 6 | 2 | 46–1500 | 4 |

**FIGURE 13–8  Packet format for Ethernet.**

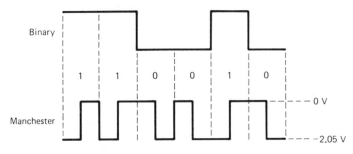

**FIGURE 13–9  Manchester code.**

*physical address* is a unique address of a single station and is denoted by a 0 as the first bit (LSB) of the destination address. A *multicast address* has two forms. If only a partial group of the total stations is addressed, the first bit is a 1. The destination address is all 1's if all of the stations in the network are to receive the transmitted packet. A delay of 9.6 μs is required between the transmission of packets.

## Hardware

Coaxial cable is the transmission medium for Ethernet. A cable segment may have a maximum distance of 500 m (1640 ft). Each segment may have up to 100 transceivers attached by way of pressure taps. Since these taps do not require cable cutting for installation, additional stations may be added without interfering with normal system operations. Interlan has developed the NT10 transceiver, in which the taps consist of two probes. One probe contacts the center conductor while the other contacts the shield. Color bands on the segment cable identify where the taps may be placed. A maximum of three segments may be connected end to end by way of repeaters to extend the system length to 1500 m (4921 ft). In Figure 13–10, this is the path between A and C. Remote repeater may be used with a maximum distance of 1000m between them. If these are used, the maximum end-to-end distance would be extended to 2.5 km (8202 ft). In Figure 13–10, if the center and right segments were interchanged, there would be a 2.5-km separation between A and B. There must always be only one signal path between any two stations. A station is connected to a controller, which is then interfaced with the transceiver to the transmission system. This is shown in Figure 13–11.

## System Operation

**Transmission.**  The data link control for Ethernet is CSMA/CD. Stations acquire access to the transmission system through contention. They are not polled, nor do they have specific time slots for transmission. A station wishing to transmit first determines if another station is currently using the transmission system. The controller accomplishes this through the transceiver by sensing the presence of a carrier on the line. The controller may be a hardware or a software function, depending

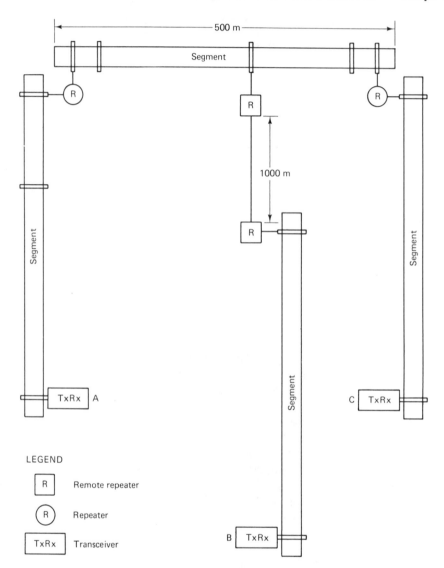

**FIGURE 13–10   Ethernet transmission system.**

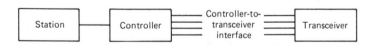

**FIGURE 13–11   Ethernet station connection to the transmission system.**

on the complexity of the station. In 1983, Xerox Corporation, in conjunction with Digital Equipment Corporation and Intel Corporation, developed and introduced a single-chip controller for Ethernet. Later, Mostek Corporation, working with Digital Equipment Corporation and Advanced Micro Devices, announced a two-chip set for this purpose. One chip is the local area network controller for Ethernet (LANCE) and the second is a Serial Interface Adapter.

The presence of a carrier is denoted by the signal transitions on the line produced by the Manchester code. If a carrier is detected, the station defers transmission until the line is quiet. After the required delay, the station sends digital data to the controller. The controller converts these data to Manchester code, inserts the CRC, adds the preamble, and places the packet on-line. The transmission of the entire packet is not yet assured. A different station may also have detected the quiet line and started to transmit its own packet. The first station monitors the line for a period called the *collision window* or the *collision interval*. This interval is a function of the end-to-end propagation delay of the line. This delay, including the delay caused by any repeaters, measured in distance, cannot exceed 2.5 km. If a data collision has not occurred in this interval, the station is said to have line acquisition and will continue to transmit the entire packet. Should a collision be detected, both of the transmitting stations will immediately abort the transmission for a random period of time and then will attempt to retransmit. A data collision may be detected by the transceiver by comparing the received signal with the transmitted signal. To make this comparison, a station must still be transmitting its packet while a previously transmitted signal has propagated to the end of the line and back. This dictates that a packet be of some minimum size. If a collision is detected, it is the controller that must take the necessary action. The controller-transceiver interface contains a line for notification of Collision Presence (10-MHz square wave). The remaining three lines of this interface are the Transmit Data, Receive Data, and power for the transceiver. Since a collision is manifested in some form of phase violation, the controller alone may detect a collision. In Ethernet, data collision is detected mainly by the transceiver. When feasible, this is supplemented by a collision detection facility in the controller. To ensure that all stations are aware of the collision, a *collision enforcement consensus procedure* is invoked. When the controller detects a collision, it transmits four to six bytes of random data. These bytes are called the *jam sequence*. If a collision has occurred, the station's random waiting time before retransmission is determined from a *binary exponential back-off algorithm*. The transmission time slot is usually set to be slightly longer than the round-trip time of the channel. The delay time is randomly selected from this interval.

**EXAMPLE**

Transmission time slot = time of 512 bits

$$\text{Maximum time of the interval} = \frac{512 \text{ bits}}{10 \text{ Mbps}} = 51.2 \ \mu s$$

Time interval = $0 - 51.2 \ \mu s$

For each succeeding collision encountered by the same packet, the time interval is doubled until a maximum interval is reached. The maximum interval is given as $2^{10} \times$ transmission time slot. After 15 unsuccessful attempts at transmitting a packet have been made, no further attempts are made and the error is reported to the station. This is the major drawback of Ethernet—it cannot guarantee packet delivery at a time of heavy transmission load.

**Reception.**    The line is monitored until the station's address is detected. The controller strips the preamble, checks the CRC, and converts the Manchester code back to digital format. If the packet contains any errors, it is discarded. The end of the packet is recognized by the absence of a carrier on the transmission line. This means that no transitions were detected for the period of 75 to 125 ns since the center of the last bit cell. The decoding is accomplished through a phase-locked loop. The phase-locked loop is initialized by the known pattern of the preamble.

# QUESTIONS

1. What is a value-added network?
2. If there were 14 terminals in a network:
   (a) How many direct connections would be required to hook up every terminal to every other terminal?
   (b) How many connections would be required if a switch was to interconnect any two terminals?
3. Select answers from the following: (a) circuit switching; (b) message switching; (c) packet switching. Which switching technique(s):
   (1) Operate in real time or near real time?
   (2) Are transparent?
   (3) Are transactional?
   (4) May encounter blocking?
   (5) Require end terminals to be compatible?
   (6) Are a form of hold-and-forward network?
   (7) Are a form of store-and-forward network?
   (8) Transmit all information in the smallest segments?
4. HDLC is the standard protocol for packet switching.   (T, F)
5. All seven levels of ISO protocols are defined also by CCITT X.25.   (T, F)
6. CCITT X.25 describes the standards between two switches.   (T, F)
7. CCITT X.25 describes packet formats.   (T, F)
8. What is the approximate end-to-end distance of most local area networks?
9. If cable is used for only a single channel, that local area network is said to be using a _____ system.
10. Data collisions are possible with token passing.   (T, F)
11. Ethernet is a form of topology.   (T, F)

12. CSMA/CD is used only with baseband systems.   (T, F)
13. If an Ethernet system was extended from 500 m to 1000 m, the minimum packet size would (*have to be increased, have to be decreased, not have to be changed*).
14. On Ethernet, how is a physical address distinguished from a multicast address?
15. The Ethernet carrier is a constant-frequency signal.   (T, F)
16. Which piece of equipment normally detects collisions on Ethernet?
17. Identify three nonrelated functions performed by the controller.
18. If an Ethernet message experienced collisions 15 times, the interval from which a back-off time was selected was doubled each time.   (T, F)
19. Draw the Manchester-coded waveform for the digital waveform shown in Figure P13–19.

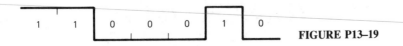

FIGURE P13–19

# FOURTEEN

# *FIBER OPTIC COMMUNICATIONS*

## INTRODUCTION

During the past 10 years, the electronic communications industry has experienced many remarkable and dramatic changes. A phenomenal increase in voice, data, and video communications has caused a corresponding increase in the demand for more economical and larger capacity communications systems. This has caused a technical revolution in the electronic communications industry. Terrestrial microwave systems have long since reached their capacity, and satellite systems can provide, at best, only a temporary relief to the ever-increasing demand. It is obvious that economical communications systems that can handle large capacities and provide high-quality service are needed.

Communications systems that use light as the carrier of information have recently received a great deal of attention. As we shall see later in this chapter, propagating light waves through the earth's atmosphere is difficult and impractical. Consequently, systems that use glass or plastic fiber cables to "contain" a light wave and guide it from a source to a destination are presently being investigated at several prominent research and development laboratories. Communications systems that carry information through a *guided fiber cable* are called *fiber optic* systems.

The *information-carrying capacity* of a communications system is directly proportional to its bandwidth; the wider the bandwidth, the greater its information-carrying capacity. For comparison purposes, it is common to express the bandwidth of a system as a percentage of its carrier frequency. For instance, a VHF radio system operating at 100 MHz has a bandwidth equal to 10 MHz (i.e., 10% of the carrier frequency). A microwave radio system operating at 6 GHz with a bandwidth equal to 10% of its carrier frequency would have a bandwidth equal to 600 MHz. Thus the higher the carrier frequency, the wider the bandwidth possible and consequently, the greater the information-carrying capacity. Light frequencies used in fiber optic systems are between $10^{14}$ and $10^{15}$ Hz (100,000 to 1,000,000 GHz). Ten percent of 1,000,000 GHz is 100,000 GHz. To meet today's communications needs or the needs of the foreseeable future, 100,000 GHz is an excessive bandwidth. However, it does illustrate the capabilities of fiber optic systems.

## HISTORY OF FIBER OPTICS

In 1880, Alexander Graham Bell experimented with an apparatus he called a *photophone*. The photophone was a device constructed from mirrors and selenium detectors that transmitted sound waves over a beam of light. The photophone was awkward, unreliable, and had no real practical application. Actually, visual light was a primary means of communicating long before electronic communications came about. Smoke signals and mirrors were used ages ago to convey short, simple messages. Bell's contraption, however, was the first attempt at using a beam of light for carrying information.

Transmission of light waves for any useful distance through the earth's atmosphere is impractical because water vapor, oxygen, and particulates in the air absorb and attenuate the ultrahigh light frequencies. Consequently, the only practical type of optical communications system is one that uses a fiber guide. In 1930, J. L. Baird, an English scientist, and C. W. Hansell, a scientist from the United States, were granted patents for scanning and transmitting television images through uncoated fiber cables. A few years later a German scientist named H. Lamm successfully transmitted images through a single glass fiber. At that time, most people considered fiber optics more of a toy or a laboratory stunt and consequently, it was not until the early 1950s that any substantial breakthrough was made in the field of fiber optics.

In 1951, A. C. S. van Heel of Holland and H. H. Hopkins and N. S. Kapany of England experimented with light transmission through *bundles* of fibers. Their studies led to the development of the *flexible fiberscope*, which is used extensively in the medical field. It was Kapany who coined the term "fiber optics" in 1956.

The *laser* (*l*ight *a*mplification by *s*timulated *e*mission of *r*adiation) was invented in 1960. The laser's relatively high output power, high frequency of operation, and capability of carrying an extremely wide bandwidth signal make it ideally suited for high-capacity communications systems. The invention of the laser greatly accelerated research efforts in fiber optic communications, although it was not until 1967 that K. C. Kao and G. A. Bockham of the Standard Telecommunications Laboratory in England proposed a new communications medium using *cladded* fiber cables.

The fiber cables available in the 1960s were extremely *lossy* (more than 1000 dB/km), which limited optical transmissions to short distances. In 1970, Kapron, Keck, and Maurer of Corning Glass Works in Corning, New York, developed an optical fiber with losses less than 20 dB/km. That was the "big" breakthrough needed to permit practical fiber optics communications systems. Since 1970, fiber optics technology has grown exponentially. Recently, Bell Laboratories successfully transmitted 1 billion bps through a fiber cable for 75 miles without a regenerator. AT&T has projected they will have a transatlantic fiber cable installed and operational by 1988.

In the late 1970s and early 1980s, the refinement of optical cables and the development of high-quality, affordable light sources and detectors have opened the door to the development of high-quality, high-capacity, and efficient fiber optics communications systems.

## FIBER OPTIC VERSUS METALLIC CABLE FACILITIES

Communications through glass or plastic fiber cables has several overwhelming advantages over communications over conventional *metallic* or *coaxial* cable facilities.

### Advantages of Fiber Systems

1. Fiber systems have a greater capacity due to the inherently larger bandwidths available with optical frequencies. Metallic cables exhibit capacitance between and inductance along their conductors. These properties cause them to act like low-pass filters which limit their bandwidths.

2. Fiber systems are immune to crosstalk between cables caused by *magnetic induction*. Glass or plastic fibers are nonconductors of electricity and therefore do not have a magnetic field associated with them. In metallic cables, the primary cause of crosstalk is magnetic induction between conductors located near each other.

3. Fiber cables are immune to *static* interference caused by lightning, electric motors, fluorescent lights, and other electrical noise sources. This immunity is also attributable to the fact that optical fibers are nonconductors of electricity. Also, fiber cables do not radiate energy and therefore cannot cause interference with other communications systems. This characteristic makes fiber systems ideally suited to military applications, where the effects of nuclear weapons (EMP—electromagnetic pulse interference) has a devastating effect on conventional communications systems.

4. Fiber cables are more resistive to environmental extremes. They operate over a larger temperature variation than their metallic counterparts, and fiber cables are affected less by corrosive liquids and gases.

5. Fiber cables are safer and easier to install and maintain. Because glass and plastic fibers are nonconductors, there are no electrical currents or voltages associated with them. Fibers can be used around volatile liquids and gases without worrying about their causing explosions or fires. Fibers

are smaller and much more lightweight than their metallic counterparts. Consequently, they are easier to work with. Also, fiber cables require less storage space and are cheaper to transport.

6. Fiber cables are more secure than their copper counterparts. It is virtually impossible to tap into a fiber cable without the user knowing about it. This is another quality attractive for military applications.

7. Although it has not yet been proven, it is projected that fiber systems will last longer than metallic facilities. This assumption is based on the higher tolerances that fiber cables have to changes in the environment.

8. The long-term cost of a fiber optic system is projected to be less than that of its metallic counterpart.

## Disadvantages of Fiber Systems

At the present time, there are few disadvantages of fiber systems. The only significant disadvantage is the higher initial cost of installing a fiber system, although in the future it is believed that the cost of installing a fiber system will be reduced dramatically. Another disadvantage of fiber systems is the fact that they are unproven; there are no systems that have been in operation for an extended period of time.

## ELECTROMAGNETIC SPECTRUM

The total electromagnetic frequency spectrum is shown in Figure 14–1. It can be seen that the frequency spectrum extends from the *subsonic* frequencies (a few hertz) to *cosmic rays* ($10^{22}$ Hz). The frequencies used for fiber optic systems extend from approximately $10^{14}$ to $10^{15}$ Hz (infrared to ultraviolet). This frequency subspectrum is called *visible light*, although it extends above and below the actual sensitivity of the human eye.

When dealing with ultrahigh-frequency electromagnetic waves, such as light,

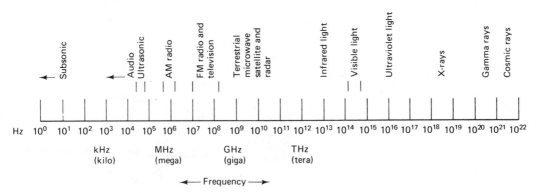

**FIGURE 14–1 Electromagnetic frequency spectrum.**

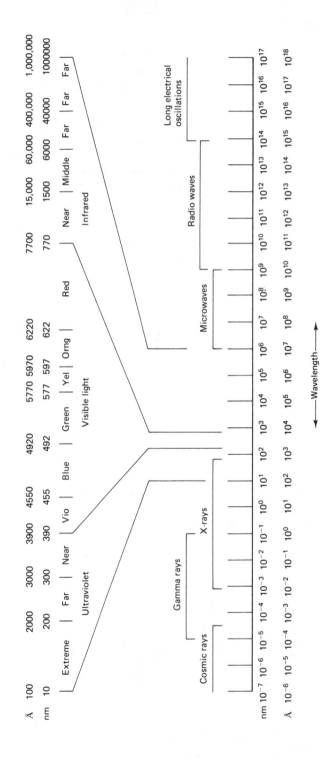

**FIGURE 14-2 Electromagnetic wavelength spectrum.**

308

it is common to use units of *wavelength* rather than frequency. The wavelengths associated with light frequencies are shown in Figure 14–2. There are two common units for wavelength: *nanometer* (nm) and *angstrom* (Å). One nanometer is $10^{-9}$ m, and 1 angstrom is $10^{-10}$ m. Therefore, 1 nanometer is equal to 10 angstroms.

# FIBER OPTIC COMMUNICATIONS SYSTEM

Figure 14–3 shows a simplified block diagram of a fiber optic communications link. The three primary building blocks of the link are the *transmitter*, the *receiver*, and the *fiber guide*. The transmitter consists of an analog or digital interface, a voltage-to-current converter, a light source, and a source-to-fiber light coupler. The fiber guide is either an ultra-pure glass or plastic cable. The receiver includes a fiber-to-light detector coupling device, a photo detector, a current-to-voltage converter, an amplifier, and an analog or digital interface.

In a fiber optic transmitter, the light source can be modulated by a digital or an analog signal. For analog modulation, the input interface matches impedances and limits the input signal amplitude. For digital modulation, the original source may already be in digital form or, if in analog form, it must be converted to a digital pulse stream. For the latter case, an analog-to-digital converter must be included in the interface.

The voltage-to-current converter serves as an electrical interface between the input circuitry and light source. The light source is either a light-emitting diode (LED) or an injection laser diode (ILD). The amount of light emitted by either an LED or an ILD is proportional to the amount of drive current. Thus the voltage-to-current converter converts an input signal voltage to a current which is used to drive the light source.

The source-to-fiber coupler is a mechanical interface. Its function is to couple the light emitted by the source into the optical fiber cable. The optical fiber consists of a glass or plastic fiber core, a cladding, and a protective jacket. The fiber-to-

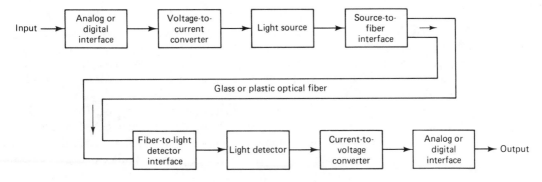

FIGURE 14–3   Fiber optic communications link.

light detector coupling device is also a mechanical coupler. Its function is to couple as much light as possible from the fiber cable into the light detector.

The light detector is very often either a PIN (*positive-intrinsic-negative*) diode or an APD (*avalanche photodiode*). Both the APD and the PIN diode convert light energy to current. Consequently, a current-to-voltage converter is required. The current-to-voltage converter transforms changes in detector current to changes in output signal voltage.

The analog or digital interface at the receiver output is also an electrical interface. If analog modulation is used, the interface matches impedances and signal levels to the output circuitry. If digital modulation is used, the interface must include a digital-to-analog converter.

## OPTICAL FIBERS

### Fiber Types

Essentially, there are three varieties of optical fibers available today. All three varieties are constructed of either glass, plastic, or a combination of glass and plastic. The three varieties are:

1. Plastic core and cladding
2. Glass core with plastic cladding (often called PCS fiber, plastic-clad silica)
3. Glass core and glass cladding (often called SCS, silica-clad silica)

Presently, Bell Laboratories is investigating the possibility of using a fourth variety that uses a *nonsilicate* substance, *zinc chloride*. Preliminary experiments have indicated that fibers made of this substance will be as much as 1000 times as efficient as glass, their silica-based counterpart.

Plastic fibers have several advantages over glass fibers. First, plastic fibers are more flexible and, consequently, more rugged than glass. They are easy to install, can better withstand stress, are less expensive, and weigh approximately 60% less than glass. The disadvantage of plastic fibers is their high attenuation characteristic; they do not propagate light as efficiently as glass. Consequently, plastic fibers are limited to relatively short runs, such as within a single building or a building complex.

Fibers with glass cores exhibit low attenuation characteristics. However, PCS fibers are slightly better than SCS fibers. Also, PCS fibers are less affected by radiation and are therefore more attractive to military applications. SCS fibers have the best propagation characteristics and they are easier to terminate than PCS fibers. Unfortunately, SCS cables are the least rugged, and they are more susceptible to increases in attenuation when exposed to radiation.

The selection of a fiber for a given application is a function of specific system

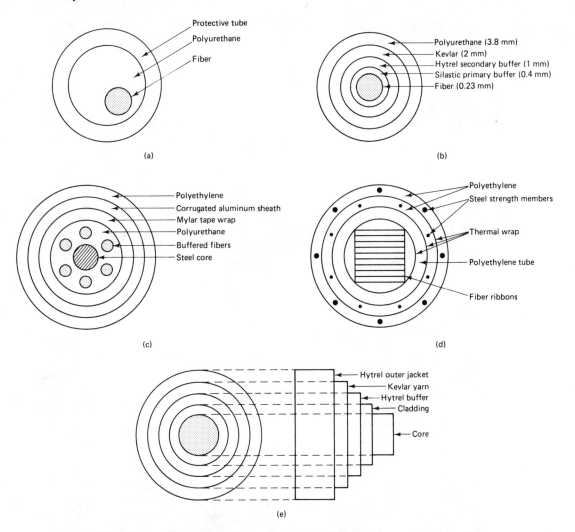

**FIGURE 14–4 Fiber optic cable configurations: (a) loose tube construction; (b) constrained fiber; (c) multiple strands; (d) telephone cable; (e) plastic-clad silica cable.**

requirements. There are always trade-offs based on the economics and logistics of a particular application.

## Fiber Construction

There are many different cable designs available today. Figure 14–4 shows examples of several fiber optic cable configurations. Depending on the configuration, the cable may include a *core*, a *cladding*, a *protective tube*, *buffers*, *strength members*, and one or more *protective jackets*.

With the *loose* tube construction [shown in Figure 14–4(a)] each fiber is contained in a protective tube. Inside the protective tube, a polyurethane compound encapsules the fiber and prevents the intrusion of water.

Figure 14–4(b) shows the construction of a *constrained* fiber cable. Surrounding the fiber cable are a primary and a secondary buffer. The buffer jackets provide protection for the fiber from external mechanical influences which could cause fiber breakage or excessive optical attenuation. Kelvar is a yarn-type material that increases the tensile strength of the cable. Again, an outer protective tube is filled with polyurethane, which prevents moisture from coming into contact with the fiber core.

Figure 14–4(c) shows a *multiple-strand* configuration. To increase the tensile strength, a steel central member and a layer of Mylar tape wrap are included in the package. Figure 14–4(d) shows a *ribbon* configuration, which is frequently seen in telephone systems using fiber optics. Figure 14–4(e) shows both the end and side views of a plastic-clad silica cable.

The type of cable construction used depends on the performance requirements of the system and both the economic and environmental constraints.


## LIGHT PROPAGATION

### The Physics of Light

Although the performance of optical fibers can be analyzed completely by application of Maxwell's equations, this is necessarily complex. For most practical applications, Maxwell's equations may be substituted by the application of *geometric ray tracing*, which will yield a sufficiently detailed analysis.

### Velocity of Propagation

Electromagnetic energy, such as light, travels at approximately 300,000,000 m/s (186,000 miles per second) in free space. Also, the velocity of propagation is the same for all light frequencies in free space. However, it has been demonstrated that in materials more dense than free space, the velocity is reduced. When the velocity of an electromagnetic wave is reduced as it passes from one medium to another medium of a denser material, the light ray is *refracted* (bent) toward the normal. Also, in materials more dense than free space, all light frequencies do not propagate at the same velocity.

### Refraction

Figure 14–5(a) shows how a light ray is refracted as it passes from a material of a given density into a less dense material. (Actually, the light ray is not bent, but rather, it changes direction at the interface.) Figure 14–5(b) shows how sunlight, which contains all light frequencies, is affected as it passes through a material

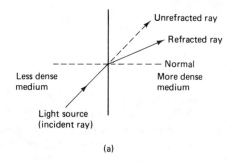

(a)

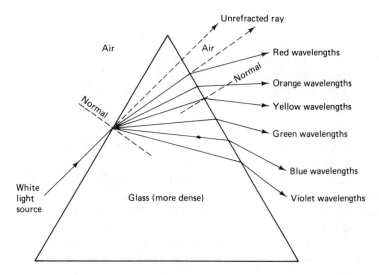

**FIGURE 14–5   Refraction of light: (a) light refraction; (b) prismatic refraction.**

more dense than free space. Refraction occurs at both air/glass interfaces. The violet wavelengths are refracted the most, and the red wavelengths are refracted the least. The spectral separation of white light in this manner is called *prismatic refraction*. It is this phenomenon that causes rainbows; water droplets in the atmosphere act like small prisms that split the white sunlight into the various wavelengths, creating a visible spectrum of color.

## Refractive Index

The amount of bending or refraction that occurs at the interface of two materials of different densities is quite predictable and depends on the *refractive index* (also called *index of refraction*) of the two materials. The refractive index is simply the ratio of the velocity of propagation of a light ray in free space to the velocity of propagation of a light ray in a given material. Mathematically, the refractive index is

$$n = \frac{c}{v}$$

where $c$ = speed of light in free space

$v$ = speed of light in a given material

Although the refractive index is also a function of frequency, the variation in most applicants is insignificant and therefore omitted from this discussion. The indexes of refraction of several common materials are given in Table 14–1.

How a light ray reacts when it meets the interface of two transmissive materials that have different indexes of refraction can be explained with *Snell's law*. Snell's law simply states:

$$n_1 \sin \theta_1 = n_2 \sin \theta_2$$

where $n_1$ = refractive index of material 1

$n_2$ = refractive index of material 2

$\theta_1$ = angle of incidence

$\theta_2$ = angle of refraction

A refractive index model for Snell's law is shown in Figure 14–6. At the interface, the incident ray may be refracted toward the normal or away from it, depending on whether $n_1$ is less than or greater than $n_2$.

Figure 14–7 shows how a light ray is refracted as it travels from a more dense (higher refractive index) material into a less dense (lower refractive index) material. It can be seen that the light ray changes direction at the interface, and the angle of refraction is greater than the angle of incidence. Consequently, when a light ray enters a less dense material, the ray bends away from the normal. The normal is simply a line drawn perpendicular to the interface at the point where the incident ray strikes the interface. Similarly, when a light ray enters a moredense material, the ray bends toward the normal.

**TABLE 14–1  Typical Indexes of Refraction**

| Medium | Index of refraction[a] |
|---|---|
| Vacuum | 1.0 |
| Air | 1.0003 (1.0) |
| Water | 1.33 |
| Ethyl alcohol | 1.36 |
| Fused quartz | 1.46 |
| Glass fiber | 1.5–1.9 |
| Diamond | 2.0–2.42 |
| Silicon | 3.4 |
| Gallium-arsenide | 3.6 |

[a] Index of refraction is based on a wavelength of light emitted from a sodium flame (5890 Å).

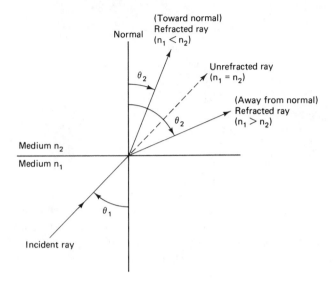

**FIGURE 14–6 Refractive model for Snell's law.**

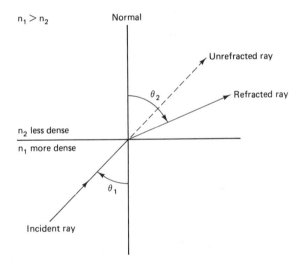

**FIGURE 14–7 Light ray refracted away from the normal.**

**EXAMPLE**

In Figure 14–7, let medium 1 be glass and medium 2 be ethyl alcohol. For an angle of incidence of 30°, determine the angle of refraction.

**Solution** From Table 14–1,

$$n_1(\text{glass}) = 1.5$$

$$n_2(\text{ethyl alcohol}) = 1.36$$

Substituting for $n_1$, $n_2$, and $\theta_1$ gives us

$$\frac{n_1}{n_2} \sin \theta_1 = \sin \theta_2$$

$$\frac{1.5}{1.36} \sin 30 = 0.5514 = \sin \theta_2$$

$$\theta_2 = \sin^{-1} 0.5514 = 33.47°$$

The result indicates that the light ray refracted (bent) or changed direction by 3.47° at the interface. Because the light was traveling from a more dense material into a less dense material, the ray bent away from the normal.

## Critical Angle

Figure 14–8 shows a condition in which an *incident ray* is at an angle such that the angle of refraction is 90° and the refracted ray is along the interface. (It is important to note that the light ray is traveling from a medium of higher refractive index to a medium with a lower refractive index.) Again, using Snell's law,

$$\sin \theta_1 = \frac{n_2}{n_1} \sin \theta_2$$

With $\theta_2 = 90°$,

$$\sin \theta_1 = \frac{n_2}{n_1} (1) \qquad \text{or} \qquad \sin \theta_1 = \frac{n_2}{n_1}$$

and

$$\sin^{-1} \frac{n_2}{n_1} = \theta_1 = \theta_c$$

where $\theta_c$ is the critical angle.

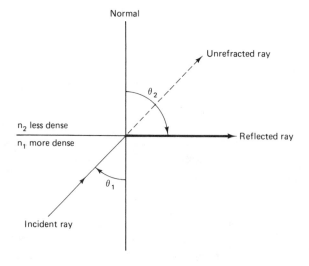

**FIGURE 14–8   Critical angle reflection.**

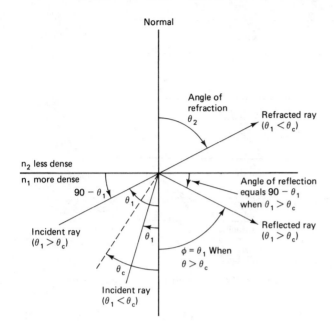

**FIGURE 14–9** **Angle of reflection and refraction.**

The *critical angle* is defined as the minimum angle of incidence at which a light ray may strike the interface of two media and result in an angle of refraction of 90° or greater. (This definition pertains only when the light ray is traveling from a more dense medium into a less dense medium.) If the angle of refraction is 90° or greater, the light ray is not allowed to penetrate the less dense material. Consequently, total reflection takes place at the interface, and the angle of reflection is equal to the angle of incidence. Figure 14–9 shows a comparison of the angle of refraction and the angle of reflection when the angle of incidence is less than or more than the critical angle.

## PROPAGATION OF LIGHT THROUGH AN OPTICAL FIBER

Light can be propagated down an optical fiber cable by either reflection or refraction. How the light is propagated depends on the *mode of propagation* and the *index profile* of the fiber.

### Mode of Propagation

In fiber optics terminology, the word *mode* simply means path. If there is only one path for light to take down the cable, it is called *single mode*. If there is more than one path, it is called *multimode*. Figure 14–10 shows single and multimode propagation of light down an optical fiber.

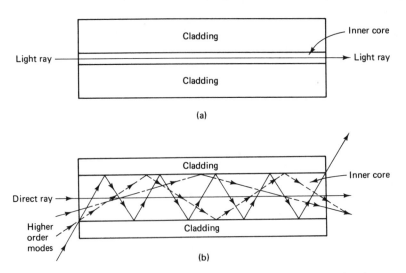

**FIGURE 14–10    Modes of propagation: (a) single mode; (b) multimode.**

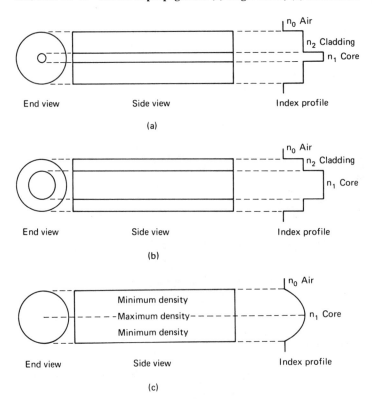

**FIGURE 14–11    Core index profiles: (a) single-mode step index; (b) multimode step index; (c) multimode graded index.**

### Index Profile

The index profile of an optical fiber is a graphical representation of the refractive index of the core. The refractive index is plotted on the horizontal axis and the radial distance from the core axis is plotted on the vertical axis. Figure 14–11 shows the core index profiles of three types of fiber cables.

There are two basic types of index profiles: step and graded. A *step-index fiber* has a central core with a uniform refractive index. The core is surrounded by an outside cladding with a uniform refractive index less than that of the central core. From Figure 14–11 it can be seen that in a step-index fiber there is an abrupt change in the refractive index at the core/cladding interface. In a *graded-index fiber* there is no cladding, and the refractive index of the core is nonuniform; it is highest at the center and decreases gradually toward the outer edge.

## OPTICAL FIBER CONFIGURATIONS

Essentially, there are three types of optical fiber configurations: single-mode step-index, multimode step-index, and multimode graded-index.

### Single-Mode Step-Index Fiber

A *single-mode step-index fiber* has a central core that is sufficiently small so that there is essentially only one path that light may take as it propagates down the cable. This type of fiber is shown in Figure 14–12. In the simplest form of single-mode step-index fiber, the outside cladding is simply air [Figure 14–12(a)]. The refractive index of the glass core ($n_1$) is approximately 1.5, and the refractive index of the air cladding ($n_0$) is 1. The large difference in the refractive indexes results in a small critical angle (approximately 42°) at the glass/air interface. Consequently, the fiber will accept light from a wide aperture. This makes it relatively easy to couple light from a source into the cable. However, this type of fiber is typically very weak and of limited practical use.

A more practical type of single-mode step-index fiber is one that has a cladding other than air [Figure 14–12(b)]. The refractive index of the cladding ($n_2$) is slightly less than that of the central core ($n_1$) and is uniform throughout the cladding. This type of cable is physically stronger than the air-clad fiber, but the critical angle is also much higher (approximately 77°). This results in a small acceptance angle and a narrow source-to-fiber aperture, making it much more difficult to couple light into the fiber from a light source.

With both types of single-mode step-index fibers, light is propagated down the fiber through reflection. Light rays that enter the fiber propagate straight down the core or, perhaps, are reflected once. Consequently, all light rays follow approximately the same path down the cable and take approximately the same amount of

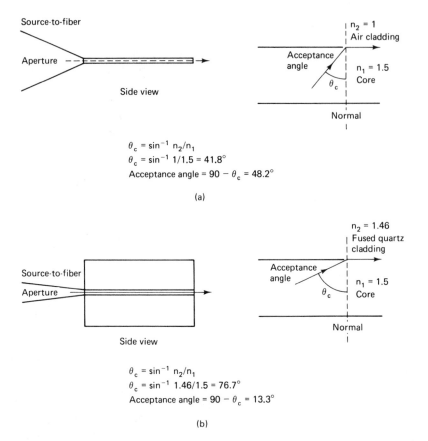

$\theta_c = \sin^{-1} n_2/n_1$

$\theta_c = \sin^{-1} 1/1.5 = 41.8°$

Acceptance angle = $90 - \theta_c = 48.2°$

(a)

$\theta_c = \sin^{-1} n_2/n_1$

$\theta_c = \sin^{-1} 1.46/1.5 = 76.7°$

Acceptance angle = $90 - \theta_c = 13.3°$

(b)

**FIGURE 14–12   Single-mode step-index fibers: (a) air cladding; (b) glass cladding.**

time to travel the length of the cable. This is one overwhelming advantage of single-mode step-index fibers and will be explained in more detail later.

## Multimode Step-Index Fiber

A *multimode step-index fiber* is shown in Figure 14–13. It is similar to the single-mode configuration except that the center core is much larger. This type of fiber has a larger light-to-fiber aperture and, consequently, allows more light to enter the cable. The light rays that strike the core/cladding interface at an angle greater than the critical angle (ray A) are propagated down the core in a zigzag fashion, continuously reflecting off the interface boundary. Light rays that strike the core/cladding interface at an angle less than the critical angle (ray B) enter the cladding and are lost. It can be seen that there are many paths that a light ray may follow as it propagates down the fiber. As a result, all light rays do not follow the same

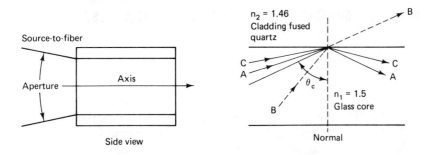

**FIGURE 14–13** Multimode step-index fiber.

path and consequently, do not take the same amount of time to travel the length of the fiber.

## Multimode Graded-Index Fiber

A *multimode graded-index fiber* is shown in Figure 14–14. A multimode graded-index fiber is characterized by a central core that has a refractive index that is nonuniform; it is maximum at the center and decreases gradually toward the outer edge. Light is propagated down this type of fiber through refraction. As a light ray propagates diagonally across the core, it is continually intersecting a less-dense-to-more dense interface. Consequently, the light rays are constantly being refracted, which results in a continuous bending of the light rays. Light enters the fiber at many different angles. As they propagate down the fiber, the light rays that travel in the outermost area of the fiber travel a greater distance than the rays traveling near the center. Because the refractive index decreases with distance from the center and the velocity is inversely proportional to the refractive index, the light rays traveling farthest from the center propagate at a higher velocity. Consequently, they take approximately the same amount of time to travel the length of the fiber.

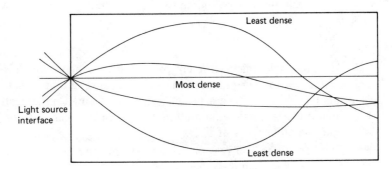

**FIGURE 14–14** Multimode graded-index fiber.

# COMPARISON OF THE THREE TYPES OF OPTICAL FIBERS

## Single-Mode Step-Index Fiber

*Advantages*

1. There is minimum dispersion. Because all rays propagating down the fiber take approximately the same path, they take approximately the same amount of time to travel down the cable. Consequently, a pulse of light entering the cable can be reproduced at the receiving end very accurately.
2. Because of the high accuracy in reproducing transmitted pulses at the receive end, larger bandwidths and higher information transmission rates are possible with single-mode step-index fibers than with the other types of fibers.

*Disadvantages*

1. Because the central core is very small, it is difficult to couple light into and out of this type of fiber. The source-to-fiber aperture is the smallest of all the fiber types.
2. Again, because of the small central core, a highly directive light source such as a laser is required to couple light into a single-mode step-index fiber.
3. Single-mode step-index fibers are expensive and difficult to manufacture.

## Multimode Step-Index Fiber

*Advantages*

1. Multimode step-index fibers are inexpensive and simple to manufacture.
2. It is easy to couple light into and out of multimode step-index fibers; they have a relatively large source-to-fiber aperture.

*Disadvantages*

1. Light rays take many different paths down the fiber, which results in large differences in their propagation times. Because of this, rays traveling down this type of fiber have a tendency to spread out. Consequently, a pulse of light propagating down a multimode step-index fiber is distorted more than with the other types of fibers.
2. The bandwidth and rate of information transfer possible with this type of cable are less than the other types.

## Multimode Graded-Index Fiber

Essentially, there are no outstanding advantages or disadvantages of this type of fiber. Multimode graded-index fibers are easier to couple light into and out of than single-mode step-index fibers but more difficult than multimode step-index fibers. Distortion due to multiple propagation paths is greater than in single-mode step-index fibers but less than in multiple step-index fibers. Graded-index fibers are easier to manufacture than single-mode step-index fibers but more difficult than multimode step-index fibers. The multimode graded-index fiber is considered an intermediate fiber compared to the other types.

## ACCEPTANCE ANGLE AND ACCEPTANCE CONE

In previous discussions, the *source-to-fiber aperture* was mentioned several times, and the *critical* and *acceptance* angles at the point where a light ray strikes the core/cladding interface were explained. The following discussion deals with the light-gathering ability of the fiber, the ability to couple light from the source into the fiber cable.

Figure 14–15 shows the source end of a fiber cable. When light rays enter the fiber, they strike the air/glass interface at normal $A$. The refractive index of air is 1 and the refractive index of the glass core is 1.5. Consequently, the light entering at the air/glass interface propagates from a less dense medium into a more dense medium. Under these conditions and according to Snell's law, the light rays will refract toward the normal. This causes the light rays to change direction and propagate diagonally down the core at an angle ($\theta_c$) which is different than the external angle

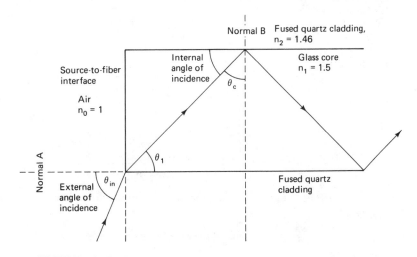

**FIGURE 14–15   Ray propagation into and down an optical fiber cable.**

of incidence at the air/glass interface ($\theta_{in}$). In order for a ray of light to propagate down the cable, it must strike the internal core/cladding interface at an angle that is greater than the critical angle ($\theta_c$).

Applying Snell's law to the external angle of incidence yields the following expression:

$$n_0 \sin \theta_{in} = n_1 \sin \theta_1$$

and

$$\theta_1 = 90 - \theta_c$$

Thus

$$\sin \theta_1 = \sin (90 - \theta_c) = \cos \theta_c$$

combining yields

$$n_0 \sin \theta_{in} = n_1 \cos \theta_c$$

Rearranging and solving for $\sin \theta_{in}$ gives us

$$\sin \theta_{in} = \frac{n_1}{n_0} \cos \theta_c$$

Figure 14–16 shows the geometric relationship.

From Figure 14–16 and using the Pythagorean theorem, we obtain

$$\cos \theta_c = \frac{\sqrt{n_1^2 - n_2^2}}{n_1}$$

Substituting for $\cos \theta_c$ yields

$$\sin \theta_{in} = \frac{n_1}{n_0} \frac{\sqrt{n_1^2 - n_2^2}}{n_1}$$

Reducing gives

$$\sin \theta_{in} = \frac{\sqrt{n_1^2 - n_2^2}}{n_0}$$

and

$$\theta_{in} = \sin^{-1} \frac{\sqrt{n_1^2 - n_2^2}}{n_0}$$

Because light rays generally enter the fiber from an air medium, $n_0$ equals 1. This simplifies to

$$\theta_{in(max)} = \sin^{-1} \sqrt{n_1^2 - n_2^2}$$

$\theta_{in}$ is called the *acceptance angle* or *acceptance cone* half-angle. It defines the maximum angle in which external light rays may strike the air/fiber interface and

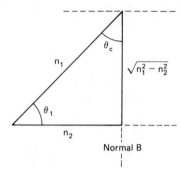

**FIGURE 14–16  Geometric relationship of Equation 10–5.**

still propagate down the fiber with a response that is no greater than 10 dB down from the peak value. Rotating the acceptance angle around the fiber axis describes the acceptance cone of the fiber. This is shown in Figure 14–17.

## Numerical Aperture

*Numerical aperture* (NA) is a figure of merit that is used to measure the light-gathering or light-collecting ability of an optical fiber. The larger the magnitude of NA, the greater the amount of light accepted by the fiber from the external light source. For a step-index fiber, numerical aperture is mathematically defined as the sin of the acceptance half-angle. Thus

$$NA = \sin \theta_{in}$$

and

$$NA = \sqrt{n_1^2 - n_2^2}$$

Also,

$$\sin^{-1} NA = \theta_{in}$$

For a graded index, NA is simply the sin of the critical angle:

$$NA = \sin \theta_c$$

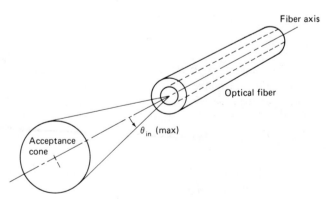

**FIGURE 14–17  Acceptance cone of a fiber cable.**

**EXAMPLE**

For this example refer to Figure 14–15. For a multimode step-index fiber within a glass core ($n_1 = 1.5$) and a fused quartz cladding ($n_2 = 1.46$), determine the critical angle ($\theta_c$), acceptance angle ($\theta_{in}$), and numerical aperture. The source-to-fiber media is air.

**Solution**   Substituting for $n_1$ and $n_2$ gives us

$$\theta_c = \sin^{-1} \frac{n_2}{n_1} = \sin^{-1} \frac{1.46}{1.5} = 76.7°$$

and

$$\theta_{in} = \sin^{-1} \sqrt{n_1^2 - n_2^2} = \sin^{-1} \sqrt{1.5^2 - 1.46^2}$$

$$= 20.2°$$

thus

$$NA = \sin \theta_{in} = \sin 20.2$$

$$= 0.344$$

## LOSSES IN OPTICAL FIBER CABLES

Transmission losses in optical fiber cables are one of the most important characteristics of the fiber. Losses in the fiber result in a reduction in the light power and thus reduce the system bandwidth, information transmission rate, efficiency, and overall system capacity. The predominant fiber losses are as follows:

1. Absorption losses
2. Material or Rayleigh scattering losses
3. Chromatic or wavelength dispersion
4. Radiation losses
5. Modal dispersion
6. Coupling losses

### Absorption Losses

*Absorption loss* in optical fibers is analogous to power dissipation in copper cables; impurities in the fiber absorb the light and convert it to heat. The ultrapure glass used to manufacture optical fibers is approximately 99.9999% pure. Still, absorption losses between 1 and 1000 dB/km are typical. Essentially, there are three factors that contribute to the absorption losses in optical fibers: ultraviolet absorption, infrared absorption, and ion resonance absorption.

*Ultraviolet absorption.* Ultraviolet absorption is caused by valence electrons in the silica material from which fibers are manufactured. Light *ionizes* the valence

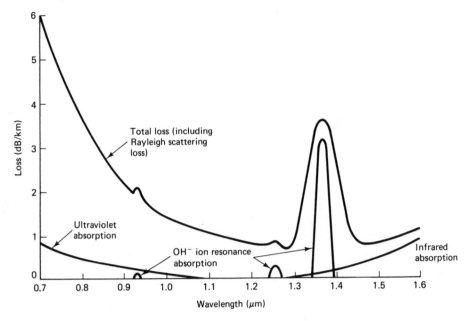

**FIGURE 14–18  Absorption losses in optical fibers.**

electrons into conduction. The ionization is equivalent to a loss in the total light field and, consequently, contributes to the transmission losses of the fiber.

*Infrared absorption.* Infrared absorption is a result of *photons* of light that are absorbed by the atoms of the glass core molecules. The absorbed photons are converted to random mechanical vibrations typical of heating.

*Ion resonance absorption.* Ion resonance absorption is caused by $OH^-$ ions in the material. The source of the $OH^-$ ions in water molecules that have been trapped in the glass during the manufacturing process. Ion absorption is also caused by iron, copper, and chromium molecules.

Figure 14–18 shows typical losses in optical fiber cables due to ultraviolet, infrared, and ion resonance absorption.

## Material or Rayleigh Scattering Losses

During the manufacturing process, glass is extruded (drawn into long fibers of very small diameter). During this process, the glass is in a plastic state (not liquid and not solid). The tension applied to the glass during this process causes the cooling glass to develop submicroscopic irregularities that are permanently formed in the fiber. When light rays that are propagating down a fiber strike one of these impurities, they are *diffracted*. Diffraction causes the light to disperse or spread out in many directions. Some of the diffracted light continues down the fiber and some of it escapes through the cladding. The light rays that escape represent a loss in light

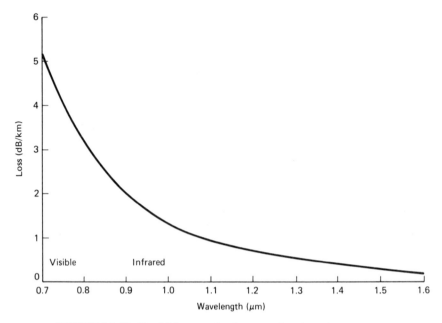

**FIGURE 14–19**   Rayleigh scattering loss as a function of wavelength.

power. This is called *Rayleigh scattering loss*. Figure 14–19 graphically shows the relationship between wavelength and Rayleigh scattering loss.

## Chromatic or Wavelength Dispersion

As stated previously, the refractive index of a material is wavelength dependent. Light-emitting diodes (LEDs) emit light that contains a combination of wavelengths. Each wavelength within the composite light signal travels at a different velocity. Consequently, light rays that are simultaneously emitted from an LED and propagated down an optical fiber do not arrive at the far end of the fiber at the same time. This results in a distorted receive signal and is called *chromatic distortion*. Chromatic distortion can be eliminated by using a monochromatic source such as an injection laser diode (ILD).

## Radiation Losses

*Radiation losses* are caused by small bends and kinks in the fiber. Essentially, there are two types of bends: microbends and constant-radius bends. *Microbending* occurs as a result of differences in the thermal contraction rates between the core and cladding material. A microbend represents a discontinuity in the fiber where Rayleigh scattering can occur. *Constant-radius bends* occur when fibers are bent during handling or installation.

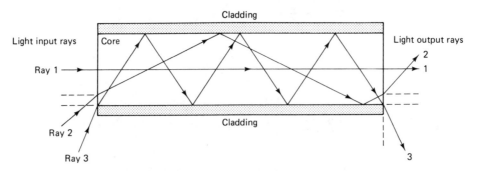

**FIGURE 14–20    Light propagation down a multimode step-index fiber.**

## Modal Dispersion

*Modal dispersion*, or *pulse spreading*, is caused by the difference in the propagation times of light rays that take different paths down a fiber. Obviously, modal dispersion can occur only in multimode fibers. It can be reduced considerably by using graded-index fibers and almost entirely eliminated by using single-mode step-index fibers.

Modal dispersion can cause a pulse of light energy to spread out as it propagates down a fiber. If the pulse spreading is sufficiently severe, one pulse may fall back on top of the next pulse (this is an example of intersymbol inference). In a multimode step-index fiber, a light ray that propagates straight down the axis of the fiber takes the least amount of time to travel the length of the fiber. A light ray that strikes the core/cladding interface at the critical angle will undergo the largest number of internal reflections and, consequently, take the longest time to travel the length of the fiber.

Figure 14–20 shows three rays of light propagating down a multimode step-index fiber. The lowest-order mode (ray 1) travels in a path parallel to the axis of the fiber. The middle-order mode (ray 2) bounces several times at the interface before traveling the length of the fiber. The highest-order mode (ray 3) makes many trips back and forth across the fiber as it propagates the entire length. It can be seen that ray 3 travels a considerably longer distance than ray 1 as it propagates down the fiber. Consequently, if the three rays of light were emitted into the fiber at the same time and represented a pulse of light energy, the three rays would reach the far end of the fiber at different times and result in a spreading out of the light energy in respect to time. This is called modal dispersion and results in a stretched pulse which is also reduced in amplitude at the output of the fiber. All three rays of light propagate through the same material at the same velocity, but ray 3 must travel a longer distance and, consequently, takes a longer period of time to propagate down the fiber.

Figure 14–21 shows light rays propagating down a single-mode step-index fiber. Because the radial dimension of the fiber is sufficiently small, there is only a single path for each of the rays to follow as they propagate down the length of the fiber. Consequently, each ray of light travels the same distance in a given period

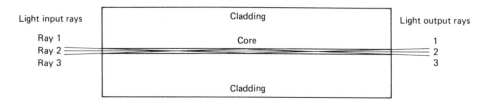

**FIGURE 14–21**  **Light propagation down a single-mode step-index fiber.**

of time and the light rays have exactly the same time relationship at the far end of the fiber as they had when they entered the cable. The result is no *modal dispersion* or *pulse stretching*.

Figure 14–22 shows light propagating down a multimode graded-index fiber. Three rays are shown traveling in three different modes. Each ray travels a different path but they all take approximately the same amount of time to propagate the length of fiber. This is because the refractive index of the fiber decreases with distance from the center, and the velocity at which a ray travels is inversely proportional to the refractive index. Consequently, the farther rays 2 and 3 travel from the center of the fiber, the faster they propagate.

Figure 14–23 shows the relative time/energy relationship of a pulse of light as it propagates down a fiber cable. It can be seen that as the pulse propagates down the fiber, the light rays that make up the pulse spread out in time, which causes a corresponding reduction in the pulse amplitude and stretching of the pulse width. It can also be seen that as light energy from one pulse falls back in time, it will interfere with the next pulse. This is called *pulse spreading* or *pulse-width dispersion* and causes errors in digital transmission.

Figure 14–24(a) shows a unipolar return-to-zero (UPRZ) digital transmission. With UPRZ transmission (assuming a very narrow pulse) if light energy from pulse A were to fall back (*spread*) one bit time ($T_b$), it would interfere with pulse B and change what was a logic 0 to a logic 1. Figure 14–24(b) shows a unipolar nonreturn-to-zero (UPNRZ) digital transmission where each pulse is equal to the bit time. With UPNRZ transmission, if energy from pulse A were to fall back one-half of a bit time, it would interfere with pulse B. Consequently, UPRZ transmissions can tolerate twice a smuch delay or spread as UPNRZ transmissions.

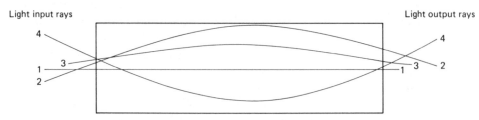

**FIGURE 14–22**  **Light propagation down a multimode graded-index fiber.**

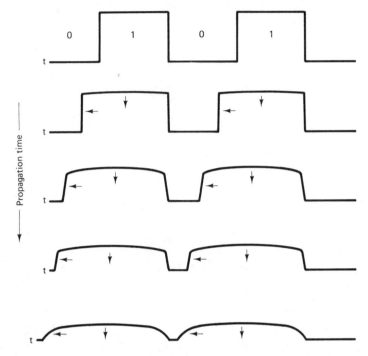

**FIGURE 14–23  Pulse-width dispersion in an optical fiber cable.**

The difference between the absolute delay times of the fastest and slowest rays of light propagating down a fiber is called the *pulse-spreading constant* ($\Delta t$) and is generally expressed in nanoseconds per kilometer (ns/km). The total pulse spread ($\Delta T$) is then equal to the pulse spreading constant ($\Delta t$) times the total fiber length ($L$). Mathematically, $\Delta T$ is

$$\Delta T \text{ (ns)} = \Delta t \left(\frac{\text{ns}}{\text{km}}\right) \times L \text{ (km)}$$

For UPRZ transmissions, the maximum data transmission rate in bits per second (bps) is expressed as

$$F_b \text{ (bps)} = \frac{1}{\Delta t \times L}$$

and for UPNRZ transmissions, the maximum transmission rate is

$$F_b \text{ (bps)} = \frac{1}{2\,\Delta t \times L}$$

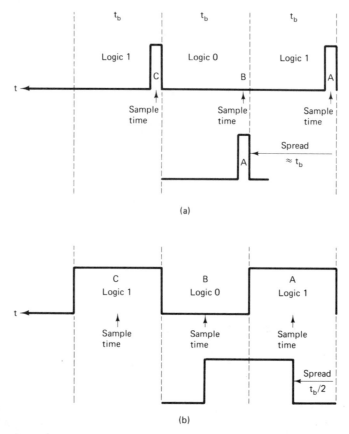

**FIGURE 14–24  Pulse spreading of digital transmissions: (a) UPRZ: (b) UPNRZ.**

**EXAMPLE**

For an optical fiber 10 km long with a pulse-spreading constant of 5 ns/km, determine the maximum digital transmission rates for (a) return-to-zero and (b) nonreturn-to-zero transmissions.

**Solution**  For return-to-zero transmission

$$F_b = \frac{1}{5 \text{ ns/km} \times 10 \text{ km}} = 20 \text{ Mbps}$$

For non-return-to-zero transmission

$$F_b = \frac{1}{(2 \times 5 \text{ ns/km}) \times 10 \text{ km}} = 10 \text{ Mbps}$$

The results indicate that the digital transmission rate possible for this optical fiber is twice as high (20 Mbps versus 10 Mbps) for UPRZ as for UPNRZ transmission.

## Coupling Losses

In fiber cables coupling losses can occur at any of the following three types of optical junctions: light source-to-fiber connections, fiber-to-fiber connections, and fiber-to-photodetector connections. Junction losses are most often caused by one of the following alignment problems: lateral misalignment, gap misalignment, angular misalignment, and imperfect surface finishes. These impairments are shown in Figure 14–25.

*Lateral misalignment.* This is shown in Figure 14–25(a) and is the lateral or axial displacement between two pieces of adjoining fiber cables. The amount of

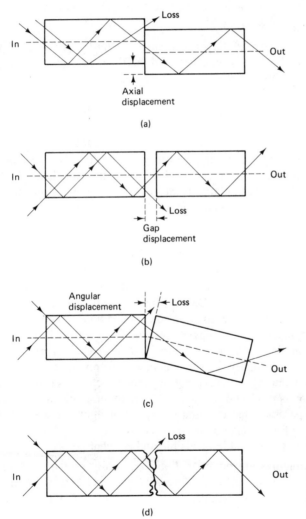

(a)

(b)

(c)

(d)

**FIGURE 14–25** Fiber alignment impairments: (a) lateral misalignment; (b) gap displacement; (c) angular misalignment; (d) surface finish.

loss can be from a couple of tenths of a decibel to several decibels. This loss is generally negligible if the fiber axes are aligned to within 5% of the smaller fiber's diameter.

*Gap misalignment.* This is shown in Figure 14–25(b) and is sometimes called *end separation.* When *splices* are made in optical fibers, the fibers should actually touch. The farther apart the fibers are, the greater the loss of light. If two fibers are joined with a connector, the ends should not touch. This is because the two ends rubbing against each other in the connector could cause damage to either or both fibers.

*Angular misalignment.* This is shown in Figure 14–25(c) and is sometimes called *angular displacement.* If the angular displacement is less than 2°, the loss will be less than 0.5 dB.

*Imperfect surface finish.* This is shown in Figure 14–25(d). The ends of the two adjoining fibers should be highly polished and fit together squarely. If the fiber ends are less than 3° off from perpendicular, the losses will be less than 0.5 dB.

# LIGHT SOURCES

Essentially, there are two devices commonly used to generate light for fiber optic communications systems: light-emitting diodes (LEDs) and injection laser diodes (ILDs). Both devices have advantages and disadvantages and selection of one device over the other is determined by system requirements.

## Light-Emitting Diodes

Essentially, a *light-emitting diode* (LED) is simply a P-N junction diode. It is usually made from a semiconductor material such as aluminum-gallium-arsenide (AlGaAs) or gallium-arsenide-phosphide (GaAsP). LEDs emit light by spontaneous emission; light is emitted as a result of the recombination of electrons and holes. When forward biased, minority carriers are injected across the *p-n* junction. Once across the junction, these minority carriers recombine with majority carriers and give up energy in the form of light. This process is essentially the same as in a conventional diode except that in LEDs certain semiconductor materials and dopants are chosen such that the process is radiative; a photon is produced. A photon is a quantum of electromagnetic wave energy. Photons are particles that travel at the speed of light but at rest have no mass. The energy gap of the material used to construct an LED determines whether the light emitted by it is invisible or visible and of what color.

The simplest LED structures are homojunction, epitaxially grown, or single-diffused devices and are shown in Figure 14–26. *Epitaxially grown LEDs* are generally constructed of silicon-doped gallium-arsenide [Figure 14–26(a)]. A typical wave-

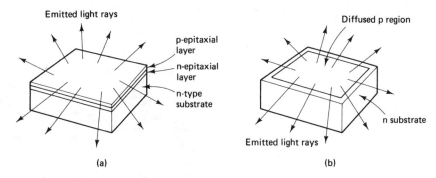

(a)                                              (b)

**FIGURE 14–26   Homojunction LED structures: (a) silicon-doped gallium arsenide; (b) planar diffused.**

length of light emitted from this construction is 940 nm, and a typical output power is approximately 3 mW at 100 mA of forward current. *Planar diffused* (*homojunction*) *LEDs* [Figure 14–26(b)] output approximately 500 μW at a wavelength of 900 nm. The primary disadvantage of homojunction LEDs is the nondirectionality of their light emission, which makes them a poor choice as a light source for fiber optic systems.

The *planar heterojunction LED* (Figure 14–27) is quite similar to the epitaxially grown LED except that the geometry is designed such that the forward current is concentrated to a very small area of the active layer. Because of this the planar heterojunction LED has several advantages over the homojunction type. They are:

1. The increase in current density generates a more brilliant light spot.
2. The smaller emitting area makes it easier to couple its emitted light into a fiber.
3. The small effective area has a smaller capacitance, which allows the planar heterojunction LED to be used at higher speeds.

**Burrus etched-well surface-emitting LED.**   For the more practical applications, such as telecommunications, data rates in excess of 100 Mbps are required.

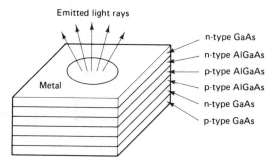

**FIGURE 14–27   Planar heterojunction LED.**

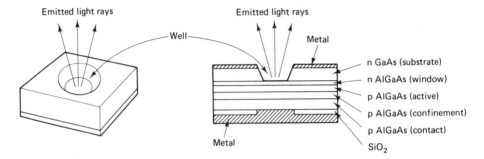

**FIGURE 14–28**    Burrus etched-well surface-emitting LED.

For these applications, the etched-well LED was developed. Burrus and Dawson of Bell laboratories developed the etched-well LED. It is a surface-emitting LED and is shown in Figure 14–28. The Burrus etched-well LED emits light in many directions. The etched well helps concentrate the emitted light to a very small area. Also, domed lenses can be placed over the emitting surface to direct the light into a smaller area. These devices are more efficient than the standard surface emitters and they allow more power to be coupled into the optical fiber, but they are also more difficult and expensive to manufacture.

**Edge-emitting LED.**    The edge-emitting LED, which was developed by RCA, is shown in Figure 14–29. These LEDs emit a more directional light pattern than do the surface-emitting LEDs. The construction is similar to the planar and Burrus diodes except that the emitting surface is a stripe rather than a confined circular area. The light is emitted from an active stripe and forms an elliptical beam. Surface-emitting LEDs are more commonly used than edge emitters because they emit more light. However, the coupling losses with surface emitters are greater and they have narrower bandwidths.

The *radiant* light power emitted from an LED is a linear function of the forward current passing through the device (Figure 14–30). It can also be seen

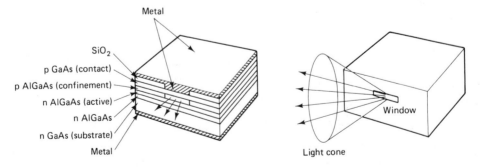

**FIGURE 14–29**    Edge-emitting LED.

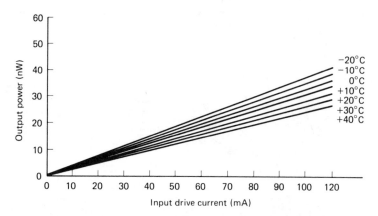

**FIGURE 14–30** Output power versus forward current and operating temperature for an LED.

that the optical output power of an LED is, in part, a function of the operating temperature.

## Injection Laser Diode

The word *laser* is an acronym for *l*ight *a*mplification by *s*timulated *e*mission of *r*adiation. Lasers are constructed from many different materials, including gases, liquids, and solids, although the type of laser used most often for fiber optic communications is the semiconductor laser.

The *injection laser diode* (ILD) is similar to the LED. In fact, below a certain threshold current, an ILD acts like an LED. Above the threshold current, an ILD oscillates; lasing occurs. As current passes through a forward-biased *p-n* junction diode, light is emitted by spontaneous emission at a frequency determined by the energy gap of the semiconductor material. When a particular current level is reached, the number of minority carriers and photons produced on either side of the *p-n* junction reaches a level where they begin to collide with already excited minority carriers. This causes an increase in the ionization energy level and makes the carriers unstable. When this happens, a typical carrier recombines with an opposite type of carrier at an energy level that is above its normal before-collision value. In the process, two photons are created; one is stimulated by another. Essentially, a gain in the number of photons is realized. For this to happen, a large forward current that can provide many carriers (holes and electrons) is required.

The construction of an ILD is similar to that of an LED (Figure 14–31) except that the ends are highly polished. The mirror-like ends trap the photons in the active region and, as they reflect back and forth, stimulate free electrons to recombine with holes at a higher-than-normal energy level. This process is called *lasing*.

The radiant output light power of a typical ILD is shown in Figure 14–32. It can be seen that very little output power is realized until the threshold current is

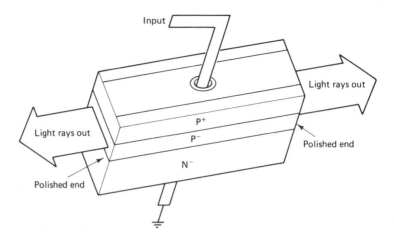

**FIGURE 14–31   Injection laser diode construction.**

reached; then lasing occurs. After lasing begins, the optical output power increases dramatically, with small increases in drive current. It can also be seen that the magnitude of the optical output power of the ILD is more dependent on operating temperature than is the LED.

Figure 14–33 shows the light radiation patterns typical of an LED and an ILD. Because light is radiated out the end of an ILD in a narrow concentrated beam, it has a more direct radiation pattern.

*Advantages of ILDs*

1. Because ILDs have a more direct radiation pattern, it is easier to couple their light into an optical fiber. This reduces the coupling losses and allows smaller fibers to be used.

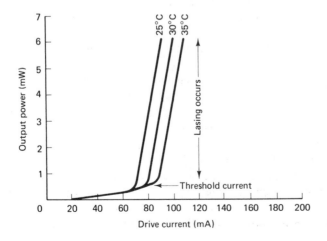

**FIGURE 14–32   Output power versus forward current and temperature for an ILD.**

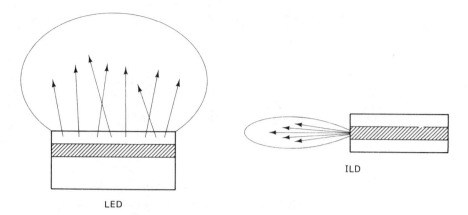

**FIGURE 14–33  LED and ILD radiation patterns.**

2. The radiant output power from an ILD is greater than that for an LED. A typical output power for an ILD is 5 mW (7 dBm) and 0.5 mW (−3 dBm) for LEDs. This allows ILDs to provide a higher drive power and to be used for systems that operate over longer distances.
3. ILDs can be used at higher bit rates than can LEDs.
4. ILDs generate monochromatic light, which reduces chromatic or wavelength dispersion.

*Disadvantages of ILDs*

1. ILDs are typically on the order of 10 times more expensive than LEDs.
2. Because ILDs operate at higher powers, they typically have a much shorter lifetime than LEDs.
3. ILDs are more temperature dependent than LEDs.

## LIGHT DETECTORS

There are two devices that are commonly used to detect light energy in fiber optic communications receivers; PIN (positive-intrinsic-negative) diodes and APD (avalanche photodiodes).

### PIN Diodes

A *PIN diode* is a *depletion-layer photodiode* and is probably the most common device used as a light detector in fiber optic communications systems. Figure 14–34 shows the basic construction of a PIN diode. A very lightly doped (almost pure or intrinsic) layer of *n*-type semiconductor material is sandwiched between the junction

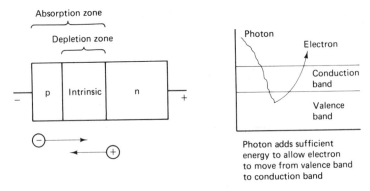

**FIGURE 14–34   PIN photodiode construction.**

of the two heavily doped *n*- and *p*-type contact areas. Light enters the device through a very small window and falls on the carrier-void intrinsic material. The intrinsic material is made thick enough so that most of the photons that enter the device are absorbed by this layer. Essentially, the PIN photodiode operates just the opposite of an LED. Most of the photons are absorbed by electrons in the valence band of the intrinsic material. When the photons are absorbed, they add sufficient energy to generate carriers in the depletion region and allow current to flow through the device.

**Photoelectric effect.**   Light entering through the window of a PIN diode is absorbed by the intrinsic material and adds enough energy to cause electrons to move from the valence band into the conduction band. The increase in the number of electrons that move into the conduction band is matched by an increase in the number of holes in the valence band. To cause current to flow in a photodiode, sufficient light must be absorbed to give valence electrons enough energy to jump the energy gap. The energy gap for silicon is 1.12 eV (electrode volts). Mathematically, the operation is as follows.

For silicon, the energy gap ($E_g$) equals 1.12 eV:

$$1 \text{ eV} = 1.6 \times 10^{-19} \text{ J}$$

Thus the energy gap for silicon is

$$E_g = (1.12 \text{ eV}) \left( 1.6 \times 10^{-19} \frac{\text{J}}{\text{eV}} \right) = 1.792 \times 10^{-19} \text{ J}$$

and

$$\text{energy } (E) = hf$$

where $h$ = Planck's constant = $6.6256 \times 10^{-34}$ J/Hz
$f$ = frequency (Hz)

Rearranging and solving for $f$ yields

$$f = \frac{E}{h}$$

For a silicon photodiode,

$$f = \frac{1.792 \times 10^{-19} \text{ J}}{6.6256 \times 10^{-34} \text{ J/Hz}}$$

$$= 2.705 \times 10^{14} \text{ Hz}$$

Converting to wavelength yields

$$\lambda = \frac{c}{f} = \frac{3 \times 10^8 \text{ m/s}}{2.705 \times 10^{14} \text{ Hz}} = 1109 \text{ nm/cycle}$$

Consequently, light wavelengths of 1109 nm or shorter, or light frequencies of $2.705 \times 10^{14}$ Hz or higher, are required to generate enough electrons to jump the energy gap of a silicon photodiode.

## Avalanche Photodiodes

Figure 14–35 shows the basic construction of an *avalanche photodiode* (APD). An APD is a *pipn* structure. Light enters the diode and is absorbed by the thin, heavily doped *n*-layer. This causes a high electric field intensity to be developed across the *i-p-n* junction. The high reverse-biased field intensity causes impact ionization to occur near the breakdown voltage of the junction. During impact ionization, a carrier can gain sufficient energy to ionize other bound electrons. These ionized carriers, in turn, cause more ionizations to occur. The process continues like an avalanche and is, effectively, equivalent to an internal gain or carrier multiplication. Consequently, APDs are more sensitive than PIN diodes and require less additional amplification. The disadvantages of APDs are relatively long transit times and additional internally generated noise due to the avalanche multiplication factor.

## Characteristics of Light Detectors

The most important characteristics of light detectors are:

*Responsitivity.* This is a measure of the conversion efficiency of a photodetector. It is the ratio of the output current of a photodiode to the input optical power and

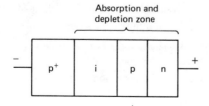

**FIGURE 14–35  Avalanche photodiode construction.**

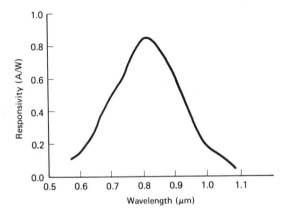

**FIGURE 14-36** **Spectral response curve.**

has the unit of amperes/watt. Responsivity is generally given for a particular wavelength or frequency.

*Dark current.* This is the leakage current that flows through a photodiode with no light input. Dark current is caused by thermally generated carriers in the diode.

*Transit time.* This is the time it takes a light-induced carrier to travel across the depletion area. This parameter determines the maximum bit rate possible with a particular photodiode.

*Spectral response.* This parameter determines the range or system length that can be achieved for a given wavelength. Generally, relative spectral response is graphed as a function of wavelength or frequency. Figure 14-36 is an illustrative example of a spectral response curve. It can be seen that this particular photodiode more efficiently absorbs energy in the range 800 to 820 nm.

# QUESTIONS

1. Define a fiber optic system.
2. What is the relationship between information capacity and bandwidth?
3. What development in 1951 was a substantial breakthrough in the field of fiber optics? In 1960? In 1970?
4. Contrast the advantages and disadvantages of fiber optic cables and metallic cables.
5. Outline the primary building blocks of a fiber optic system.
6. Contrast glass and plastic fiber cables.
7. Briefly describe the construction of a fiber optic cable.
8. Define the following terms: velocity of propagation, refraction, and refractive index.

9. State Snell's law for refraction and outline its significance in fiber optic cables.

10. Define *critical angle*.

11. Describe what is meant by mode of operation; by index profile.

12. Describe a step-index fiber cable; a graded-index cable.

13. Contrast the advantages and disadvantages of step-index, graded-index, single-mode propagation, and multimode propagation.

14. Why is single-mode propagation impossible with graded-index fibers?

15. Describe the source-to-fiber aperture.

16. What are the acceptance angle and the acceptance cone from a fiber cable?

17. Define *numerical aperture*.

18. List and briefly describe the losses associated with fiber cables.

19. What is *pulse spreading*?

20. Define *pulse spreading constant*.

21. List and briefly describe the various coupling losses.

22. Briefly describe the operation of a light-emitting diode.

23. What are the two primary types of LEDs?

24. Briefly describe the operation of an injection laser diode.

25. What is lasing?

26. Contrast the advantages and disadvantages of ILDs and LEDs.

27. Briefly describe the function of a photodiode.

28. Describe the photoelectric effect.

29. Explain the difference between a PIN diode and an APD.

30. List and describe the primary characteristics of light detectors.

# PROBLEMS

1. Determine the wavelength in nanometers and angstroms for the following light frequencies.
   (a) $3.45 \times 10^{14}$ Hz
   (b) $3.62 \times 10^{14}$ Hz
   (c) $3.21 \times 10^{14}$ Hz

2. Detemine the light frequency for the following wavelengths.
   (a) 670 nm
   (b) 7800 Å
   (c) 710 nm

3. For a glass ($n = 1.5$)/quartz ($n = 1.38$) interface and an angle of incidence of 35°, determine the angle of refraction.

4. Determine the critical angle for the fiber described in Problem 3.

5. Determine the acceptance angle for the cable described in Problem 3.

6. Determine the numerical aperture for the cable described in Problem 3.

7. Determine the maximum bit rate for RZ and NRZ encoding for the following pulse-spreading constants and cable lengths.
   (a) $\Delta t = 10$ ns/m, $L = 100$ m
   (b) $\Delta t = 20$ ns/m, $L = 1000$ m
   (c) $\Delta t = 2000$ ns/km, $L = 2$ km
8. Determine the lowest light frequency that can be detected by a photodiode with an energy gap $= 1.2$ eV.

# INDEX

## A

Absolute phase delay, 29
Acceptance angle, 323
Acceptance cone, 324
A-channel signaling, 174
ACIA, 67
Acoustic couplers, 124
Adaptive delta modulator (ADM), 185
Adaptive equalizer, 116, 121, 123
Added channel framing, 188
Added digit framing, 187
Advanced data communication control procedures (ADCCP), 292
Aliasing, 159
Alphanumeric characters, 48, 51
Alternate mark inversion (AMI), 180
Amplitude distortion, 144, 208
Amplitude regulation, 141
Amplitude shift keying (ASK), 88, 95
Angle of refraction, 314
Angstrom, 309
Answer frequencies, 96
Aperture distortion, 159
Aperture time, 159
Application programs, 16

Area code, 5
ARQ-code, 53–54
ASCII, 48, 50, 225
Asynchronous data, 19
Attention identifiers (AIDS), 242
Attenuation distortion, 23, 26, 28
Attribute characters, 234, 240
Automatic data access arrangement (ADAA), 126
Avalanche photodiode (APD), 310, 341

## B

Balanced modulator, 98
Balanced transmission line, 81
Baluns, 81
Bandwidth efficiency, 104
Bandwidth parameters, 25–26
Baseband, 146, 296
Baud, 105
Baudot code, 48–49
B-channel signaling, 174
Bessel function chart, 92
Binary six-zero substitution (B6ZS), 182
Binary synchronous communications (BSC), 225

Binary three-zero substitution (B3ZS), 182
Bipolar, 179
Bit, 20
Bit error rate (BER), 120–121
Bit interleaving, 189
Bit-oriented protocol, 225–266
Bit synchronization, 19
Block check character (BCC), 56–57, 232
Block check sequence (BCS), 56–57
Blocking, 288
Branch area exchange, 2
Brickwall filter (see Nyquist filter)
Broadband transmission, 297
Broadcast address, 215, 267
Burst mode, 195

## C

Call directing code (CDC), 214
Character-oriented protocol, 225
Character synchronization, 19
Carrier recovery, 113, 116, 124
Carrier synchronization, 19
Carrier-to-noise ratio, 121
Carterfone decision, 16
CCITT, 83, 188
CCITT TDM, 178
CCITT V.24, 83
CCITT V.28, 83
CCITT V.41, 279
CCITT X.1, 290
CCITT X.4, 291
CCITT X.20, 83
CCITT X.25, 291–93
CCITT X.28, 291
CCITT X.29, 291
Channel access for LAN, 297
Channel banks:
   A-Type, 135
   D-Type, 167, 184, 187
Character codes, 48
Character language, 48
Character sets, 48
Character synchronization, 19
Circuit switching, 288
Cladded fiber, 305
Clock recovery, 116, 124
C-message noise, 34

C-message weighting, 33
C-notched noise, 35
Codecs, 12, 175, 189
Coherent detection (demodulation), 38,
   113, 115
Collision enforcement consensus procedure,
   301
Collision interval, 301
Collision window, 301
Combo chip, 190
Commercial television terminal, 147
Companding:
   Analog, 165
   Digital, 167
Compromise equalizers, 119, 123
Conditioning (see Line conditioning)
Configure command/response, 278
Constellation diagram, 110
Contention, 13
Continuous phase FSK, 97
Continuous variable-slope delta modulation
   (CVSD), 185
Copy command, 252
Copy control characters (CCC), 234, 252
Costas loop, 114
Couplers (CBT, CBS), 126, 127
Critical angle, 316–17
Cross-connects, 175
Crosshairs, 209
CSMA/CD, 297
Cyclic redundancy check (CRC), 57, 267, 298

## D

Dark current, 342
Data above video (DAV), 150
Data above voice (DAV), 150
Data access arrangement (DAA), 126
Data communications, 12
Data communications equipment (DCE), 18
Datagram, 292–93
Data in voice (DIV), 152
Data level point (DLP), 43
Data link control characters, 48, 50, 219, 225
Dataphone digital service, 154–55
Datascopes, 56
Dataset, 18, 88
Data terminal, 148

Data terminal equipment (DTE), 18, 88
Data transition jitter, 210
Data under voice (DUV), 149, 155
dB, 40
dBm, 41
dBmO, 41
dBrn, 44
dBrnc, 43
dBrncO, 43
Dedicated facilities, 9
Dedicated loop, 6 (see also Local loop)
Delta modulation, 183
Descrambler, 116, 118, 124
Deviation ratio, 93
Dial pulsing, 3
Dial switch, 2
Dibit, 101
Differential gain, 26
Diffraction, 327
Digital communications, 12, 154
Digital data service, 10, 150, 154
Digital hierarchy, 175
Digital line signals, 179
Digital radio, 154
Digital transmission advantages, 156
Direct distance dialing (DDD), 1, 2, 94, 124, 126
Disconnect mode, 271
Distributed parameters, 24
Double buffering, 68
Double sideband suppressed carrier, 100
Double-sided Nyquist bandwidth, 104, 206
Dropouts, 37
Duplex (see Full duplex)
Duty cycle, 179
Dynamic range, 164, 166

**E**

EBCDIC, 48, 51–52, 57, 225, 266
Echoplex, 125
Echo suppressor, 7
Elastic store, 149
Electromagnetic frequency spectrum, 307
Electromagnetic wavelength spectrum, 308
Electronic Industries Association (EIA), 18, 71, 79, 83
Electronic switching system (ESS), 7

Envelope delay, 26, 29
distortion (EDD), 30
Equalizers, 25, 94, 116, 120, 206
Equipment busy, 5
Error detection, 53, 181
Ethernet, 297–98
Exact-count encoding, 53
Eye patterns, 208

**F**

Facility, 23
Facility parameters, 25, 33
Federal Communications Commission (FCC), 26
Fiber construction, 311
Fiber guide, 309
Fiber optic system, 304
Fiber types, 310
Fixed data rate mode, 195
Folded binary code, 169
Foldover distortion, 159
Formants, 172
Forward error correction, 60
Frame, 173, 266
Frame bit, 173, 187
Frame check character (FCC), 57
Frame check sequence (FCS), 57, 267
Frame error, 67
Frame synchronization, 187
Frequency deviation, 89, 92
Frequency division multiplexing (FDM), 95, 134
Frequency modulation (FM), 89, 92
Frequency response, 26
Frequency shift, 38
Frequency shift keying (FSK), 18, 89
Front-end processor (FEP), 18
Full duplex, 15
Full/full duplex, 15

**G**

Gain hits, 37
Go-ahead sequence, 277
Graded index, 319
Granular noise, 184
Graphic control characters, 48, 51

Group, 135
Group address, 215, 267
Group separator, 234
Guard band, 135
Guided fiber cable, 304

## H

Half duplex, 15
Hamming code, 60
Hardwire coupling, 124, 127
Harmonic distortion, 39
Hartley's law, 20
HDLC, 279
h-factor, 89
High speed modems, 116
Hold and forward network, 288
Holding tone, 35
Hold time, 160
Host, 14
Hybrids, 7–8
Hybrid data, 149

## I

Idle channel noise, 164
Idle line 1's, 68, 72
IEEE 802, 298
Impulse noise, 36
Incident ray, 316
Index profile, 317–19
Information carrying capacity, 20,
    304
Information format, 268
Injection laser diode (ILD), 309, 337
Interface parameters, 25, 32
Intermodulation distortion, 39
Intersymbol interference, 206
Invert-on-zero coding, 276
ISO, 279
    protocol hierarchy, 289–90
Isochronous transmission, 20

## J

Jam sequence, 301
Jumbogroup, 137

## L

Leased facility, 9
Light detectors, 337
Light emitting diode (LED), 309, 334
Light sources, 334
Linear PCM codes, 163
Line conditioning, 26
Line control unit (LCU), 18, 214
Line drivers, 71
Loading coil, 24, 156
Local area network (LAN), 295
Local dial switch, 2
Local loop, 2, 10, 23
Localnet, 297
Longitudinal current, 81
Longitudinal redundancy check (LRC), 56
Loop start, 2
Lower sideband, 135
Low speed modems, 116

## M

Manchester code, 298, 301
Mark frequency, 89
Mark hold, 72
M-ary, 112
Mastergroup, 136
Mastergroup terminal, 147
Master station, 13
Medium-speed modems, 116
Mesh network, 296
Message channel, 23
Message switching, 288
Metallic circuit current, 81
Microwave radio channel, 136
Microwave transmission, 144
Midrise quantization, 164
Midtread quantization, 164
Minimum shift keying (MSK), 97
Modems, 18
    asynchronous, 19, 88, 116
    dial-up, 125
    synchronous, 9, 88, 116, 122
    103, 96
    202S, 94, 214
    202T, 89, 214
Modem synchronization, 115

Modes of propagation through fiber:
   single mode, 317
   multimode, 317
Modified data, 249
Modulation index, 91
Morse code, 16, 48
µ-law, 165
Muldem, 177
Multicast address, 299
Multidrop, 14
Multifrequency signaling, 7
Multiplexer, 16, 175
Multiplexing, 95, 134
Multipoint system, 13

**N**

Nanometer, 309
Noise, 33
Noise immunity, 156
Noise weighting, 33
Nonlinear distortion, 39
Nonlinear encoding, 164
Nonlinear PCM codes, 163
Non-return-to-zero transmission (NRZ),
   179–81
Nonuniform encoding, 164
Normal response mode, 271
NRZI, 276
Null, 246
Numerical aperture, 325
Nyquist filter, 104
Nyquist sampling theorem, 157

**O**

1004 Hz deviation, 26, 33
Off-hook, 2
On-hook, 3
Optical fiber losses:
   absorption, 326
   chromatic dispersion, 328
   coupling losses, 323
   material or Rayleigh scattering, 327
   modal dispersion, 329
   pulse spreading, 330
   radiation, 328

Order of the arrow, 55
Originate frequency, 96

**P**

Packet, 288
Packet switching, 286, 288
Pad, 124
Parity, 53
   checker, 55
   generator, 54
   error, 67
Peak-to-average ratio, 40
Percentage error for digital companding, 170
Permanent virtual circuit, 292–93
Phase delay, 29
Phase distortion, 23–24, 208
Phase hits, 37
Phase intercept distortion, 38
Phase jitter, 37, 140
Phase locked loops, 94, 113
Phase shift keying (PSK), 18, 89
   BPSK, 98
   differential, 114
   8-PSK, 107
   Offset QPSK, 107
   OKQPSK, 107
   QPSK, 102
   16-PSK, 110
Photoelectric effect, 340
Photophone, 305
Physical address, 299
Picturephone terminal, 147
Pilot carrier, 140
Pilot frequency, 144
PIN diode, 310, 339
Poll, 214
   general, 229
   specific, 230
Poll/final bit, 268
Post equalization, 94, 120
Pre-equalization, 119
Prefix, 4
Primary station, 14
Prismatic refraction, 313
Private branch exchange (PBX), 27, 189
Private line service, 1, 9
   foreign exchange (FX), 10

Private line service (*cont.*)
  full data (FD), 10
  full period (FP), 10
Private switched network, 27
Probability of error (P(e)), 120
Propagation time, 29
Protocol:
  asynchronous, 214
  bit-oriented, 225
  character-oriented, 225
  chips, 281
  line, 20, 214
  message, 20
  synchronous, 225
Public data network, 286
Public telephone network (PTN), 1, 6, 88
Pulse amplitude modulation (PAM), 157
Pulse code modulation (PCM), 147, 157
Pulse duration modulation (PDM), 157
  differential PCM, 186
Pulse position modulation (PPM), 157
Pulse spreading constant, 331
Pulse stretching, 330
Pulse transmission, 205
Pulse width dispersion, 331
Pulse width modulation (PWM), 157

**Q**

Quadrature amplitude modulation (QAM), 18,
  89, 109–110
Quadrature loop, 114
Quantization error (see Quantization noise)
Quantization noise, 160

**R**

Receiver overrun, 67
Redundancy, 53
Reference noise, 43
Refraction, 312
Refractive index, 313
Regenerative repeaters, 156
Remote stations, 14
Remote terminal equipment, 215
Resolution, 162
Responsivity, 341
Return-to-zero transmission (RZ), 179–81

Reverse interrupt, 250
Ring modulator, 99
Ring network, 14, 296
Robbed digit framing, 187–88
RS 232C, 19, 64, 71, 93, 125
RS 422A, 79
RS 423A, 79
RS 449, 79
RZ AMI, 180–81

**S**

Sample and hold, 157, 159
Sampling rate, 157
Scramblers, 116, 118, 124
SDLC, 225, 266
  frame, 266
  loop, 276
Secondary channel, 95
Secondary station, 14
Selection, 214, 231
Serial interface, 71
Setup time, 288
Shannon limit for information capacity, 21
Shift register mode, 200
Side frequency, 92–93
Signaling functions, 167
Signal quality, 121, 123
Signal-to-C-notched noise, 26
Signal-to-quantization noise ratio (SQR), 162
Simplex, 15
Single channel terminals, 146
Single frequency interference, 38
Single frequency tone (SF), 7
Single sideband suppressed carrier, 135
(sin x)/x function, 180
Slave station, 13
Slope overload, 184
Snell's law, 314, 324
Space, 246
Space frequency, 89
Spectral null, 206
Spectral response, 342
Squaring loop, 113
Star network, 13, 296
Start bit, 68
Start bit validation, 68
Station, 2

Station busy, 3
Station controller (STACO), 18, 215
Statistical framing, 188
Status and sense message, 230, 250
Steering circuitry, 66
Step index, 319
Store and forward network, 189, 288
Subscriber, 2
Supergroup, 135
Supervisory frame, 268
Supervisory signaling, 2, 201
Switched carrier, 72
Switched network, 2
Switcher's switch, 4
Switching hierarchy, 5
Switching techniques, 286
Symbol, 104
    rate, 104
    substitution, 55
Synchronization, 19
Synchronization time, 187, 188
Synchronous data, 19–20

**T**

Talk battery, 2
Tandem switch, 4
T-carrier, 12, 154, 157, 174, 180, 182–83, 187
Telco, 1
Telecommunications, 12
Telegraph, 16
Telephone circuit, 23

Telephone number, 2
Telephone prefix, 2, 4
Teletype, 83
Telex code, 49
Telset, 2
Terminating set, 7
Ternary signals, 208
Time division multiplexing (TDM), 173
Timing recovery, 180
Tip and ring, 2, 124
Token passing, 297
Toll call, 4
Touch tone signaling, 2
Training sequence, 115–116
Transactional switch, 289
Transit time, 342
Transmission codes, 48
Transmission level point (TLP), 41
Transmission parameters, 25
Transmission voltages, 179
Transmit start code (TSC), 214
Transparent switch, 289
Transparent text, 256
Tribit, 107
Tributaries, 14
Trunk circuits, 2, 10
    common usage, 7
    direct, 4
    interoffice, 4
    intertoll, 6
    tandem, 4
    toll-connecting, 5